WITHDRAWN
UTSA LIBRARIES

Neural Networks and Animal Behavior

MONOGRAPHS IN
BEHAVIOR AND ECOLOGY

Edited by John R. Krebs
and Tim Clutton-Brock

Neural Networks and Animal Behavior,
by Magnus Enquist and Stefano Ghirlanda

The Evolution of Animal Communication: Reliability and Deception in Signaling Systems, by William A. Searcy and Stephen Nowicki

Hormones and Animal Social Behavior,
by Elizabeth Adkins-Regan

Sexual Conflict, by Göran Arnqvist and Locke Rowe

Mating Systems and Strategies, by Stephen M. Shuster and Michael J. Wade

The African Wild Dog: Behavior, Ecology, and Conservation, by Scott Creel and Nancy Marusha Creel

Sperm Competition and Its Evolutionary Consequences in the Insects, by Leigh W. Simmons

Model Systems in Behavioral Ecology: Integrating Conceptual, Theoretical, and Empirical Approaches, edited by Lee Alan Dugatkin

Social Foraging Theory, by Luc-Alain Giraldeau and Thomas Caraco

Levels of Selection in Evolution, edited by Laurent Keller

Parasites in Social Insects, by Paul Schmid-Hempel

Foundations of Social Evolution, by Steven A. Frank

Sex, Color, and Mate Choice in Guppies,
by Anne E. Houde

Female Control: Sexual Selection by Cryptic Female Choice, by William G. Eberhard

Social Evolution in Ants, by Andrew F. G. Bourke and Nigel R. Franks

Leks, by Jacob Höglund and Rauno V. Alatalo

Polygyny and Sexual Selection in Red-Winged Blackbirds, by William A. Searcy and Ken Yasukawa

Sexual Selection, by Malte Andersson

Parasitoids: Behavioral and Evolutionary Ecology,
by H. C. J. Godfray

The Evolution of Parental Care, by T. H. Clutton-Brock

Dynamic Modeling in Behavioral Ecology,
by Marc Mangel and Colin W. Clark

Foraging Theory, by David W. Stephens
and John R. Krebs

Neural Networks and Animal Behavior

Magnus Enquist
Stefano Ghirlanda

PRINCETON UNIVERSITY PRESS
PRINCETON AND OXFORD

Published by Princeton University Press, 41 William Street,
Princeton, New Jersey 08540

In the United Kingdom: Princeton University Press,
3 Market Place, Woodstock, Oxfordshire OX20 1SY

Library of Congress Cataloging-in-Publication Data

Enquist, Magnus.
Neural networks and animal behavior / Magnus Enquist, Stefano Ghirlanda.
p. cm.—(Monographs in behavior and ecology)
Includes bibliographical references (p.) and index.
ISBN-13: 978-0-691-09632-2 (alk. paper)—ISBN-13: 978-0-691-09633-9 (pbk. : alk. paper)
ISBN-10: 0-691-09632-5 (alk. paper)—ISBN-10: 0-691-09633-3 (pbk. : alk. paper)
1. Animal behavior—Computer simulation. 2. Neural networks (Computer science). I. Ghirlanda, Stefano, 1972– II. Title. III. Series.

QL751.65.D37 E57 2005
591.5—dc22 2005040561

British Library Cataloging-in-Publication Data is available

This book has been composed in Times Roman in LaTeX

The publisher would like to acknowledge the authors of this volume for providing the camera-ready copy from which this book was printed.

Printed on acid-free paper. ∞

pup.princeton.edu

Printed in the United States of America

10 9 8 7 6 5 4 3 2 1

Contents

Preface

The aim of this book is to explore the potential of neural network models to explain animal behavior at the motivational, ontogenetic and evolutionary levels. Our focus is on understanding behavior based on principles of operation of nervous systems. Ethology, psychology and behavioral ecology have produced a large body of theory of behavior, but it is generally not formulated in terms of operation of the nervous system. The latter can be achieved, we will argue, with neural network models. We will focus on general principles rather than analyzing in detail specific nervous systems or neural structures. Thus, although nervous systems exhibit bewildering complexities, we will avoid them for the most part. The reason is that we want to explore how much understanding and predictive power can be gained by introducing general and basic consideration of nervous systems into models of behavior. There are also practical reasons for this choice. First, while empirical knowledge of nervous systems is increasingly detailed, we often do not know whether a given detail reflects a general and important principle or a less important contingency. Hence we prefer not to include too much detail before we have understood more general issues. Second, it is unlikely that we will ever know in detail the nervous system of more than a few species. Thus a useful theory of behavior should abstract as much as possible from features of particular nervous systems. Indeed, nervous systems differ in many respects across taxa (say, insects and mammals). If general principles of behavior exist, they must emerge from something other than either the gross architecture of the nervous system or its fine details. It is possible that our hopes will not be fulfilled and that simple neural networks may not prove satisfactory models of behavior. We keep an open mind on this possibility. This book will not decide this matter, but hopefully it will show that further exploration of neural network models of behavior is worthwhile.

Our models will try to include (usually in simplified form) features that are common to most, if not all, nervous systems. The foremost such feature is, of course, that nervous systems are made of neurons wired together. We also will consider main facts of neuron physiology, how neurons interact, the properties of receptors and their organization in sense organs, and how the nervous system interfaces with the body. Using this knowledge to build models of behavior has not been a very common research strategy. Behaviorists, for instance, have held that inquiries into the mechanisms of behavior are just speculation. Cognitive psychologists, although in most ways opposite to behaviorists, have made an effectively similar claim that physiology can be ignored in the analysis of "intelligent behavior" and "cognition." In contrast, ethologists and some animal psychologists have considered behavior as the product of concrete biological mechanisms that can be investigated fruitfully.

One thread in this book is that some of this work anticipates important aspects of neural network models of behavior. Surprisingly, however, this has not led to widespread use of neural networks in animal behavior theory. We will try to remedy this situation and show the potentials of these models in many areas of animal behavior, from mechanisms of behavior to the evolution of behavior.

In writing this book we had to explore some areas of research with which we had little experience. This has been very rewarding and has taught us about both animal behavior and neural networks. However, we cannot present a mature theory of behavior in terms of neural network models, nor can we cover in its entirety the existing literature. It is likely that we have neglected important work, and for this we apologize to the reader and the authors.

WHO SHOULD READ THIS BOOK?

The reader we primarily have in mind is a student of animal behavior, but we hope that this book also will interest cognitive scientists, engineers and people working with neural networks in general who have an interest in animal behavior. This book, in fact, presents many concepts of animal behavior theory that are not widely known outside ethology and animal psychology, showing their connection with neural network models.

HOW TO READ THIS BOOK

This book is organized in six chapters. Readers less interested in the mathematical aspects of neural network models (or very familiar with them) can skip most of Chapter 2. We advise them, however, to read at least the first section (excluding Section 2.1.3). Here follows an outline of the chapters.

Chapter 1 starts by asking what it means to explain behavior, discussing the differences between motivational, ontogenetic and evolutionary explanations. We list some basic requirements for models of behavior and then introduce the reader to different modeling traditions. The difference between operational (black-box) models and physiological models is introduced. Lastly, we informally present artificial neural networks and outline their potentials as models of animal behavior.

In Chapter 2 we offer a technical description of the most widely used kinds of neural networks. We consider models of individual nerve cells, feedforward networks and recurrent networks, as well as the methods developed to have networks perform specific tasks and to organize networks in response to inputs. Concrete examples of what networks can achieve are provided, as well as simple mathematical analysis of key properties. A final section introduces the reader to computer simulation of neural networks.

In Chapter 3 we describe a general approach to modeling behavior and use it to develop neural network models of reactions to stimuli, motivation, decision making, behavior sequences and motor control. The approach includes realistic modeling of network inputs, based on knowledge of sensory physiology, and a conceptual boundary separating the brain from both the body and the external world (rather than a boundary between the individual and the external world, as often done). We also compare networks with other approaches such as ethological and psychological theory and control theory.

Chapter 4 considers some general questions about ontogeny, such as the role of genes in building brain networks and the material basis of memory. Mechanisms of learning are discussed, and various methods for updating network memory are considered. We present some applications of neural networks to learning and to ontogenetic phenomena such as classical and instrumental conditioning, perceptual learning and imprinting.

Chapter 5 considers the evolution of behavior. Most theoretical studies of behavioral evolution assume few constraints on the phenotype. We discuss what neural networks can teach us about possible constraints on nervous systems and how such constraints may affect evolution. We present some concrete examples concerning the evolution of behavior used in animal communication.

The final chapter, Chapter 6, summarizes what the exploration in this book has taught us about behavior and neural networks as a tool for understanding behavior. We compare neural network models with other approaches to behavior and consider some challenges that future research must address.

ACKNOWLEDGMENTS

We are grateful to the publisher for inviting us to write this book and for editorial support during its realization. We thank the Department of Zoology of Stockholm University, where most of the work for this book has been carried out. Our research has been funded by the Swedish Research Council, the Tercentenary Fund of the Bank of Sweden, Marianne och Marcus Wallenbergs Stiftelse, and the Italian Ministry for Education, University and Research (through grants to Emmanuele A. Jannini, Department of Experimental Medicine, University of L'Aquila, and Vincenzo Di Lazzaro, Institute of Neurology, Catholic University of Rome). SG has enjoyed encouragement and financial support from his family and girlfriend, as well as warm hospitality at the Institute for the Science and Technologies of Cognition, CNR, Rome. We have greatly benefited from tools developed by the free software community, including the GNU/Linux operating system, many GNU project tools and LaTeX.

Our understanding of the topics covered in this book has benefited greatly from discussions with many colleagues, especially Anthony Arak, Gianluca Baldassarre, Jerry Hogan, Liselotte Jansson, Björn Merker, Stefano Nolfi, Daniel Osorio, Domenico Parisi and Peter Wallén. Michael Ryan and an anonymous reviewer provided useful comments on an earlier draft of the book.

Neural Networks and Animal Behavior

Chapter One

Understanding Animal Behavior

The subject of this book is animal behavior. What is animal behavior, and what does it mean to understand animal behavior? As we shall see in this chapter, there are several answers to these questions, depending on research tradition and what one intends to understand. At the same time, different theories of behavior have much the same scope:

- They deal with how the animal as a whole interacts with its physical, ecological and social environment, in particular through reception of sensory stimulation and behavioral actions such as motor patterns, pheromone release, change in body coloration and so forth.
- They want to explain, predict or control what animals do.
- They consider situations in which internal factors such as memory and physiological states are not easily accessible.

Besides our theoretical interests, there are great practical demands for knowledge of behavior. People who work with animals, such as zookeepers, farmers, animal trainers, veterinarians and conservationists, constantly need such knowledge.

Ethology and comparative psychology are two major research traditions dealing with behavior. Within these disciplines, it has been and still is important to explore behavior as a function of external stimuli and readily observable factors such as species, age and sex. Internal factors such as physiological states and memory, on the other hand, are studied indirectly or inferred from observations of behavior, history of events and the passage of time. The reason for this difference is that in almost all situations in which we encounter animal behavior, it is relatively easy to monitor and control the external situation and to record behavior, but the access to internal factors is usually limited. However, ignoring internal factors undoubtedly has shortcomings. First, it is clear that external factors are not alone in causing behavior. An animal can react quite differently to the same piece of food, e.g., depending on hunger and memory of experiences with similar food items. Second, it is difficult to reconstruct behavior mechanisms from pure observations of behavior, and physiological knowledge about internal mechanisms may greatly facilitate the development of behavior models.

But what is the proper compromise between the complexity of nervous system and body physiology and the need for understanding at the behavioral level? The role of internal factors in models of behavior has been and still is a matter of much discussion. The most famous of these debates is the confrontation between behaviorists, who attempted to avoid internal factors altogether, and cognitive psychologists, who instead encouraged theorizing about internal processes (Leahey

Figure 1.1 A newly hatched cuckoo chick dumps the eggs of its foster parents. The proximate causes of this behavior are not the same as its ultimate causes. The proximate explanation is that touch-sensitive receptors on the back of the chick elicit a behavior sequence that eventually results in eviction of the eggs. The ultimate, or evolutionary explanation is that chicks whose genes coded for this behavior survived better than chicks without such genes, since the former did not share the foster parents' efforts with foster siblings. Reproduced from Davies (1992) by permission of artist David Quinn.

2004; Skinner 1985; Staddon 2001; see Section 1.3.5). We will discuss this issue at length in Chapter 6. In short, our position is that internal factors are vital for understanding behavior, but at the same time we side with behaviorists (and, of course, ethologists) in their focus on behavior and regard their contribution to the understanding of behavior as very significant.

In this chapter we first consider the different kinds of explanations that have been considered for behavior. We then introduce the reader to major theories of behavior, focusing on their structure and the causal factors invoked to explain behavior. Finally, we introduce neural networks models, which we will explore in this book as a potential framework for understanding behavior.

1.1 THE CAUSES OF BEHAVIOR

In biology, two types of causal explanations are generally recognized: proximate and ultimate explanations (Baker 1938; Mayr 1961). Proximate explanations appeal to motivational variables, experiences and genotype as the cause of behavior. Ultimate explanations refer to selection pressures and other factors that cause the evolution of behavior. These two kinds of causal explanations are independent and complementary, serving different purposes: one cannot replace the other. To clarify this concept, Lorenz (1981; §1.6) offers the following example. The ultimate cause of cars is, of course, traveling. If the engine breaks, however, the ultimate cause cannot start it again: we need to know how the engine works (proximate causes). Figure 1.1 gives another example of this distinction, in the context of behavior. Note that, in principle, we can learn independently about proximate and ultimate causes. However, the two are also related (since behavior is a result of evolution and influences evolution), and we probably will learn faster about one by also considering the other.

A more detailed set of explanations was suggested Niko Tinbergen in defining the scope of ethology (Tinbergen 1963). Rather than distinguishing just between proximate and ultimate causes, he argued that four questions must be answered to understand behavior. These are usually summarized as follows:

1. What causes a behavior to appear at a given moment and how does the behavioral machinery work?
2. How does behavior develop during an individual's lifetime?
3. What is the evolutionary history of the behavior?
4. How does the behavior contribute to survival and reproduction?

The first two questions are about the proximate explanation. Splitting it in two allows us to study behavior mechanisms at one time separately from how they change with time, an advantage on which we will capitalize. Tinbergen's third question aims at a description of evolutionary change and does not refer to any causal explanation. The fourth question, as it is expressed, does not strictly refer to a causal explanation but to so-called final or functional explanations. These, with a leap of logic, invoke the *effects* of a behavior (contribution to survival and reproduction) as the *cause* of the behavior itself. In practice, however, the fourth question often covers true causal studies of the outcome of evolution, i.e., studies of how natural selection and other factors can modify behavior in an evolving population. Additional discussion of explanations of animal behavior can be found in Alcock and Sherman (1994), Dewsbury (1992, 1999), and Hogan (1994a). In this book we consider three kinds of causal explanations closely related to Tinbergen's:

1. Motivation
2. Ontogeny
3. Evolution

To us, understanding behavior means having answers to all of these questions, and this book is organized after this classification. The three explanations consider the causation of different phenomena and invoke different causes as follows.

A *motivational explanation* refers to individual behavior, and the goal is to predict behavior from variables such as external stimulation and internal physiological states and to understand the behavior mechanism (Hogan 1994a). The term *motivation* refers to generally reversible and often short-term changes in behavior. An example is how animals use behavior to regulate food and water intake. Other examples include mechanisms of perception, decision making, motor control, etc. These topics are covered in Chapter 3.

Ontogeny refers to the development of the behavior mechanism during an individual's lifetime. The causes of ontogeny are the genotype and all the experi-

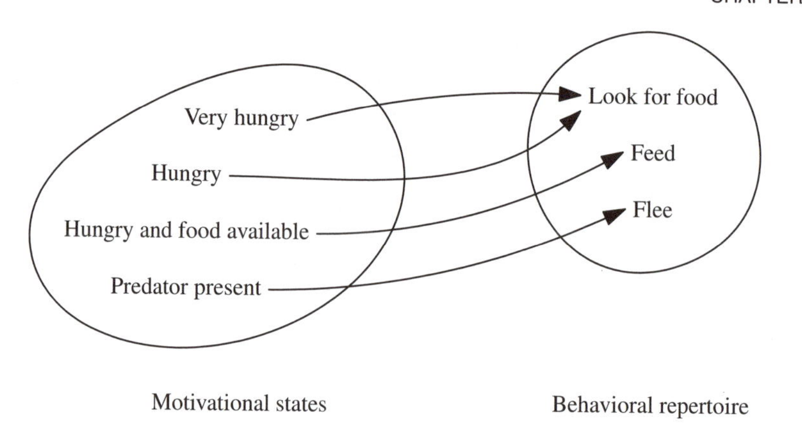

Figure 1.2 An idealized behavior map associating motivational state (external and internal factors) with behavioral responses.

ences the individual has. The changes that occur include changes in the structure of the nervous system and memory changes, leading to generally long-term and less reversible changes in behavior (Hogan 1994a). Particular phenomena include learning, maturation, genetic predispositions and the development of the nervous system. Chapter 4 is dedicated to learning and other ontogenetic phenomena.

Lastly, *evolution of behavior* (or any other trait) occurs in a population of individuals and is manifested by changes in the population's gene pool (Futuyma 1998). Causes of genetic changes include mutations, recombination, natural selection and chance. Selection may stem from the abiotic environment, from other species and from the behavior of conspecifics. These are the topics of Chapter 5. In species with culture, additional factors would have to be considered, but this is beyond the scope of this book.

1.2 A FRAMEWORK FOR MODELS OF BEHAVIOR

1.2.1 The behavior map: From motivational state to response

Central to this book and a starting point for most of our discussions is the assumption that behavior can be described and predicted based on knowledge about relevant *motivational variables*. These are simply the variables that enter motivational explanations of behavior. For instance, feeding behavior may be caused by the presence of food stimuli in combination with hunger, as illustrated in Figure 1.2. Research on behavior is partly about identifying motivational variables, and today, many are known, including physiological variables such as hormone levels and external variables such as the shape and color of objects, the structure of mating calls, etc. The collection of all motivational variables at a given time is the animal's *motivational state*. A complete description of behavior assigns a behavioral response to each possible motivational state. Mathematically, we can express this notion

Table 1.1 How the behavior map enters different explanations of behavior.

	Level of explanation		
	Motivation	**Ontogeny**	**Evolution**
Question	What are the properties of the behavior map?	How is the behavior map determined?	How does evolution change behavior maps?
Causes	External stimuli and internal states	Genes, experiences	Environment (physical, biological, social), mutations, chance
Effects	A behavioral response	A behavior map	Genetic code to develop a behavior map
Important states	Internal states of the body and nervous system	Memory and nervous system connectivity	Gene frequencies and genotypes

with a function m that establishes a mapping from the set of motivational states X to the set of responses R. The latter contains the behavioral repertoire and other responses, such as hormone secretion. If we use $\mathbf{x}$ and $\mathbf{r}$ to refer to a single state and response, respectively, we can write

$$\mathbf{r} = m(\mathbf{x}) \tag{1.1}$$

We will refer to m as the *behavior map*. This expression may sound technical, but it actually refers to something familiar to all students of behavior. For instance, the concepts of stimulus-response relationship, decision rule or response gradient are all examples of behavior maps. That is, they can be all written in the form of equation (1.1) with appropriate choices of the input and output spaces and of the map function m. We use the expression behavior map for several reasons. First, it is not linked to any modeling tradition, but we can identify a behavior map in all theories of behavior. Second, the concept of a behavior map provides a sharp definition of what we mean by "understanding behavior" at different levels of explanation (Table 1.1). Third, our book explores neural networks as behavior maps.

The behavior map may be stochastic rather than deterministic. This means that a probability is assigned to each possible behavioral response rather than predicting exactly which one will occur. Whether true stochasticity occurs in behavior is a largely an unresolved problem (see, e.g., Dawkins & Dawkins 1974). Rather than attempting to solve this difficult issue, we note that stochastic factors may be included in the formalism above. For instance, we can include motivational variables that vary at random (Chapter 3).

1.2.2 The state-transition equation: Changes in motivational state

In addition to knowledge of the behavior map, if we want to predict sequences of behavior, we also need to know how the motivational state changes with time. Such changes may reflect changes in the environment (e.g., in the availability of a given food source) or changes within the animal (e.g., altered hormone levels). This book studies the impact of both external and internal factors. However, with respect to state changes, we are mainly interested in changes within the animal: first, because they are caused by processes in the animal whereas most changes to external states occur independent of the individual; and, second, because internal states allow the animal to organize its behavior in time and to detect temporal patterns in sensory input (Chapter 3).

Formally, changes in state can be described by means of equations of the form

$$\mathbf{x}(t+1) = M(\mathbf{x}(t), \mathbf{r}(t)) \tag{1.2}$$

This equation simply says that the state at time $t+1$ is a function of the state at time t and of the response of the system at time t (discrete rather than continuous time is considered for simplicity). In systems theory (Metz 1977; Minsky 1969), equation (1.2) is known as a *state-transition equation*, whereas equation (1.1) is the system's *response function*, or *output function*. Examples of state-transition equations will be seen below, as well as in later chapters.

1.2.3 The behavior map in ontogeny and evolution

In summary, motivational explanations consider the properties of a particular behavior map, i.e., how behavioral responses are caused by motivational states. A full understanding of motivational process also requires an understanding of motivational state transitions. Ontogenetic and evolutionary explanations of behavior deal with different causes and effects.

In ontogenetic explanations, the effect is a behavior map, and the causes are an individual's genes and experiences. Thus ontogeny considers how the map develops and changes during an individuals life. It also covers the nature-nurture issue. One can express mathematically the behavior map as a function of the genotype g and the history of experiences h:

$$m = f(g, h) \tag{1.3}$$

However, we prefer an approach based on state. An animal does not store its entire history; instead, its experiences result in a change to state variables W that determine the properties of the behavior map. Formally, we can write

$$m = f(W) \tag{1.4}$$

Most models of ontogeny and learning, including those based on neural networks, are of this kind. Note that the state variables of ontogenetic process are different from those of motivational process. The most obvious state variable in ontogeny is memory, and learning is a key mechanism for memory changes (memory state transitions). The state transition equation may be complex and depend on current state, genotype, motivational state and response.

Finally, to understand the genetic evolution of behavior, we have to consider evolutionary dynamics and a new class of state variables: the genes. Changes in gene frequencies and genotypes g are caused by the physical, ecological and social environment, as well as by mutations and other mechanisms that generate new genotypes. Evolution does not change the behavior map directly: the effect of genetic evolution is rather a genetic program for developing a behavior map.

1.2.4 Requirements on models of behavior

In summary, to understand behavior we need ways of describing behavior maps and state-transition equations. Ideally, models of behavior should fulfill the following requirements:

- **Versatility:** We observe great diversity in behavior both between species and, on a less dramatic scale, within species. The basic structure of a general model of behavior should allow for a diversity of behavior maps to be formed.
- **Robustness:** Regardless what explanation we seek, the mechanisms should display some robustness; otherwise, behavior would be vulnerable to any genetic or environmental disturbance. Thus small disturbances should not cause any major changes in performance.
- **Learning:** Learning from experience is a general feature of animal behavior. The model should allow learning, which should integrate realistically with the behavior map.
- **Ontogeny:** The behavior system of an individual develops in a sequence of events in which genes, the developing individual and the environment interact. The structure of the model should allow for gradual development of a behavior map from scratch to an adult form.
- **Evolution:** For evolution to occur, genetic variation must exist that affects the development of behavior mechanisms. In a model, it should be possible to specify how genes control features of behavior maps and learning mechanisms. In addition, a model should allow the evolution of nervous systems from very simple forms to the complexities seen in, e.g., birds and mammals.

1.3 THE STRUCTURE OF BEHAVIOR MODELS

A diversity of behavior models exists, often developed for particular purposes such as the study of learning or perception. Important contributions to animal behavior theory come from the ethological tradition (McFarland 1974a; McFarland & Houston 1981; Simmons & Young 1999) and from comparative psychology (Mackintosh 1994) but also from neuroscience (Cacioppo et al. 2000; Gazzaniga 2000; Kandel et al. 2000) and computer science (Wilson & Keil 1999). Here we consider different modeling traditions from the point of view of model structure. By *structure* we simply mean the basic machinery of the model, e.g., how a motivational model generates the response from motivational variables. So far we have not provided any structures to the models we have considered. Adding structure to a model of behavior includes making assumptions about features of real animals, such as sense organs, central processes, memory, motor control and what are the available responses. What this means in practice will become clear in the following.

1.3.1 Operational and physiological models

What structure we need depends, of course, on our aims. We may distinguish roughly between operational and physiological models. An *operational model* aims to describe behavior realistically, but its structure is not intended to resemble the internal structure of animals. Such models are often referred to as *black-box models* to indicate lack of concern about underlying mechanisms. A *physiological model*, on the other hand, attempts to take into account more of the physiology that produces behavior, e.g., body and nervous system physiology.

As will become clear in the following, the majority of models, both of human and of animal behavior, are operational models. Operational models are easier to build because only input-output relationships need to be described accurately and because we have fewer constraints on how to achieve such relationships. For instance, we may use the latest computer algorithm for pattern recognition instead of figuring out how a bunch of interconnected neurons can recognize anything. This is legitimate so long as we do not claim that agreement with observations implies identity of internal structure. Another factor favoring operational models is that physiology is more difficult to observe or control compared with behavior.

A crucial difference between physiological and operational models, and indeed the whole point in distinguishing them, is that the structure of physiological models is not inferred solely from behavioral observations but is at least partly based on knowledge of physiology. There are several motives for doing this. One is to understand the neurophysiology of behavior. Another is the belief that a physiological model can be a more accurate behavioral model because its structure more closely mirrors the internal structure of animals.

In the following we consider the structure of behavior models from major research traditions, i.e., what behavior maps and state equations are used. We also discuss the extent to which models are operational or physiological. This will allow us, throughout this book, to compare neural networks with other modeling traditions, highlighting differences and similarities.

1.3.2 Models with little or no structure

Some models of behavior make few assumptions about internal structure, allowing any input-output relationship to be formed with equal ease. Such unconstrained models have been used mainly for descriptive purposes and in studies of the evolution of behavior. A simple way of creating an unconstrained behavior map is to arrange motivational factors and behavioral responses in a *look-up table*, where each entry represents a particular input, and the content of the entry represents the behavioral output (Figure 1.3). It is possible to consider histories of events by including one table entry for each possible history (including the current situation). Thus a look-up table can cover learning, but not in its usual sense of *changing* stimulus-response relationships or memory. Rather, the effect of a particular individual history is to read out from the table one response instead of another.

One way to describe more explicitly sequences of events is to arrange them in a *tree structure*. As time unfolds, one progresses along the tree, following one branch or another depending on what decisions are taken. Such tree models can describe histories of experiences and allow for a unique responses to each history. A particular kind of tree model has been developed within game theory, referred to as the *extensive-form description* of a game (Figure 1.4; Fudenberg & Tirole

Motivational state	Behavior response
Hungry	Look for food
Very hungry	Look for food
Hungry + food available	Feed
Predator present	Flee
...	...

Figure 1.3 A look-up table. By finding a particular motivational state in the table, the behavioral response is found. This particular example corresponds to the behavior map in Figure 1.2.

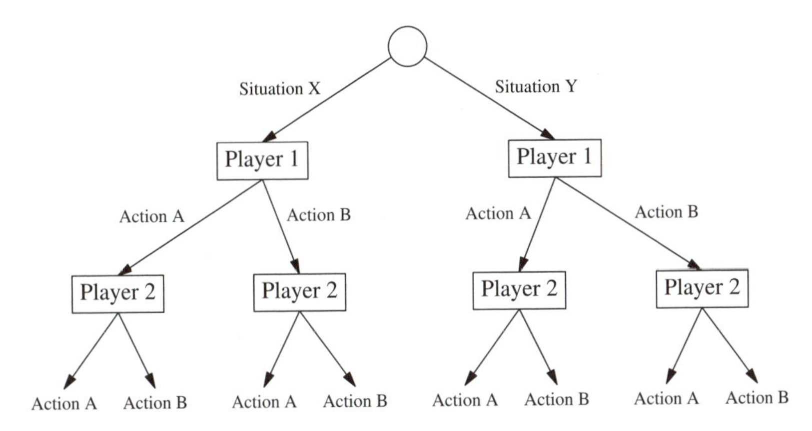

Figure 1.4 An extensive-form description of a game between two individuals ("players"). Each player has two possible actions, labeled A and B. The top node indicates that the game can start in situation X or Y, which is outside the players' control. Based on what situation occurs, player 1 decides whether to do action A or B. Based on player 1's action, player 2 decides in turn. The extensive-form representation of games can also illustrate cases in which players have imperfect information about external situations or each other's actions.

1992). The extensive form is more refined than a simple tree because it can take into account both events observed by the player and those not observed. Both extensive-form descriptions and look-up tables can be extended to continuous variables. In the case of look-up tables, for instance, the table becomes a function that translates a continuous input into outputs.

Look-up tables dominate behavioral modeling in evolutionary biology, where the aim is to seek a behavior map that maximizes fitness (or some other currency such as rate of food intake) under given ecological or social conditions (Grafen 1991). The table is more commonly called a *strategy*, but it retains the meaning of a prescription of how to behave in each possible situation. Constraints on what behavior maps can be implemented are often weak and are introduced in mainly two ways. First, behaviors may differ in the cost that is paid to perform them. Second, the set of available behaviors is often limited to obtain more sensible results (e.g., a maximum running speed of preys and predators may be assumed). However, it is still true that all behavior maps, among those allowed, are assumed equally easy to form (e.g., Kamil 1998).

Models with no structure offer complete flexibility but have also drawbacks. First, they offer no insight into mechanisms of behavior. Second, each response is set independently of others. Thus there is no a priori reason why responses to similar inputs should yield similar outputs. For the same reason, responding in novel situations is undetermined, because the corresponding entry does not exist in the look-up table. In contrast, animal behavior exhibits clear regularities as a function of similarity between stimuli or situations (Section 3.3). Models without structure are unrealistic also because storing the response to each possible history of events requires an enormous memory. Of course, animals do not recall their entire history when deciding what to do. Instead, their behavior mechanism goes through successive changes of state, and the current state depends partly on the history of events. A last drawback of these models is that each table entry must be programmed genetically because there is no place for learning. We will return to these issues several times in this book, particularly in Chapter 5 on evolution.

1.3.3 Behavior as a function of motivational state

In this section we consider models that attempt to predict behavior based on a limited number of motivational variables. The majority of motivational models are of this kind (e.g., Bolles 1975; McFarland 1971; McFarland & Houston 1981; Metz 1977). Motivational variables generally are categorized as either external stimuli (input to the system) or internal factors (system state variables). External stimulation can vary in many ways: different sense organs may be stimulated and each in many different ways. Examples of internal factors are water balance, hormone levels and memory. We shall elaborate on this distinction in Chapter 3.

Relevant motivational variables typically are identified from observations of behavior, often combined with functional considerations (e.g., that body temperature must be maintained within limits for life to continue). The motivation state may also be identified from the animal's recent history (e.g., hunger increases with the time since the last meal; Bolles 1975). Sometimes causal factors are induced from statistical analysis of observed data (Heiligenberg 1974; Wiepkema 1961) or by invoking physiological considerations. For instance, models of feeding and drinking may consider stretch receptors in the stomach, water uptake rates from the stomach and receptors that monitor cellular water levels in the blood or elsewhere in the body (McFarland & Baher 1968; Toates 1986, 1998). The concept of state, including motivational state, today plays an important role also in evolutionary theory of animal behavior (Houston & McNamara 1999).

Given that relevant variables have been identified, the next question is how to combine them in a model of behavior. This is the issue of structure. A simple and classical example of a mathematically defined behavior map is given by summation models of stimulus control. These attempt to predict the reaction to a compound stimulus from knowledge of reactions to its components in isolation. For a compound made up of three stimuli, for instance, the model can be formalized as

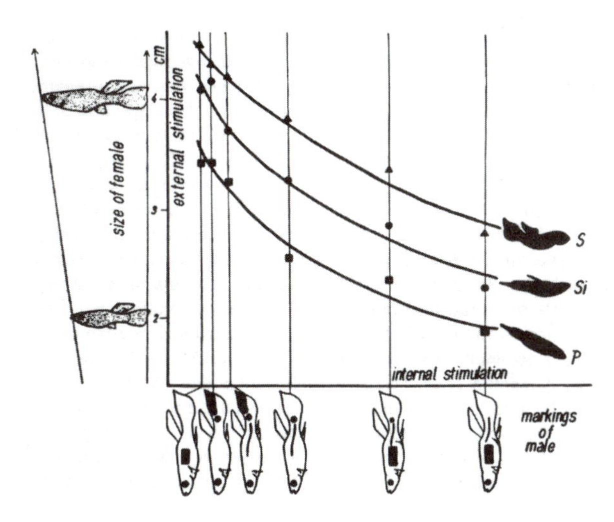

Figure 1.5 Interaction of internal and external factors in eliciting courtship behavior in male guppies (*Lebistes reticulatus*). The external factor is the size of a female dummy (vertical axis), whereas the internal factor is readiness to mate, as inferred from male coloration. The three curves represent the induction of three different components of courtship (pursuit of the female and two intensities of bending the body). From Baerends et al. (1955).

follows. To describe which stimuli are present, we introduce the motivational state $\mathbf{x} = (x_1, x_2, x_3)$, with $x_i = 1$ if stimulus i is present and 0 otherwise. We then define responding to the compound stimulus as the sum of the effects of its component:

$$r = W_1 x_1 + W_2 x_2 + W_3 x_3 \tag{1.5}$$

where W_i is the response to stimulus i. Only present stimuli influence responding, since $x_i = 0$ for absent ones. Note the separation between state variables, which describe which stimuli are present, and the model structure, which describes how to obtain the response from the state, i.e., through a weighted sum. Models with this structure are the so-called law of stimulus summation (Leong 1969; Lorenz 1981; Seitz 1940–1941, 1943; Tinbergen 1951) and psychological models such as Rescorla and Wagner (1972) and Blough (1975). Note that some of the x_i's could be made to represent internal factors. In the latter case, external and internal factors would combine additively to produce the response. Of course, there are other ways in which factors can be combined (Figure 1.5; Krantz & Tversky 1971; McFarland & Houston 1981), and individual x_i's may not just take values of 0 and 1.

Another important theoretical concept is that of thresholds. Consider first the issue of whether to perform a given behavior or not based on the value of a single motivational variable, e.g., whether to eat or not for a given hunger level. A common idea is that the behavior is performed if the motivational variable exceeds a threshold value T. Thresholds can also be applied to a combination of state variables such as the weighted sum in equation (1.5). That is, a response to a given stimulus situation is assumed to occur if

$$W_1x_1 + W_2x_2 + W_3x_3 > T \tag{1.6}$$

A more complex situation occurs when incompatible responses are motivated simultaneously (Ludlow 1980, 1976; McFarland 1989; McFarland & Sibly 1975). For instance, a hungry and thirsty animal cannot drink and eat at the same time. Usually, different behaviors or behavioral subsystems are assumed to "compete," which means that the animal will show the behavior corresponding to the highest motivation (McFarland 1999).

The concept of thresholds and decision making can be generalized to *decision boundaries* in motivational state space. The motivational state **x** of the organism is represented as a point in a multidimensional state space, and to each point one response is assigned. Different regions in the space then are defined by setting decision boundaries. States within the same region yield the same behavior, but crossing a decision boundary leads to a change in behavior (McFarland 1999; McFarland & Houston 1981).

To understand how behavior sequences are generated, we also need to consider how motivational state changes because the updated state will determine the subsequent responses (McFarland 1971). This is a complex problem that operates at various levels and time scales. Early theory of motivational state transitions was based on the drive concept (Hinde 1970). Each behavior was assumed to be controlled by a single internal factor ("drive") that builds up with time and is reset to a low value each time the behavior is elicited. Since then, a diversity of state-transition mechanisms has been identified, and pure drives are probably rare. Modeling of motivational state-transitions and behavior sequences have benefited largely from applications of control theory and systems theory, yielding the most refined approach to date for modeling motivational processes (McFarland 1971; Metz 1977; Toates 1998; Toates & Halliday 1980). An important concept is that of feedback mechanisms (McFarland 1971; Toates 1998). Many behavioral processes can be regarded naturally as controlling internal as well as external variables important for survival and reproduction. For instance, a lizard may use behavior to regulate its body temperature by moving between warmer and colder places within its territory. Feedback allows systems to respond to state changes brought about by previous actions or the environment. Figure 1.6 depicts two block diagrams of mechanisms for motor control, one without a feed back loop and one with such a loop.

However, control of behavior is not limited to feedback from consequences of behavior. Within animals, many state variables have been identified that participate in the generation of behavior sequences. How such states change depends very much on their nature and may involve processes intrinsic to the nervous system (e.g., internal clocks), physiological states and processes outside the nervous system (e.g., hormones, water and energy status) and perceived external events and circumstances (e.g., social stimuli) or any combinations of these categories.

1.3.3.1 Summary

The concept of motivational state has proved useful, and models based on such a concept have been remarkably successful. An important contribution has been the

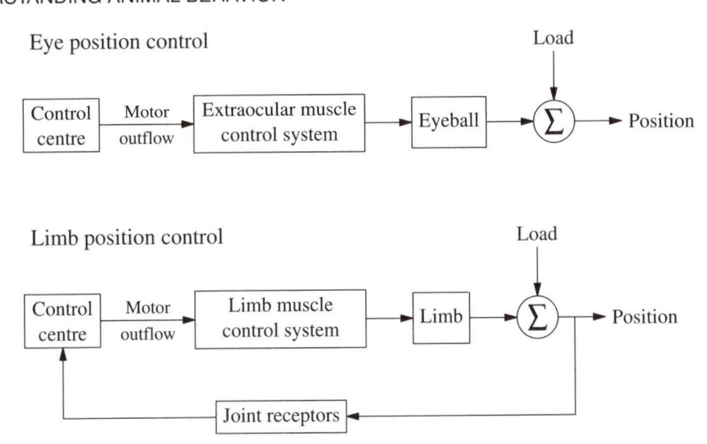

Figure 1.6 Outline of the eyeball and limb position-control systems illustrating the two principles of open-loop and closed-loop control. The feedback loop in the limb control system allows correction of any deviation of the limb from the intended position. Reprinted by permission from McFarland (1971).

establishment that behavior systems can be regarded as dynamical systems (McFarland 1971; McFarland & Houston 1981; Metz 1977) by complementing the equation that describes the behavior map (output function) with state-transition equations that update the motivational state (Houston & McNamara 1999; McFarland & Houston 1981). At least as they are applied today, however, models based on motivational state also have some weaknesses. First, applications of control theory to animal behavior have seldom considered learning (although learning to control is possible, see Sutton & Barto 1998; Widrow & Stearns 1985). Second, motivational factors are inferred from behavioral observations and sometimes from knowledge about body physiology, but the structure of the models is inferred from behavioral observations only. The concept of response threshold, for instance, has been criticized, and it remains unclear whether it can describe animal decision making satisfactorily (Section 3.7). Third, in some cases it is unclear what insight is gained by viewing a given aspect of nervous system operation as a control problem (e.g., stimulus recognition).

1.3.4 Animal learning theory

Behavior-level models of learning have been developed mainly within psychology (Dickinson 1980; Mowrer & Klein 2001; Pearce 1997). These models tend to focus on changes in memory, which is assumed to consist of "associative strengths" between events, usually between one or more stimuli and one behavioral response (Dickinson 1980; Pearce 1997; Rescorla & Wagner 1972). In our words, these models study how the behavior map changes through learning. The associative strength between a stimulus and the response is usually written V. An important class of learning models predicts the change in associative strengths ΔV during repeated experimental trials where a stimulus is associated with an event such as the delivery of food (classical or instrumental conditioning; Chapter 4). Under such conditions, V is assumed to change according to the following equation (or similar ones; Blough 1975; Bush & Mosteller 1951; Rescorla & Wagner 1972):

$$\Delta V = \eta(\lambda - V) \tag{1.7}$$

where η regulates the speed of change, and λ is the maximum value that V can reach in the given experiment (influenced by such variables as stimulus intensity and the nature of the event paired with the stimulus, see Section 4.5). In systems theory terminology, equation (1.7) is a state-transition equation with associative strength as the state variable. To translate state into behavior, it is further assumed that the likelihood of responding to the stimulus is an increasing function of V:

$$\Pr(\text{response}) = f(V) \tag{1.8}$$

In other words, V can be regarded as a tendency to respond. Note that this is the same problem of linking state to behavior that we discussed earlier in connection with motivational states.

An interesting application of these models deals with the relative influence of several different stimuli on behavior. If stimulus 1, stimulus 2 and so on are present simultaneously, their total associative strength is assumed to be the sum of the associative strengths of the individual stimuli:

$$V_{\text{TOT}} = \sum_i V_i \tag{1.9}$$

where V_i is the associative strength of stimulus i, and the sum extends over all stimuli present on a given experimental trial. V_i changes according to

$$\Delta V_i = \eta(\lambda - V_{\text{TOT}}) \tag{1.10}$$

To see more clearly how the model relates to the behavior map formalism, we proceed as above (see equation 1.5) and introduce the variables x_1, x_2, etc. such that $x_i = 1$ if stimulus i is present, and $x_i = 0$ if it is absent. We then can write the full model as a pair of equations:

$$\begin{cases} \Pr(\text{response}) = f\left(\sum_i V_i x_i\right) \\ \Delta V_i = \eta\left(\lambda - \sum_i V_i x_i\right) x_i \end{cases} \tag{1.11}$$

The first equation is the behavior map, and the second is the state-transition equation describing how the behavior map changes as a consequence of experiences. Note that the equation to calculate total associative strength is the same as the equation for calculating response in the summation models considered earlier (equation 1.5), although the notation differs slightly. The focus on learning means that here the weights, or associative strengths, are included among state variables, i.e., those variables which can change. In equation (1.5) they were instead fixed parameters because the focus was on responding to different combinations of stimuli rather than learning.

Learning models from animal learning theory are operational models. Their structure is inferred from behavioral observations. They focus on changes in associative strengths rather than behavior but make also predictions about behavior (otherwise, it would be impossible to test them). However, detailed assumptions on the response function f are seldom provided, and complex motivational situations are not studied.

1.3.5 Cognitive models

By *cognitive models* we mean models in the style of cognitive psychology and computer science (together often referred to as *cognitive science*) as it emerged from the late 1950s until today (Crevier 1993; Leahey 2004; Wilson & Keil 1999). The main feature of cognitive models is a strong focus on the representation and processing of information. Traditional cognitive science was aimed at humans, but today there is considerable research under the heading of animal cognition, including a journal by this name (Balda et al. 1998; Bekoff et al. 2002; Dukas 1998a; Gallistel 1990; Mackintosh 1994; Pearce 1997; Shettleworth 1998). The classical research program of cognitive psychology argued that any intelligent organism or machine could be fully understood in terms of a program operating on stored information. Thus one could ignore what hardware is used practically, whether a computer or a brain (Neisser 1967).

While the primary aim of cognitive models is to understand how the brain represents and processes information, they also predict behavior and are evaluated based on such predictions. In this respect, cognitive modeling is similar to animal learning theory (it could be argued that since the 1970s, the two fields have come progressively closer). The structure of cognitive models specifies both how memory is updated based on experiences (the state-transition equation) and how responses are decided on based on information stored in memory (the behavior map, or response function). Typically, memory is described in functional terms as internal representation of knowledge about the world, e.g., knowledge of locations, angles, velocities, conditional probabilities (e.g., Gallistel 1990). Decisions and memory updates are "computational" in the specific sense of manipulation of symbols, very much in the sense of formal mathematics. The following two examples illustrate these typical aspects.

The first example is about the ability of desert ants, *Cataglyphis bicolor*, to navigate in the absence of clear landmarks (Wehner & Flatt 1972; Wehner & Srinivasan 1981). After long and tortuous wanderings searching for food, these ants are capable of aiming for their nest following an approximately straight path. Gallistel (1990) has proposed that ants achieve this by constantly calculating the direction and distance to the nest. Such calculations are based on solar heading, obtained from the visual system; the speed whereby the ant moves, possibly obtained from the brain's motor command; and the sun's current azimuth. The latter is computed with the aid of an internal "solar ephemeris function" that describes the sun's course and is based on an internal clock. These inputs allow the animal to compute its position in a coordinate system from which direction and distance to the nest subsequently are computed. The model is illustrated in Figure 1.7 (further discussion in Chapter 6).

A second example illustrates another cognitive approach, which considers that organisms live in an uncertain world and can learn about the world through observations. Many empirical studies have shown that animals and humans are sensitive to temporal correlations or contingencies apparent in their interaction with the outer world (Dickinson 1980; Mackintosh 1994; Shanks 1995). This has spawned the idea that the central nervous system operates as a "statistical machine" (Shanks

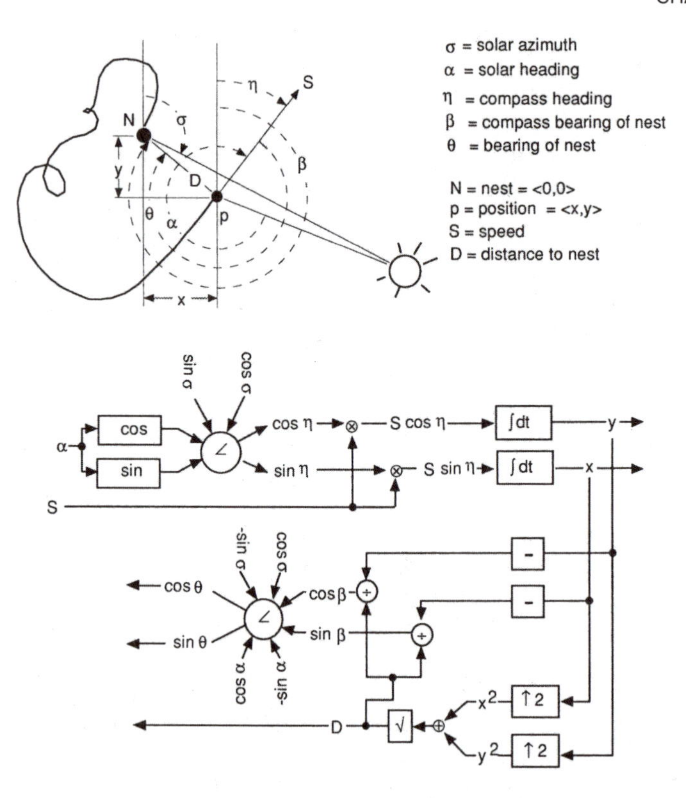

Figure 1.7 A cognitive model of navigation in the desert ant, *Cataglyphis bicolor*. Coordinates x and y are internal variables describing the position of the ant relative to the nest (origin of the coordinate system). External inputs are the angle between the ant direction and the sun (α) and the ant speed S. An internal function calculates the solar azimuth σ as a function of time of day. Output of the model is the bearing and distance from, home which are assumed to guide behavior. Reprinted from Gallistel (1990) by permission of C. R. Gallistel.

1995). In such a system, memory would consist of conditional probabilities that are used in statistical decision making and are updated as a result of experiences.

It should be clear from these examples that the elements of cognitive models are postulated because they are assumed to serve a function. Representations of velocities, distances and conditional probabilities are there because the animal needs to navigate, find food, avoid predators, etc. This prominence of functional considerations has two consequences for typical cognitive models. First, it blurs the distinction between proximate and ultimate explanations (Figure 1.1). However, that the function of memory is to store information is not an explanation of how memory works. Second, emphasis on function and symbolic information processing discourages thinking of how nervous systems actually implement the proposed functions and computations (Chapter 6). Third, emphasis on function leads naturally to the idea that what is computed is computed *correctly*. For these reasons, some behavioral phenomena have so far benefited little from a cognitive approach either because it is difficult to know what is functional, because the computational content of a behavior is unclear (e.g., sleep) or because behavior is, in some conditions, not functional. Examples of the latter include biases in responding to stimuli and other features of decision making (Chapter 3).

Cognitive models typically are intermediate between operational and physiological models. They aim at describing real mechanisms within the animal, at the level of symbolic information processing, but they are inferred from behavioral observations and functional arguments (Gallistel 1990; Leahey 2004). The extent to which this is legitimate is the subject of enduring debate. In contrast to the cognitive approach, most classical behaviorists held an extreme negative view and argued that assumptions about internal or mental variable were speculative and unscientific (Skinner 1985). For a critical examination of both positions, see Staddon (2001).

1.3.6 Neuroethology

Our overview of models of animal behavior would not be complete without mentioning neuroethology and similar research efforts (Ewert 1980; Ewert et al. 1983; Simmons & Young 1999). Neuroethology was born out of classical ethology as an effort to understand behavior based on detailed knowledge about the anatomy and physiology of nervous systems. It focuses on animals in general and usually on much simpler nervous systems than the human brain. It is hard to overrate the success of this research program. A number of behavior systems have been studied thoroughly. One example is studies of sensory processing and decision making in the frog retina and brain (Ewert 1980, 1985). Examples of other success stories are the motor control of swimming in the lamprey (Grillner et al. 1995) and pheromone searching in moths (Kennedy 1983).

The purpose of neurophysiological models is not only to predict behavior but also to provide an understanding of how nervous systems operate. Pure neurophysiological models of behavior represent the opposite of pure black-box models. They are important because they give the real picture of behavior mechanisms. They are crucial for the neural network approach explored in this book by providing an important reference, independent of behavioral observations, of how to build models of behavior. To the student of behavior, the downside of neuroethological models lies in their complexity. Each model embodies many particular aspects of specific nervous systems, building on detailed knowledge that we are unlikely to ever have but for a few species. Complexity also makes these models impractical, in the sense that a considerable expertise is needed to understand each of them. Lastly, even the most detailed physiological and anatomical knowledge may not by itself reveal principles of operation. Today there is promising cooperation between neuroethology and neural network research. Research on visuomotor coordination in frogs and toads is a good example (Cervantes-Péres 2003). Neural network research has produced tools that allow us to analyze findings from empirical neuroethology, including theory, modeling techniques and computer simulation techniques. These, however, have seldom been used to develop simplified models that provide intuitive understanding of basic principles of behavior.

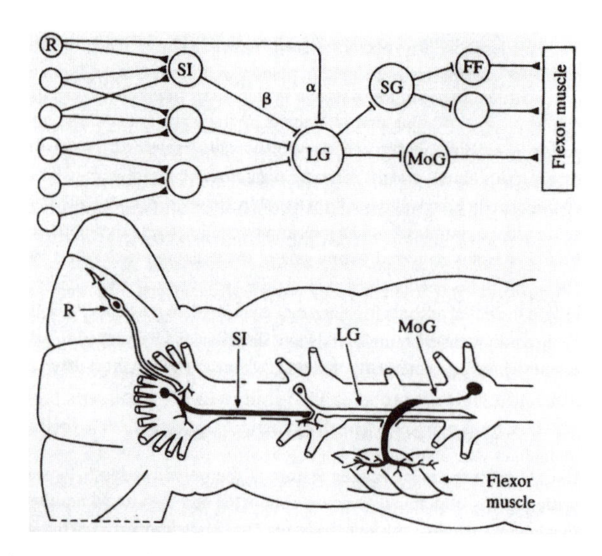

Figure 1.8 Functional scheme (top) and biological localization (bottom) of the neural network responsible for a flight response in the crayfish. (Based on Wine & Krasne 1982.) The network includes, at one end, touch receptors in the tail and, at the other end, motoneurons connecting to muscles. Interneurons process information from receptors and, if a tactile stimulus is strong enough, send to motoneurons a command that allows the crayfish to rapidly move away from the stimulus. Reprinted by permission from Simmons and Young (1999), *Nerve Cells and Animal Behaviour* (Cambridge University Press).

1.4 NEURAL NETWORK MODELS

This book explores neural networks as models of behavior, also known as *artificial neural networks*, *connectionist networks* or *models*, *parallel distributed processing models* and *neurocomputers*. We call them neural network models or sometimes just neural networks or networks when the meaning is clear. To avoid confusion between model and reality, we talk of biological neural networks or nervous systems in connection with real animals. In this section we offer a brief introduction to neural network models; the next chapter covers them more thoroughly.

The basic feature of neural network models is that they are inspired by biological neural networks. As an illustrative example, Figure 1.8 portrays part of the nervous system of a crayfish. It consists of neurons connected together, often in a way that is consistent across individuals, forming a network of interacting cells. Some neurons are *receptors*; i.e., they transform physical stimuli such as oscillating air pressure (sound) into electrical or chemical signals that are communicated to other neurons. Other neurons are *effectors*; i.e., they represent the output of the nervous system to the body. Some effectors are connected to muscles (motoneurons), whereas others secrete chemicals that affect other cells in the body. Neurons that are neither receptors nor effectors are generically called *interneurons*. Obviously, their organization is of paramount importance for nervous system operation.

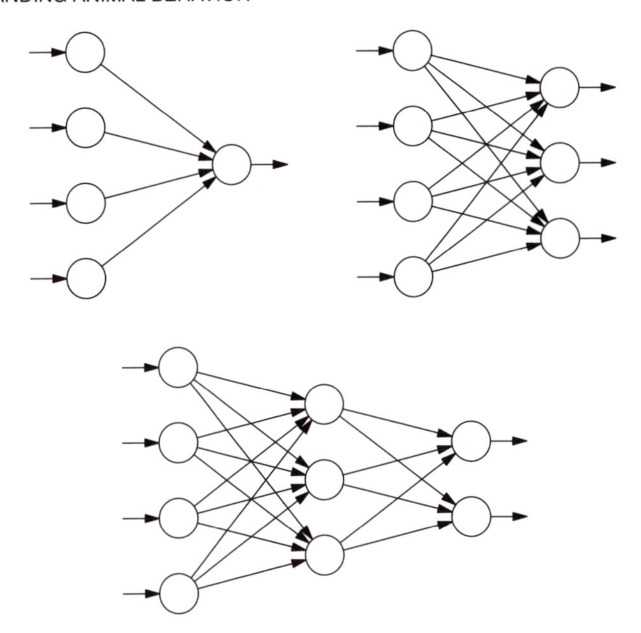

Figure 1.9 Examples of feedforward artificial neural networks. Top left: A simple architecture where input nodes connect directly to a single output node. Top right: Network with multiple output nodes. Bottom: Network with an additional layer of nodes between input and output ("hidden nodes").

A neural network model can include all the features we see in Figure 1.8. It mimics the architecture of nervous systems by connecting elementary neuron-like units, referred to as *nodes* or *units*, into networks. Each node typically has links or connections with many other nodes. Nodes can stimulate or inhibit each other's activity like neurons. Stimulation is entered into the network via artificial receptors and sense organs, and the activity of output units represents behavior patterns or muscle contractions (depending on the level of analysis). Although nervous systems are made up of many cells (about 1000 in nematode worms, many billions in birds and mammals), their modeling is simplified because their operation can often be described as the mass action of many essentially identical neurons with regular patterns of connectivity.

Neural network models can have different architectures, working principles and aims. One popular architecture that will be studied many times in this book is the feedforward network consisting of a layer of input nodes, zero or more intermediate layers and a layer of output nodes (Figure 1.9, see also Figure 1.8). An important property of neural networks is that in addition to having an interpretation as model nervous systems, they are also mathematically defined models. Typically, the activity of each node is a number, either continuous (e.g., between 0 and 1) or discrete (e.g., 1 for "active" and 0 for "inactive"). We write as z_i the activity of node i. The index i runs from 1 to N, the total number of nodes, and serves to identify each node. Node activity is assumed to be an increasing function f of the total input that

Table 1.2 A behavior map, interpreted as a simple model of feeding, that can be realized by a network with two input nodes and one output node.

Input		Desired output	
Values	**Interpretation**	**Value**	**Interpretation**
(0,0)	Not hungry, no food	0	Don't feed
(0,1)	Not hungry, food	0	Don't feed
(1,0)	Hungry, no food	0	Don't feed
(1,1)	Hungry, food	1	Feed

the node receives from other nodes in the network, written y_i. Thus

$$z_i = f(y_i) \tag{1.12}$$

The total input, in turn, is computed as a weighted sum of the activities of other nodes. Formally, this is expressed as

$$y_i = \sum_i W_{ij} z_j \tag{1.13}$$

The number W_{ij}, called a *weight*, is the strength of the connection from node j to node i, akin to the strength of biological synapses. Absence of a connection between two nodes is indicated by $W_{ij} = 0$. Equations (1.12) and (1.13) are the basic building blocks of most neural network models. All nodes in the network operate based on such equations, possibly with different choices of the function f. One important exception is nodes that model receptors, whose activity should reflect the activity of the corresponding receptor in the stimulus conditions we want to model. When a node models a receptor, we often write its activity as x rather than z. Likewise, the activity of output nodes is often written as r. This agrees with the notation used earlier for behavior maps in general.

As a simple example of behavioral modeling with a neural network, consider an animal that decides to feed or not based on hunger and availability of food. More precisely, the animal feeds when (1) it is hungry, and (2) food is available, but not when only one or none of these conditions is met. Our network model has two input nodes connected directly to one output node. The activity of input node i is x_i ($i = 1,2$), and the corresponding weight to the output node is W_i. We assume that the output node is active (which is interpreted as feeding) if the total input $W_1 x_1 + W_2 x_2$ overcomes a threshold θ. For simplicity, we assume that nodes are either active ($x_i = 1$) or inactive ($x_i = 0$). The first input node is active when the animal needs energy ("hunger" node); the second, when food is available ("food" node). There are thus four possible input patterns, and the network should react only to one (Table 1.2). This must be accomplished by tuning the three parameters W_1, W_2 and θ. The solution is simple: one of the weights must be less than $\theta/2$, and the other must be greater than $\theta/2$, with $\theta > 0$. With such a choice, when both inputs are active, the output node receives an input of $W_1 + W_2$, which is larger than θ, as required to feed. If the food terminates, or if hunger ceases, one of the x_i's falls to 0, and feeding stops. The network thus can decide whether to feed or not, integrating information about hunger and food availability. The model obviously is simplistic, and will be developed further in Chapter 3. Note, for instance, that the assumption of a "food" node does not explain why some stimuli are treated as food and others are not. A "hunger" node is more realistic because it may correspond to neurons sensitive to blood sugar levels, located, e.g., in the hypothalamus and liver of mammals (Toates 2001).

The preceding example uses the simplest kind of network, featuring a number of input nodes connected directly to one output node (Figure 1.9, top left). Such a network is described fully by the single equation

$$r = f\left(\sum_i W_i x_i\right) \tag{1.14}$$

which is very similar (or exactly the same, depending on f) to some of the ethological and psychological models surveyed earlier. We will consider such a basic network numerous times. Usually, however, neural network models are a bit more complex. Additional output cells allow for more than one type of response, and the ability to form input-output relationships is enhanced significantly by adding one or more intermediate layers of nodes between the input and output nodes. Finally, recurrent connections (feedback loops) allow the network to handle time, i.e., to respond to temporal sequences of inputs and to organize output in time.

1.4.1 Features of neural network models

Neural network models have many features particularly appealing to students of behavior that we will try to communicate in this book. Here we provide a brief summary of such features, to be justified more fully in the following chapters. We first consider the list of requirements on page 7:

- **Versatility:** Neural network models can implement practically any behavior map (Haykin 1999). A particular network may be fitted to many requirements, and additional flexibility is achieved by considering different network architectures.
- **Robustness:** The operation of a neural network is seldom affected in major ways by damage to a small fraction of nodes or connections. The performance of each task is divided among many nodes, and memories are encoded over many connection weights (of course, this does not hold for very small networks or nervous systems).
- **Learning:** Learning has been a prominent part of neural network research since its beginnings. Networks can learn by means of procedures that change connection weights either autonomously or under external guidance. From a physiological point of view, most training procedures have several or many unrealistic features, but progress in this area has been steady over the last few decades.
- **Ontogeny:** Neural network models to date have not been applied systematically to problems of behavioral ontogeny. Some applications have been developed, with encouraging results (Chapter 4).

- **Evolution:** Various features of neural network models, such as architecture, learning rules and properties of connections and nodes, can be assumed to arise from different "genes." Computer simulations of behavioral evolution can be set up to investigate what networks evolve to solve particular tasks. Such investigations are still at their beginnings. One obstacle is that we still ignore a lot about how genes control nervous system development.

Other features of neural network models are also relevant to modeling behavior:

- **Structural constraints:** The structure of these models essentially is based on knowledge of the nervous system: it is not inferred from observations of behavior or from functional considerations. In this sense, neural network models are unique among models of behavior. Such constraints, when used properly, can greatly diminish the danger of introducing processes that cannot have a counterpart in biological nervous systems.
- **Parallel processing:** Processing in neural network is mainly parallel rather than sequential, as in digital computers. This means that the processes from reception of stimuli to responses can occur in much fewer cycles of operation, each consisting of many thousands or millions of simultaneous computations. This is an important step toward matching the abilities of nervous systems to react very quickly in a large range of conditions.
- **Generalization:** In new situations, animals show systematic generalization based on similar past conditions, and this is a crucial part of their ability to confront the world. As a consequence of their structure, many network models generalize naturally. This is a great advantage compared with models that either ignore generalization (Section 1.3.2) or simply assume that it occurs (Section 3.3.3).
- **Definiteness and accessibility:** Since neural network models are specified formally, they can be investigated in all details of operation. They can be studied with tools such as the theory of dynamical systems and computer simulations or in much the same ways as neurophysiologists study nervous systems. For instance, one can study what happens when a specific part of the network is damaged or removed. This can bring insight into network operation and can also be compared with lesion studies in nervous systems.
- **Unifying power:** Neural network models can be applied to all aspects of behavior: processing of stimuli, central processing and motor control. Neural networks can learn, and the substrate of memory (connection weights) is clear. Most approaches to behavior, on the other hand, are geared toward specific aspects (e.g., perception or cognition) and may have problems at producing a unified theory of behavior.

1.4.2 A brief history

The history of neural network models is rooted in attempts to understand the nervous system and behavior, starting with the discovery by Santiago Ramòn y Cajal, Camillo Golgi and others at the end of the 19th century that the nervous system is composed of an intricate network of cells. Developments of neural network mod-

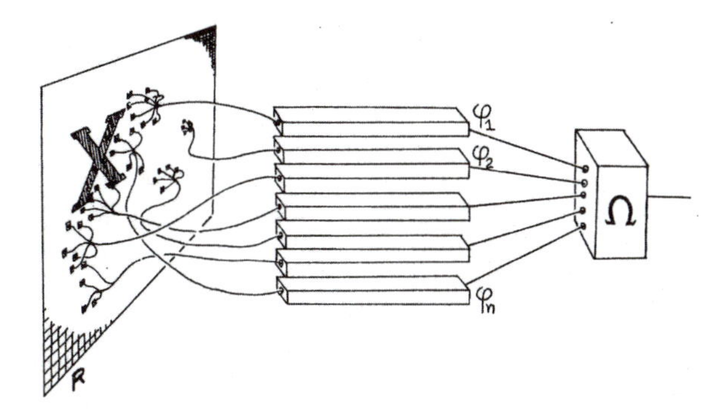

Figure 1.10 Rosenblatt's (1958; 1962) perceptron, as depicted by Minsky & Papert (1969). Each φ is assumed to be 0 or 1 depending on the result of a simple computation based on the state of a limited number of receptors. The summation unit Ω performs a weighted sum of the φ's. If the sum is larger than a threshold value, the perceptron response Ψ is 1; if the sum is smaller than the threshold, the response is 0. Reprinted by permission from Minsky and Papert (1969). *Perceptrons*. The MIT Press.

els, however, are also rooted in mathematics, theories for intelligent machines and philosophy (Crevier 1993; Minsky 1969; Wang 1995).

The development of neural network models started in earnest in 1943 when Warren McCulloch and Walter Pitts described how arbitrary logical operations could be carried out by networks of nodes (so-called formal neurons) that could either respond or not respond to an input, computed as the weighted sum of the activation state of other nodes in the network (McCulloch & Pitts 1943). This way of computing the total input to a node has stayed in practically all later neural network models. An important aspect of McCulloch and Pitts' work was that neural processing could be formulated mathematically. Six years later, Donald Hebb published his famous book, *The Organization of Behavior*, which, among other things, includes a theory of how connection strengths between nerve cells may act as memory and how such memory may change as a consequences of experiences (Hebb 1949). Perhaps the first full-blown neural network model was designed and simulated on computers by Frank Rosenblatt (1958, 1962). His "perceptron" is depicted in Figure 1.10. An important aspect of Rosenblatt's work was providing the perceptron with a learning algorithm that allowed it to solve a wide range of classification problems, whereby each of many input patterns should be assigned to one of two categories. Bernard Widrow and Marcian Hoff further developed learning algorithms introducing a general procedure for training two-layer networks (Widrow & Hoff 1960), to be known later as the δ rule (Section 2.3.2).

The early enthusiasm for neural network models was subdued in 1969 by Minsky and Papert's book, *Perceptrons*, pointing out some severe limitations of perceptrons. Perceptrons cannot solve all categorization problems, and some that are solvable in principle are very hard to solve in practice (Chapter 2). The authors also suggested that multilayer networks would suffer from similar limitations, but this turned out to be incorrect. Minsky and Papert also acknowledged that recurrent networks (with feedback loops) could be much more capable (Chapters 2 and 3) but did not discuss them for several reasons. One was a desire to compare parallel computers (i.e., neural networks) with traditional serial computers; recurrent networks have both serial and parallel elements and thus were set aside. The perceptron was also chosen for its relative mathematical simplicity, which made a comprehensive analysis feasible. The book is nevertheless very interesting and contains many insights about neural networks that may also apply to biological nervous systems.

The 1980s saw a series of important publications that inspired a new wave of research into neural network theory and applications. Efficient learning algorithms for multilayer networks were made widely known, such as the now-celebrated back-propagation algorithm, and it was shown that multilayer feedforward networks could overcome some limitations of perceptrons (Ackley et al. 1985; Haykin 1999; Rumelhart et al. 1986). Recurrent networks were also studied, and it was described how information could be stored in recurrent networks (Amit 1989; Cohen & Grossberg 1983; Hopfield 1982). Kohonen (1982) also published his results on self-organizing maps, showing how simple learning rules could organize networks based on experience without external guidance. Crucial to the diffusion of neural networks into cognitive psychology was the two-volume book, *Parallel Distributed Processing: Explorations in the Microstructures of Cognition* edited by James McClelland and David Rumelhart (1986; Rumelhart & McClelland 1986b).

To date, the study of artificial neural network has generated an extensive body of theory (e.g., Arbib 2003; Haykin 1999). Neural networks have proved to be a powerful explanatory tool applied to a wide variety of phenomena such as perception, concept learning, the development of motor skills, language acquisition in humans and studies of amnesia and brain damage (Arbib 2003; Churchland 1995; Churchland & Sejnowski 1992; McClelland & Rumelhart 1986). Recent progress has occurred along several lines. One is the study of more powerful and/or biologically realistic network architectures, e.g., recurrent networks that can handle time (Elman 1990; Jordan 1986). Fundamental areas of research are how networks can be trained to solve a specific task and how they can learn from experiences without explicit supervision. From the present perspective, an important issue is the development of more biologically realistic learning mechanisms. Many methods for training networks are not realistic because they are external to the network rather than built into the network itself, but significant improvements have been achieved (Chapters 2 and 4). This is one area that needs more research, and it is important to recognize that besides all the progress, a number of issues remain to be resolved. Other areas in need of further research are, for instance, the genetic control of network development and how unrealistic retroactive interference (new learning destroying earlier memories) can be tackled (French 1999).

The preceding sketch of the history of neural network models is, of course, far from complete. The development of neural network models, for instance, has also benefited from the gains in our understanding of neurophysiology. Examples are the discovery of receptive fields in visual perception (Hubel & Wiesel 1962) and the study of learning in invertebrates (Kandel et al. 2000; Morris et al. 1988). There are also engineering applications that are of potentially great interest to biology and behavior, such as research into robotics (McFarland & Bösser 1993; Nolfi & Floreano 2000; Webb 2001). Designing functional robots based on neural networks forces the engineer to think about design issues similar to the ones evolution has tackled in animals. This is often revealing for biologists because it highlights problems of design that otherwise may be ignored or considered trivial. For instance, biologists and psychologists often step over the problem of recognizing stimuli based on retinal images. This is a far from trivial task that people working with robots are forced to consider. Readers who are interested in a more complete history of neural network models are referred to Arbib (2003) and Haykin (1999).

1.4.3 Neural network models in animal behavior research

It may seem obvious that students of animal behavior should have embraced neural network models, but this has not been the case. The situation is rather the opposite. With some exceptions, neural network models have been ignored among ethologists, behavioral ecologists and animal psychologists. For instance, they are absent from most textbooks on animal behavior. The most important exception, of course, is neuroethologists (Cliff 2003; Simmons & Young 1999), but their focus is not primarily on general theories of behavior in the sense discussed in this chapter. Neural networks are now increasingly popular in neuroscience (e.g., Dayan & Abbott 2001), where recent and important advances on our understanding of learning rely extensively on neural network models (Chapter 4).

Of course, students of animal behavior have thought regularly about the neural machinery underlying behavior. For example, both Ivan Pavlov and Edward Thorndike speculated on the nervous structures behind their behavioral observations. The concept of connectionism, now often used as a label for neural network research, can also be traced back to these pioneers (Kandel et al. 2000; Mowrer & Klein 2001). Less often recognized are a number of network-like models by ethologists and animal psychologists (Baerends 1971; Blough 1975; Fentress 1976; Hinde 1970; Horn 1967; Sutherland 1959; Thompson 1965). Figure 1.11 shows three such models. Baerend's (1971) model at top left is similar to Rosenblatt's perceptron (Figure 1.10) and continues the tradition of ethological summation models (Section 1.3.3). The response arises from the weighted sum of signals from "evaluation units" (E), which in turn analyze signals from receptors (R). The model by Blough (1975; discussed in Chapter 3) has a similar structure and also rediscovers Widrow and Hoff's learning rule first published in 1960. Figure 1.11 also shows Thompson's model from 1965. Its aim was to illustrate how generalization (similar responses to different stimuli) can arise from the fact that similar stimuli are processed by partially overlapping sets of neurons. This is a classical idea, present in Pavlov (1927), as well as in current neural network models (Chapter 3). The last model in the figure is Sutherland's intriguing model from as early as 1959. It depicts a multilayer network with adjustable connections around the same time as Rosenblatt published his work on perceptrons (Sutherland 1959, 1964; Sutherland & Mackintosh 1971). It is interesting that these early efforts by biologist and psychologists developed with little or no contact with that part of the artificial intelligence community that in the same years actively pursued neural networks.

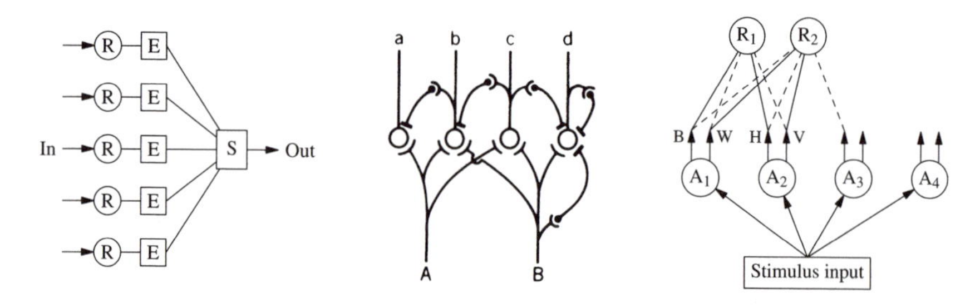

Figure 1.11 Early network-like models. Left: Barends's (1971) model (redrawn); R is a receptor in the receptive field of unit E, which evaluates the input received; S sums signals from the E units. Middle: Thompson's (1965) suggestion of neural circuitry underlying generalization, whereby physically distinct stimuli (A and B) may activate overlapping sets of neurons, thus eliciting similar behavior (smaller circles are inhibitory neurons). Reprinted by permission from Mostofsky, ed., *Stimulus Generalization*. © 1965 by the Board of Trustees of the Leland Stanford Jr. University. Right: Sutherland's model of learning (Sutherland 1959, 1964; Sutherland & Mackintosh 1971). A_i is an analyzer that examines stimuli along a stimulus dimension. R is a response unit that can be attached to different analyzers. Solid lines are existing response attachments, dashed lines represent further possible attachments. In this figure, A_1 analyzes brightness and discriminates between black (B) and white (W), whereas A_2 discriminates between horizontal (H) and vertical (V) orientation. Learning is assumed to consist of (1) learning to use analyzers relevant for the discrimination and (2) learning to attach the right response to the relevant analyzers.

So why are neural network models often absent from current animal behavior theory? One reason could be that neural network research emerged mainly from the artificial intelligence community, with a strong focus on humans rather than animals in general. To date, much more neural network research is concerned with humans than with other animals. For instance, James McClelland, one the most productive scientist in the field, writes

> Connectionist cognitive modeling is an approach to understanding the mechanisms of *human cognition* through the use of simulated networks of simple, neuronlike processing units. (McClelland 1999; emphasis added)

Without a look at history, the focus on humans does not seem motivated because it easier to understand the neural basis of behavior in organisms less complex than humans. Differences between research traditions may also have hindered the development of neural network models of animal behavior. Most work in animal behavior is empirical, whereas neural network models emerged from a theoretical research community. Among students of animal behavior, the computer simulation techniques and formal mathematics that are so helpful in exploiting neural network models were not widely available. Terminological barriers might also have contributed. For instance, the term *adaptive* in biology is used strictly to indicate a trait that evolved because it increases individual reproduction (fitness). In engineering and cognitive science the expression *adaptive system* has a broader scope and refers to any system that can change to better serve its purpose. Engineers thus may speak of *adaptation* when biologists would speak of *learning*. Another example is neural network theory referring to *self-organization* where ethologists would talk of *learning* or *development* and experimental psychologists of *perceptual learning*. Important terms such as *reinforcement* and *association* are also used in slightly different ways.

In conclusion, traditional ethological and animal psychological thinking contains a number of seeds to neural network models, but the development of neural network theory has occurred mainly outside these disciplines. Neural network models still have not gained broad popularity in animal behavior research, but there is today a growing interest in neural networks, particularly among animal psychologists.

1.4.4 The diversity of network models: Our approach

To apply neural networks to animal behavior, we need to decide which particular models to use. This is not always easy. Neural network models come in many forms and purposes. There are biologically or psychologically motivated models with aims ranging from the very details of neural mechanisms to the highest mental functions in humans. There are models aimed primarily at machine intelligence and engineering applications with little concern for biological realism. The latter nevertheless can be of interest to modeling animal behavior for practical and theoretical reasons. In this book we try to use simple networks as long as they can account for the behavior we investigate. There are a number of reasons for this approach. First, complexity should not be introduced prematurely and without reason. We could try to include from the start all known details of nervous systems, but this would result in a model as difficult to understand as the real system. The only way to understand a complex structure is to start by abstracting pieces from it or study simplified models that are more readily analyzed. The crucial test of a simplified model is the extent to which it captures essential features of the behavior of real animals. If it does, then the fact that the model is neurally inspired rather than based on some other metaphor should be considered a bonus rather than a hindrance.

Second, we need models of animal behavior that are easy to understand and practical to use. We also need for general models as opposed to specific to a particular behavior and a particular species. The subject of animal behavior covers species with a nervous system made up of a few cells to the complexity of primate brains, and we know that many behavioral phenomena are surprisingly general.

Third, the differences between network models should not be exaggerated. Starting from a simple model such as Rosenblatt's perceptron, we can add complexity gradually and in several ways. For instance, we can add more layers of nodes, recurrent connections and more complex node dynamics. We thus can obtain complex models gradually from simple ones and study how their properties change in the process. This is also relevant when studying the evolution from simple to complex behavior (Chapter 5).

In practice, our approach means that multilayer feedforward networks operating in discrete time are our first choice of model. Indeed, most applications of neural networks are of this kind. If such models prove insufficient to analyze a given behavioral finding, we consider additional elements such as recurrent connections or dynamics in continuous time. Note that we do not deny the importance of detailed models. If neural network models are to fulfill their promises, the future will contain simple, intuitive, general and practical models, as well as detailed models closer to real nervous systems.

CHAPTER SUMMARY

- "Behavior" is a legitimate level of analysis. It has theoretical importance in ethology, psychology, behavioral ecology and evolutionary biology and practical importance to many people working and living with animals in zoos, veterinary clinics, farms, etc.
- Explaining behavior includes understanding how motivational factors control responding, how behavior develops during an individual's lifetime and how evolution shapes behavior. Understanding of actual internal processes may also be important.
- Most attempts to understand animal behavior have inferred models from behavioral observations and functional considerations. This is true for most of ethology, experimental psychology and cognitive psychology. A major exception is neuroethology, which studies how nervous systems generate behavior.
- Neural network models offer a potential to develop models of behavior that are informed not only by behavioral observations but also by the actual structure of nervous systems. Can this result in increased knowledge about behavior? Our aim is to explore this question.
- We have also touched on some fundamental issues about how to model behavior, from neurophysiology to "cognition." On important issue is how to deal with animals' internal states. We will return to these issues throughout this book and in the concluding chapter.

FURTHER READING

Ethological and biological theory of behavior:

Eibl-Eibesfeldt I, 1975. *Ethology: The Biology of Behavior.* New York: Holt, Rinehart & Winston.

Hinde RA, 1970. *Animal Behaviour.* Tokyo: McGraw-Hill Kogakusha, 2 edition.

Krebs JR, Davies NB, 1987. *An Introduction to Behavioural Ecology.* London: Blackwell, 2 edition.

McFarland DJ, 1999. *Animal Behaviour: Psychobiology, Ethology and Evolution.* Harlow, England: Longman, 3 edition.

Neuroethology:

Ewert JP, 1985. Concepts in vertebrate neuroethology. *Animal Behaviour* 33, 1–29.

Simmons PJ, Young D, 1999. *Nerve Cells and Animal Behaviour.* Cambridge, England: Cambridge University Press, 2 edition.

Animal learning theory, animal cognition, human cognition:

Klein SB, 2002. *Learning: Principles and Applications.* New York: McGraw-Hill, 4 edition.

Mackintosh NJ, editor, 1994. *Animal Learning and Cognition.* New York: Academic Press.

Pearce JM, 1997. *Animal Learning and Cognition.* Hove, East Sussex: Psychology Press, 2 edition.

Shanks DS, 1995. *The Psychology of Associative Learning.* Cambridge, England: Cambridge University Press.

For the discussion between behaviorism and cognitive psychology:

Leahey TH, 2004. *A History of Psychology.* Englewood Cliffs, NJ: Prentice-Hall, 6 edition.

Staddon JER, 2001. *The New Behaviorism: Mind, Mechanism and Society.* Hove, East Sussex: Psychology Press.

Neural network models:

Arbib MA, 2003. *The Handbook of Brain Theory and Neural Networks.* Cambridge, MA: MIT Press, 2 edition.

Churchland PM, 1995. *The Engine of Reason, the Seat of the Soul: A Philosophical Journey into the Brain.* Cambridge, MA: MIT Press.

Churchland PS, Sejnowski T, 1992. *The Computational Brain.* Cambridge, MA: MIT Press.

Dayan P, Abbott LF, 2001. *Theoretical Neuroscience: Computational and Mathematical Modeling of Neural Systems.* Cambridge, MA: MIT Press.

Haykin S, 1999. *Neural Networks: A Comprehensive Foundation.* New York: Macmillan, 2 edition.

Chapter Two

Fundamentals of Neural Network Models

This chapter contains a technical presentation of neural network models and of how they are rooted in neurophysiology. The aim is to provide the reader with a basic understanding of how neural network models work and how to build them. We begin with basic concepts of neurophysiology, which are then summarized in a simple mathematical model of a network node. We continue showing different ways to connect nodes into networks and discuss what different networks can do. Then we consider how to set network connections so that the network behaves as desired. We conclude with a section on computer simulation of neural networks.

2.1 NETWORK NODES

In Chapter 1 we saw that neural network models consist of interconnected nodes that are inspired by biological neurons. Nodes in a neural network model are typically much simpler than neurons, including only features that appear important for the collective operation of the network. A minimal set of such features may be summarized as follows:

- Nodes can be in different states, including different levels of activity and possibly different values of internal variables.
- Nodes are connected to each other, and through these connections, they influence each other.
- The influence of a node on others depends on the node's own activity and on properties of the connection (excitatory or inhibitory, weak or strong, etc.).

Below we provide a basic sketch of how neurons operate (referring to Dayan & Abbott 2001; Kandel et al. 2000; and Simmons & Young 1999 for further details) and then we introduce the simple node model that we will use throughout this book.

2.1.1 Neurons and synapses

Neurons can be of many different types and can interact in different ways (Figure 2.1; Gibson & Connors 2003; Kandel et al. 2000; Simmons & Young 1999; Toledo-Rodriguez et al. 2003). Neurons have output pathways by which they influence other neurons and input pathways through which they are influenced. We begin by reviewing the operation of spiking neurons and their interactions through chemical synapses. Then we consider nonspiking neurons and interactions through electrical synapses and neuromodulators.

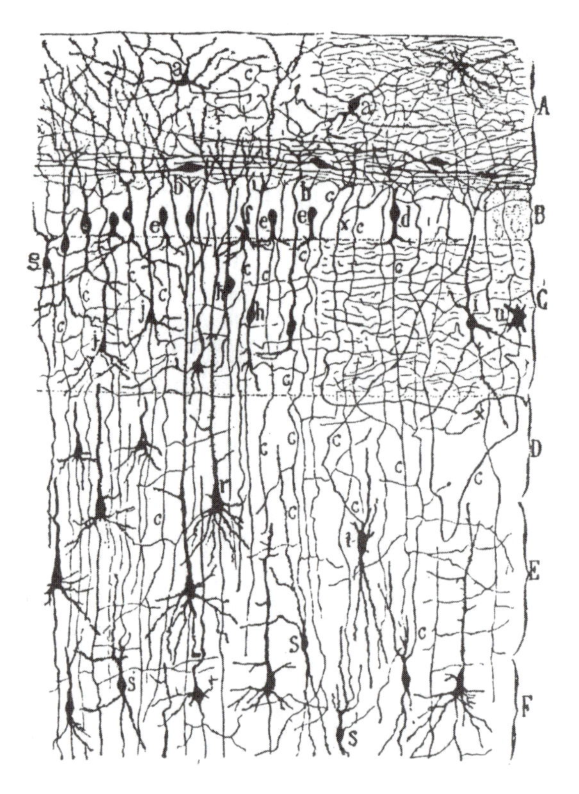

Figure 2.1 Part of the neural network of the inferior occipital cortex of a rabbit of eight days. Uppercase letters label anatomically distinguishable cortical layers, lowercase letters label different neuron types. Drawing by Ramón y Cajal, reprinted from De Felipe & Jones (1989).

The distinguishing feature of a *spiking neuron* is that it can generate electrical pulses called *spikes* or *action potentials*. These can travel along the neuron's axon, an elongated branching structure that grows out of the cell body. Axons can reach out for distances between a few millimeters and up to tens of centimeters, thus enabling both short- and long-distance interaction between neurons. The numbers of these interactions are often staggering. For instance, each cubic millimeter of mouse cortex contains about 10^5 neurons and an estimated 1 to 3.5 km of axons, whereby each neuron connects to about 8000 others (Braitenberg & Schüz 1998; Schütz 2003).

Neurons connect to each other by structures called *synapses*. At a synapse, the axon of one neuron (called *presynpatic*) meets a *dendrite* of another neuron (called *postsynaptic*). Dendrites form a structure similar to the axon but usually shorter and with more branches. While axons are output pathways, dendrites are input pathways. When a spike traveling along the axon of the presynaptic neuron reaches a chemical synapse, it causes the release of chemicals (*neurotransmitters*) that can excite or inhibit the postsynaptic neuron, i.e., favor or oppose the generation of a spike by altering the electrical potential of the neuron. The magnitude of this effect depends on the amount of neurotransmitter released by the presynaptic neuron, as well as on the sensitivity of the postsynaptic neuron to the neurotransmitter. Altogether, these factors determine the "strength" of a synapse, i.e., the extent to which the presynaptic neuron is capable of influencing the postsynaptic one. Fundamental work has related basic learning phenomena such as habituation and classical conditioning to changes in the number and efficacy of synapses rather than to changes in patterns of connections between neurons (Chapter 4, Kandel et al. 2000).

The excitatory or inhibitory effect of the neurotransmitters released by the presynaptic neuron is transient. A simple picture of spiking neurons is that they keep track of excitatory and inhibitory inputs that arrive within a short time window, extending up to a few tens of milliseconds in the past. Whenever the balance of excitatory and inhibitory input within this time window overcomes a threshold, a spike is generated. Thus a neuron generates more spikes per unit time the larger the difference between excitatory and inhibitory input. Neuron activity is often expressed in terms of spike frequency, measured in Hertz (1 Hz = 1 spike per second). Physiology limits the spike frequency of most neurons to a maximum of a few hundred Hertz, although rates close to 1000 Hz can be observed in some neurons.

It is often of great importance to know how neuron activity changes when input varies. This holds also for receptor cells that convert light, chemical concentrations, sound waves, etc. into signals that are passed on to other neurons. In general, neurons respond nonlinearly to change in input; i.e., increasing the input in fixed steps does not increase output by the same amount for every step (Figure 2.2; Simmons & Young 1999; Torre et al. 1995; and Simmons & Young 1999 for reviews).

As mentioned earlier, *nonspiking neurons* also exist. Their influence on other neurons depends continuously on the neuron's electrical potential, e.g., through continuous changes in the rate of neurotransmitter release (Burrows 1979; Simmons & Young 1999). Likewise, chemical synapses are not the only way neurons interact. For instance, neurons may be joined by electrical synapses, or *gap junctions*. These allow neurons to exchange electric currents, thereby altering each other's electrical potential without the occurrence of spikes. Electrical synapses are often found in invertebrate nervous systems or in sense organs in vertebrates (Simmons & Young 1999). Recently, they have been discovered in the cortex of mammals (Gibson & Connors 2003). Neurons also influence each other by releasing relatively long-lived substances, called *neuromodulators*, into the extracellular fluid. These can change the properties of neurons over large regions of the nervous system, sometimes for hours. For instance, some neurons may become more excitable and others less so (Dickinson 2003; Harris-Warrick & Marder 1991; Hasselmo et al. 2003).

The activity of neurons depends on internal factors as well as external input. For instance, neurons respond to changes in input with a characteristic delay (Figure 2.3, left). Furthermore, the response to a constant input is often not constant at all (Wang & Rinzel 2003). A neuron may *habituate*, i.e., decrease its output to constant input (Figure 2.3, right), or it may produce oscillating activity. Neurons may also be spontaneously active in the absence of any input, as will be considered in Section 3.8 relative to the generation of motor patterns.

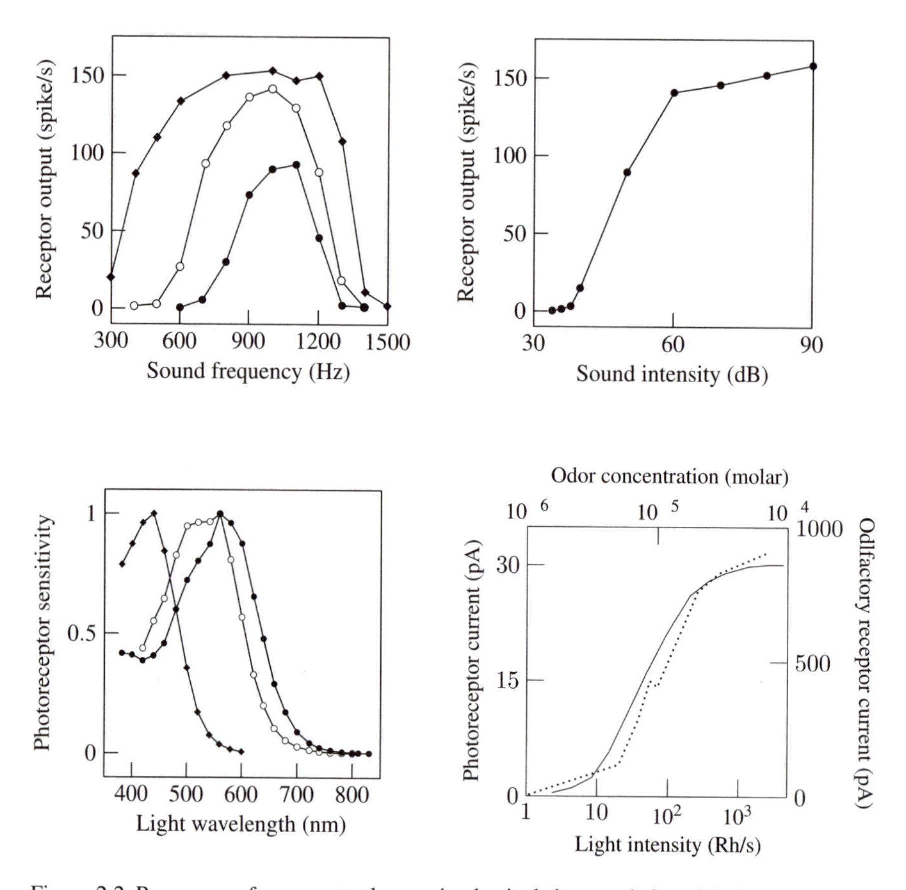

Figure 2.2 Responses of neurons to changes in physical characteristics of the input. Top left: Response of a ganglion cell in the ear of a squirrel monkey to sounds varying in frequency for three different intensities: 50 dB (closed circles), 60 dB (open circles) and 80 dB (diamonds). Top right: Response of the same cell to sounds of varying intensity and frequency of 1100 Hz (data from Rose et al. 1971). Bottom left: Relative sensitivity to lights of different wavelengths of the three kinds of cones in the retina of *Macaca fascicularis* (data from Baylor et al. 1987). Bottom right: Output of a newt (*Triturus cristatus*) photoreceptor to light varying in intensity (continuous line, left and bottom scales; data from Forti et al. 1989), and output of an olfactory receptor to different odorant concentrations (dotted line, scales at top and right; data from Menini et al. 1995). Note the similar shape of the two curves and of the curve in the panel above.

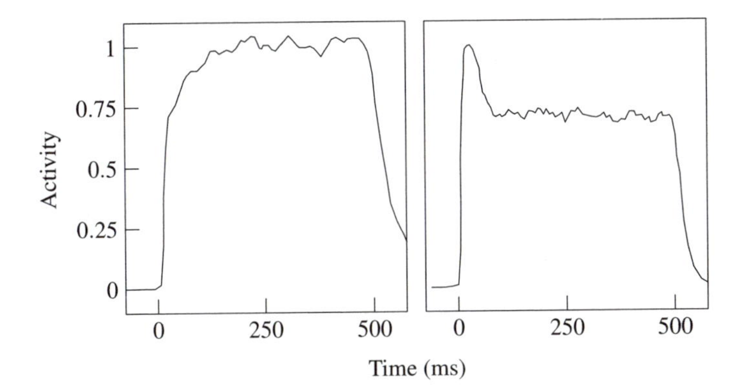

Figure 2.3 Responses of dark-adapted photoreceptors to a flash of light of 500 ms. Left: A cranefly (*Tipula*) photoreceptor approaches a steady-state activity rather slowly (activity is expressed as proportion of peak activity). Right: A fleshfly (*Sarcophaga*) photoreceptor reacts with a fast activity peak followed by lower steady-state activity (activity is expressed as proportion of maximum activity). Data from Laughlin (1981).

2.1.2 The basic node model

We now define mathematically network nodes and connections. Nodes in a network are indexed by integer numbers; a generic node is indicated by a lowercase letter such as i or j. The activity of node i is a (nonnegative) number written z_i. Its biological counterpart is the short-term average spiking frequency of a neuron or, for nonspiking neurons, the difference in electrical potential between the outside and inside of the neuron (see also Dayan & Abbott 2001; §1.2). The activities of all network nodes can be collected in an *activity vector* $\mathbf{z}$:

$$\mathbf{z} = \begin{pmatrix} z_1 \\ z_2 \\ \vdots \\ z_N \end{pmatrix} \tag{2.1}$$

where N is the number of nodes. Nodes are connected to each other via weights, continuous numbers that correspond to the strength of biological synapses. The weight that connects node j to node i is written W_{ij}. To indicate compactly all weights in the network, we introduce the *weight matrix* $\mathbf{W}$, which can be represented as an array with N rows and N columns:

$$\mathbf{W} = \begin{pmatrix} W_{11} & W_{12} & \cdots & W_{1N} \\ W_{21} & W_{22} & \cdots & W_{2N} \\ \vdots & \vdots & \ddots & \vdots \\ W_{N1} & W_{N2} & \cdots & W_{NN} \end{pmatrix} \tag{2.2}$$

A first key assumption of neural network models is that node j conveys to node i an input equal to the product $W_{ij}z_j$. This models the biological fact that a presynaptic neuron influences a postsynaptic one in a way that depends jointly on the presynaptic neuron's activity and on the strength of the synapse. The total input received by node i is written y_i and is calculated as the sum of all the inputs received:

$$y_i = \sum_j W_{ij} z_j \tag{2.3}$$

where the index j runs over all nodes in the network. Absence of a particular connection is expressed simply as $W_{ij} = 0$. Equation (2.3) is written more compactly as a relationship between vectors:

$$\mathbf{y} = \mathbf{W} \cdot \mathbf{z} \tag{2.4}$$

where $\mathbf{y}$ is the vector of all inputs, and $\mathbf{W} \cdot \mathbf{z}$ is the *matrix product* between $\mathbf{W}$ and $\mathbf{z}$. That is, $\mathbf{W} \cdot \mathbf{z}$ is simply a short way of writing the vector whose element i is $\sum_j W_{ij} z_j$ (the application of matrix and vector algebra to neural networks is detailed in Dayan & Abbott 2001 and Haykin 1999).

A second key assumption regards how input received by a node causes node activity. The function f that expresses the relationship between node input and node activity is called the *transfer function*:

$$z_i = f(y_i) \tag{2.5}$$

From a biological perspective, the transfer function embodies hypotheses about how neurons respond to input. In practice, however, the transfer function should also be easy to work with. The simplest transfer function is the *identity function*, by which node output equals node input:

$$f(y) = y \tag{2.6}$$

This function is used in mathematical analyses of neural network models because of its simplicity, but it has disadvantages if we recall that node activity should model the spiking rates or electrical potentials of neurons. Both these quantities are constrained within a range, and moreover, spiking rates cannot be negative. This results in curvilinear relationships between input magnitude and neuron activity (Figure 2.2) that are not captured by equation (2.6). It is also important that a network using only the identity transfer function can realize only a restricted set of input-output mappings (see Sections 2.2.1.1 and 2.3.1.1).

A simple solution is to use *logistic functions*. Some examples are shown in Figure 2.4. The output of a logistic function is always positive and less than 1; the latter can be interpreted as corresponding to the maximum activity of a neuron. Logistic functions have the form

$$f(y) = \frac{1}{1 + e^{-a(y-b)}} \tag{2.7}$$

where the parameters a and b may be set to approximate empirical transfer functions. By tuning them, it is possible to regulate node activity to null input and how steeply activity rises as input increases (Figure 2.4, see also Figure 2.2). It is clear from Figure 2.2, however, that not even logistic functions satisfactorily describe all kinds of neurons and that especially to model receptors, different functions are needed. This issue will be taken up in Chapter 3.

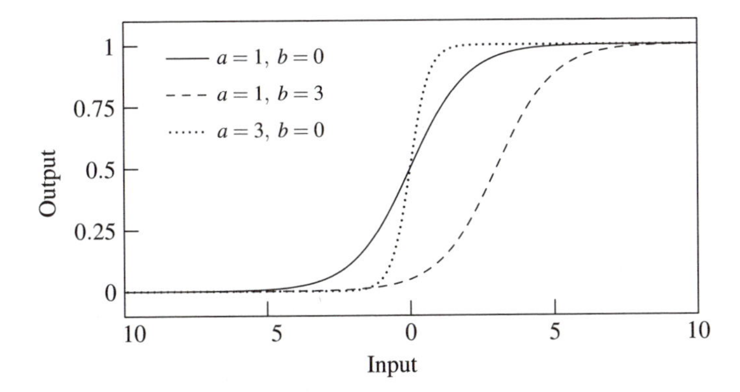

Figure 2.4 The logistic function, equation (2.7), for several choices of parameters. Note the similarity with the curves in the right part of Figure 2.2.

The node model introduced so far is summarized in Figure 2.5 and formally by

$$z_i = f\Big(\sum_j W_{ij} z_j\Big) \tag{2.8}$$

Accordingly, the operation of a whole network can be expressed symbolically as $\mathbf{z} = f(\mathbf{W} \cdot \mathbf{z})$. Obviously, this model is much simpler than neurons and synapses. Yet it has proven very useful to build models of behavior and nervous systems. In neural network studies, both simpler and more complex nodes are used, depending on the precise aim of the study. Simpler nodes sometimes are used to simplify mathematical analysis. An instance is a node having only two states, "active" and "inactive" (Amit 1989; McCulloch & Pitts 1943). More complex nodes have many uses, among which to investigate neuron physiology. For instance, Bush & Sejnowski (1991) reconstructed from photographs the branching structure of a few neurons, and Lytton & Sejnowski (1991) simulated them as complex electrical networks, accurately reproducing the observed spiking patterns. A network with such complex nodes, however, would be very difficult to build and to understand. Moreover, we are not guaranteed to get a better model of behavior simply by including more detail. We still ignore which features of neurons are crucial for nervous system operation and which are less important (Graham & Kado 2003). Thus we will start with the basic node described earlier, and add complexity only when needed to model behavioral phenomena.

2.1.3 More advanced nodes

This section shows how the basic node can be developed to include additional features of neurons. It can be skipped at first reading because most of our network models will use the basic node described earlier. We start by asking how to calculate node activity in response to an input $y(t)$ that varies with time. The simplest possibility is to construe time as a succession of discrete *time steps* and to compute node activity at time $t+1$ from node input at time t using equation (2.8):

$$z(t+1) = f(y(t)) \tag{2.9}$$

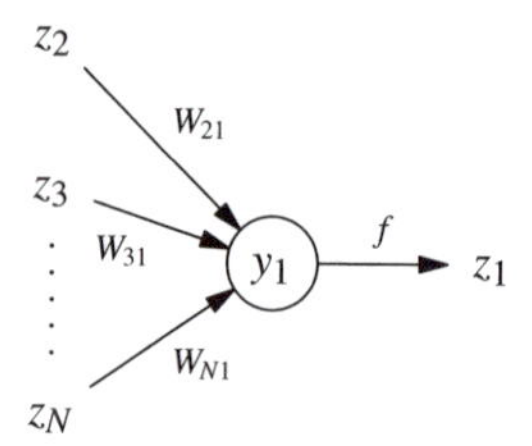

Figure 2.5 The basic node model. Node 1 receives input from nodes $2,\ldots,N$ through the weights $W_{12},\ldots,W_{1N}$. The total input received is $y_1 = \sum_{j=2}^{N} W_{1j}z_j$; it is passed through the transfer function f to yield node activity z_1.

The key feature of this equation is that node activity is determined solely by node input. The activity of neurons, on the other hand, may depend on factors such as the recent history of activity. In other words, neurons cannot instantly follow changes in input. For example, Figure 2.3 shows that a cranefly photoreceptor reacts with some delay to the onset of a light stimulus. This can be mimicked by letting $z(t+1)$ be a weighted average of $z(t)$ and $f(y(t))$:

$$z(t+1) = kf(y(t)) + (1-k)z(t) \tag{2.10}$$

Thus node activity in the next time step depends on current activity, as well as on current input, in a proportion given by the parameter k ($0 < k < 1$). If k is small, $z(t+1)$ tends to be close to $z(t)$; hence the node responds slowly to changes in input. Equation (2.10) is often written in terms of the *change* in activity from one time step to the next, $\Delta z(t) = z(t+1) - z(t)$. If we subtract $z(t)$ from both sides and write $1/k = \tau$, we get

$$\tau\Delta z(t) = f(y(t)) - z(t) \tag{2.11}$$

The parameter τ is called the *time constant* of the node: a smaller τ means that $z(t)$ can change more quickly. Figure 2.6 shows that the cranefly photoreceptor is well described by equation (2.11), with $\tau = 35\,\text{ms}$. Equation (2.11) is also an approximation to simple models of node dynamics in continuous time (the main difference is that, in continuous time, the increment $\Delta z(t)$ is replaced by the time derivative $\mathrm{d}z(t)/\mathrm{d}t$).

How is equation (2.10) related to equation (2.9)? If the input is constant, the value $z(t)$ calculated from equation (2.11) eventually reaches a steady state equal to the one predicted by the latter (which is the same as equation 2.5). The steady state is obtained by setting $\Delta z(t) = 0$ (no change), in which case equation (2.11) becomes identical to equation (2.9). The steady state is approached at a rate given by τ; hence, if input changes on a time scale much longer than τ, the two equations are equivalent. This justifies using equation (2.9) when we are not interested in short-term responses.

We now show how further properties of neurons can be modeled by adding internal states. Our example is sensory habituation (or adaptation), shown in Figure 2.6. We introduce a single state variable h that represents the level of habituation, and assume that the node receives an inhibitory input of $-h$:

$$\tau\Delta z(t) = f\big(y(t) - h(t)\big) - z(t) \tag{2.12}$$

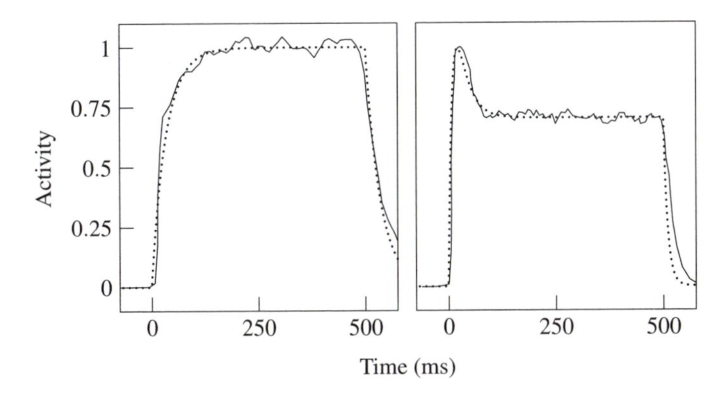

Figure 2.6 Two node models fitted to the data in Figure 2.3. Left: A cranefly (*Tipula*) photoreceptor is well described by equation (2.11), with $\tau = 35\,\text{ms}$. Right: A fleshfly (*Sarcophaga*) photoreceptor is well described by equations (2.12) and (2.13), with $\tau = 10\,\text{ms}$, $\text{T} = 25\,\text{ms}$, $a = 0.5$.

Intuitively, h should increase when the node is strongly active and decrease when node activity is low. This is accomplished by the following equation:

$$\text{T}\Delta h(t) = ay(t) - h(t) \tag{2.13}$$

where T is a second time constant, and a regulates the effect of habituation. The latter can be seen by calculating steady-state node activity arising from a constant input y. Setting $\Delta z(t) = 0$ and $\Delta h(t) = 0$ in the preceding equations, we obtain two equations for the steady-state values z and h:

$$\begin{aligned} z &= f(y - h) \\ h &= ay \end{aligned} \tag{2.14}$$

which imply $z = f((1-a)y)$. Hence the steady state of a habituating node is the same as that of a nonhabituating node whose input is reduced by a factor $1-a$. Figure 2.6 shows that this simple model can accurately describe the habituating photoreceptor of a fleshfly. Benda & Herz (2003) show that more complex kinds of habituation can be described with simple models. Models with a few internal variables have been developed also for other characteristics of neurons, such as spontaneous activity (Wang & Rinzel 2003).

2.2 NETWORK ARCHITECTURES

The pattern of connections among nodes defines the *network architecture* that is responsible for much of a network's properties. In the following we present a number of architectures relevant to behavioral modeling. We will first consider *feedforward networks*, so named because node activity flows unidirectionally from input nodes to output nodes, without any feedback connections. *Recurrent networks*, i.e., networks with feedback connections, will be considered later.

In describing neural networks, we retain notations used for behavior maps in general (Chapter 1). Input and output spaces can be defined by choosing some nodes as *input nodes* and others as *output nodes*. For instance, the former may model receptors and the latter motoneurons (connected to muscles) or neurons that secrete hormones or other chemicals (Chapter 3). Nodes that are neither input nodes nor output nodes traditionally are called *hidden nodes*. We write network input vectors as $\mathbf{x}_1$, $\mathbf{x}_2$, etc. Greek indices refer to different inputs and Latin indices to particular nodes. Thus $x_{\alpha i}$ is the activity of input node i elicited by the input $\mathbf{x}_\alpha$. Likewise, $r_{\alpha j}$ is the activity of output node j in response to input $\mathbf{x}_\alpha$. The letters y and z will be still used, as above, for the input and output of a generic node, respectively.

2.2.1 Feedforward networks

Figure 1.9 on page 19 shows some feedforward networks. Stimuli external to the network are assumed to activate the network's input nodes. This activation pattern is processed in a number of steps corresponding to node "layers." The activity in each layer is propagated to the next layer via a matrix of connections. Any number of layers may be present, although two- and three-layer networks are most common. Feedforward processing is an important organizational principle in nervous systems, e.g., in the way information from receptors reaches the brain. Although purely feedforward processing is rare, we shall see that these networks are helpful models of behavior. Moreover, feedforward networks are an important component of more complex architectures.

2.2.1.1 The perceptron

The first network architecture we consider is the so-called perceptron (Minsky & Papert 1969; Rosenblatt 1958, 1962). It consists of a single node receiving a number of inputs via a weight vector $\mathbf{W}$, as shown in Figure 1.9 on page 19. Owing to its simplicity, the perceptron is fully specified by a single equation:

$$r = f\Big(\sum_j W_j x_j\Big) = f(\mathbf{W}\cdot\mathbf{x}) \tag{2.15}$$

(see equation 2.8, and note that in this simple case we need only one index to label weights). The perceptron can be interpreted in several ways. It can model a very simple nervous system, where a number of receptors are connected directly to an output neuron, or it can be a building block for larger systems. In his seminal work, Rosenblatt considered the x_j's as the outputs of simple *feature detectors*, e.g., reporting about particular aspects of a visual scene (see the "analyzers" in Figure 1.11). The perceptron had to combine information about the presence or absence of individual features to reach yes-no decisions about the whole input pattern, e.g., about whether it belonged to some category. To this end, Rosenblatt used a transfer function assuming only the values 0 and 1 as follows:

$$f(y) = \begin{cases} 1 & \text{if } y > \theta \\ 0 & \text{otherwise} \end{cases} \tag{2.16}$$

Table 2.1 A behavior map, interpreted as a simple model of feeding, that cannot be realized by a perceptron with two input nodes and one output node. The first two columns, each split in two, provide the mathematical description and interpretation of inputs and outputs. The third column shows how to choose perceptron parameters to realize single input-output mappings. However, there is no set of parameters that can realize all four mappings.

Input		**Desired output**		**Desired output obtained if:**
Values	**Interpretation**	**Value**	**Interpretation**	
(0,0)	All black	0	Don't eat	$\theta > 0$
(0,1)	Black and white	1	Eat	$W_2 > \theta$
(1,0)	White and black	1	Eat	$W_1 > \theta$
(1,1)	All white	0	Don't eat	$W_1 + W_2 < \theta$

where θ is a number serving as a *threshold*: if the total input $y = \mathbf{W} \cdot \mathbf{x}$ is above threshold, the output will be 1; otherwise, it will be 0. Rosenblatt also devised an algorithm to adjust the weights to correctly classify a set of input patterns into two categories. For instance, we could use this algorithm to solve the feeding control problem discussed on page 20.

Perceptrons, however, cannot realize all behavior maps. Consider a perceptron with two input nodes that model photoreceptors and an output node whose activity is interpreted as eating a prey. Each receptor reacts with 0 when no or little light is captured and with 1 when a lot of light is captured, e.g., when light reflected from a white object reaches the receptor. The model organism meets two kinds of potential prey. All-white prey is unpalatable and should be ignored. We assume that the presence of white prey reflects enough light to the retina to always activate both receptors. The second kind of prey instead should be eaten. Its presence activates only one receptor (e.g., because it is smaller), resulting in two possible input patterns: (0,1) and (1,0). The required behavior map is shown in Table 2.1 and is known as the *exclusive-or problem* (XOR). The last column in the table shows how the three perceptron parameters, W_1, W_2 and θ, must be chosen to realize each input-output mapping. For instance, the perceptron reacts to $(0,1)$ if $W_1 \times 0 + W_2 \times 1 = W_2$ is larger than θ (second table row). The problem is that the four conditions cannot all be satisfied by the same set of parameters: the sum $W_1 + W_2$ cannot be smaller than the positive number θ if each of W_1 and W_2 alone is bigger! The limits of perceptrons will be further discussed below (see also Minsky & Papert 1969).

2.2.1.2 *From perceptrons to behavior*

A few modifications make perceptrons more suitable for behavioral modeling, although their computational limitations are not removed. Rosenblatt's threshold mechanism (equation 2.16) implies that the same input always yields the same output. In contrast, an animal's reaction to a given stimulus typically is variable

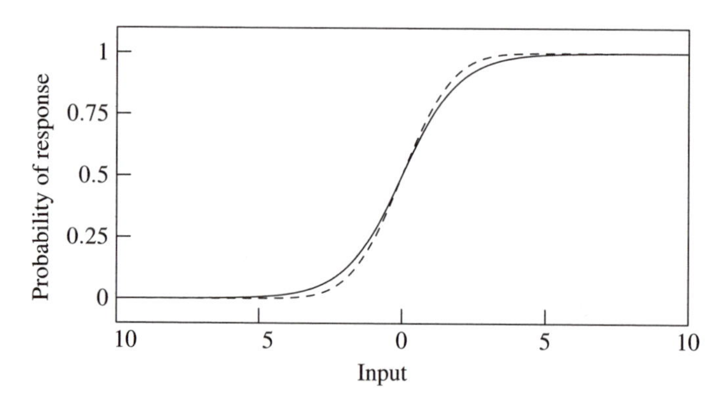

Figure 2.7 The probability that a node responds to an input for two different choices of the transfer function. Solid line: Equation (2.7), with $a = 1$ and $b = 0$. Dashed line: Equation (2.18), with $\sigma = 0.5$ and ε drawn from a standard normal distribution.

because the stimulus is not the only factor that influences behavior. We can remedy this shortcoming by following two slightly different routes. By the first route, we can use a transfer function f that rises smoothly from 0 to 1 (Figure 2.4) and interpret node activity as the probability that the animal responds to the stimulus:

$$\Pr(\text{reaction to } \mathbf{x}) = f(\mathbf{W} \cdot \mathbf{x}) \tag{2.17}$$

By the second route, the threshold mechanism is retained, but it is applied after adding to y a random number ε, e.g., drawn from a standard normal distribution. Formally, we replace equation (2.16) with

$$r = \begin{cases} 1 & \text{if } y + \sigma\varepsilon > \theta \\ 0 & \text{otherwise} \end{cases} \tag{2.18}$$

where σ is a positive number that regulates the importance of randomness. Output to $\mathbf{x}$ is now variable because ε is drawn anew each time $\mathbf{x}$ is presented. This, however, does not mean that behavior is wholly random because, according to equation (2.18), the probability of responding to $\mathbf{x}$ increases steadily as the input $y = \mathbf{W} \cdot \mathbf{x}$ increases. As shown in Figure 2.7, there can be very little difference in average behavior using either equation (2.17) or equation (2.18). In the former case, perceptron output is interpreted directly as the probability of responding. This is handy in theoretical analyses and to compare a model with data about average behavior. The latter transfer function can be used to generate individual responses to each input presentation, as needed in simulations of behavior (Section 2.3.4).

A very simple model of reaction to stimuli is obtained by considering equation (2.17) and equation (2.6) together:

$$\Pr(\text{reaction to } \mathbf{x}) = \mathbf{W} \cdot \mathbf{x} \tag{2.19}$$

with the further assumption that the probability is 0 if $\mathbf{W} \cdot \mathbf{x} < 0$ and 1 if $\mathbf{W} \cdot \mathbf{x} > 1$. In this way we can interpret the result of a straightforward computation as the probability that an animal reacts to a stimulus. We will see below that this simple model is very helpful to understand some important working principles of neural networks. In Chapter 3 we will apply equation (2.19) to many behavioral phenomena.

2.2.1.3 *General feedforward networks*

It is possible to build feedforward networks more powerful than perceptrons. For instance, we can add output nodes (Figure 1.9 on page 19). In studies of motor control, the output nodes may model motoneurons that control individual muscles. In this case, the sequence of output patterns generated by the network would correspond to sequences of muscle contractions (Section 3.8). When modeling behavior, it is often more convenient to interpret the activity of each output node as the tendency to perform a particular behavior pattern. One way to decide which behavior is actually performed is to assume that the most active node always takes control. Alternatively, we may interpret the activity of a node as the (relative) probability that the corresponding behavior be performed. We will see in Section 3.7 that both rules can emerge from reciprocal inhibition between the output nodes (possibly in the presence of random factors).

Feedforward networks can also be expanded with additional layers (Figure 1.9 on page 19). If $\mathbf{z}_i$ is the ith layer of nodes, and $\mathbf{W}^{(i)}$ is the weight matrix from layer i to layer $i+1$, the operation of a general feedforward network may be summarized by $\mathbf{z}_{i+1} = f(\mathbf{W}^{(i)} \cdot \mathbf{z}_i)$. Such networks, often called *multilayer perceptrons*, can solve a wider range of problems than networks with only an input and an output layer. In fact, multilayer perceptrons are able to realize arbitrary input-output maps given enough hidden units and provided the node transfer function is nonlinear (Haykin 1999). For instance, the behavior map in Table 2.1, which is impossible for perceptrons, can easily be realized by simple multilayer networks such as those shown in Figure 2.8. If the transfer function of hidden nodes is linear, it can be shown that a multilayer network is equivalent to one without hidden layers. The reason is that a single weight matrix is enough to perform any linear transformation between the input and output spaces. We will see in Section 3.4 that multilayer linear networks are nevertheless interesting to study. In Chapter 3 we will also use some nonlinear multilayer networks to model behavior that is too complex for linear networks.

2.2.2 Recurrent networks

Feedforward networks can realize any map with fixed input-output relationships but not maps in which the output to a given input depends on previous inputs. For instance, feedforward networks cannot distinguish the input sequences $\{\mathbf{x}_1, \mathbf{x}_2\}$ and $\{\mathbf{x}_2, \mathbf{x}_2\}$. The output to $\mathbf{x}_2$ at the second time step will be the same in both cases because the output of a feedforward network is determined solely by current input. Hence the time-processing abilities of feedforward networks are very limited. While some improvement is possible (Section 2.3.5), recurrent networks are a superior alternative to analyze input sequences and generate output sequences. The key architectural element in recurrent networks is *feedback loops*, or *recurrent connections*. This means that the activity of a node at a given time can be influenced by the activity of the same node at previous times. Usually this influence is mediated by the activity of other nodes at intervening times. In other words, while in feedforward networks only the weights serve as memory of the past, in recurrent networks

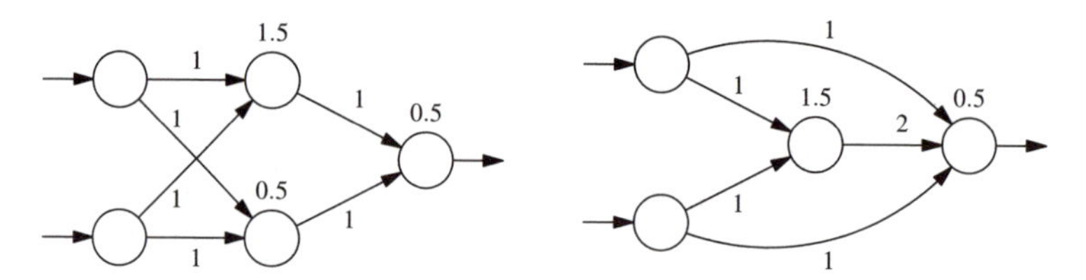

Figure 2.8 Two networks that realize the input-output map in Table 2.1 (exclusive-or problem, or XOR). Node transfer function is equation (2.16); numbers above each node are threshold values. The reader can check that among the input patterns (0,0), (1,1), (0,1) and (1,0), only the latter two activate the output node.

the activity of nodes is also a source of memory. Recurrency is a common feature of nervous systems and underlies their more sophisticated abilities. For instance, feedback is very common in the cerebral cortex, both within and between different cortical areas.

The behavior of recurrent networks can be very diverse (Amit 1989; Dayan & Abbott 2001; Hertz et al. 1991). In response to a constant input, a recurrent network may settle into a persistent pattern of activity (possibly a different one for different inputs), it may cycle repeatedly through a sequence of states, or it may exhibit irregular behavior. When the input varies in time, the network generally will not settle into one pattern of activity but will generate a sequence of output patterns, in principle of arbitrary complexity. Thus recurrent networks appear a promising tool to model time structures in animal behavior but also a tool that is not trivial to handle. In Section 2.3.5 we will show a few recurrent architectures and how to build a recurrent network with desired properties. In the following chapters we will apply recurrent networks to behavioral modeling.

We conclude with a technical note that can be skipped at first reading but should be considered when using recurrent networks. To work with recurrent networks, we need a way to calculate node activity as time proceeds, i.e., a dynamics for the network. The simplest possibility is to consider discrete time steps, and to use equation (2.8) to calculate the activity of all nodes at time $t+1$ based on activities at time t:

$$z_i(t+1) = f\Big(\sum_j W_{ij} z_j(t)\Big) \tag{2.20}$$

This simple dynamics is adequate in some cases but in general has drawbacks. Its main feature is that all nodes are updated simultaneously, as if a "global clock" synchronized the network (*synchronous dynamics*). By contrast, neurons in a nervous system can change their activity at any time, independently of what other neurons are doing. A change in one neuron then affects other neurons in the form of changed input (with some delay owing to spike travel times and synaptic transmission times). Given the same weights, synchronous dynamics can produce different results than more realistic continuous-time dynamics. Moreover, small changes to the update procedure or small amounts of noise can alter network behavior (Amit 1989). An obvious way to overcome these shortcomings is to use continuous time, but in many cases it is enough to modify equation (2.20) slightly. For instance, we can use equation (2.11), with $y(t) = \sum_j W_{ij} z_j(t)$ (see Section 2.1.3 for details). The latter is in fact a discrete-time approximation of a simple continuous time model (since digital computers operate in time steps, such an approximation is always necessary in computer simulations). Another possibility is as follows. At each time step we choose a random ordering of the nodes. We then update the first chosen node. When we update the second node, we use the updated activity of the first node, and so on for all nodes. Since the random order is different at each time step, the procedure mimics the uncoordinated dynamics in real nervous systems and thus is called *asynchronous dynamics* (Amit 1989). Any result obtained in this way obviously is resistant to changes in the order of update operations.

2.3 ACHIEVING SPECIFIC INPUT-OUTPUT MAPPINGS

The input-output map realized by a network depends, of course, on the values of the weights. In practice, much of the work with neural networks is concerned with setting weights. Up to now we have considered problems in which suitable weights could be found either by intuition or by using simple mathematics. In this section we describe some powerful weight-setting methods that can deal with more complex problems. In the next section we describe procedures that organize network weights in response to incoming input rather than seeking to achieve a particular input-output map.

Weight-setting methods can be viewed either as models of real-world phenomena, i.e., how synapses change in nervous systems, or simply as a tool to build networks that handle a given task, without any additional claim. Here the focus is on the latter view, whereas Chapter 4 covers biological learning. Our goal is to build a network that produces a number μ of assigned input-output mappings:

$$m(\mathbf{x}_\alpha, \mathbf{W}) = \mathbf{d}_\alpha \qquad \forall \alpha = 1, \ldots, \mu \tag{2.21}$$

where m is the input-output map realized by the network, $\mathbf{W}$ denotes all network weights, and $\mathbf{d}_\alpha$ is the desired output pattern to input $\mathbf{x}_\alpha$. Equation (2.21) is a system of μ equations where the unknowns are network weights. The direct analysis of these equations is typically very challenging or plainly unfeasible. The only case that can be treated thoroughly with simple mathematics is linear networks, i.e., networks in which all nodes have a linear transfer function, such as equation (2.6). We discuss this case (with a final note on networks with nonlinear output nodes) before presenting more general weight setting techniques.

2.3.1 Setting weights in linear networks

If the transfer function of all nodes is linear, weights satisfying equation (2.21) can be found by formal methods, or it can be proved that they do not exist. We start with a linear perceptron, equation (2.19), and with a single input-output mapping. This corresponds to equation (2.21) with $\mu = 1$ and $m(\mathbf{x}_1, \mathbf{W}) = \mathbf{W} \cdot \mathbf{x}_1$. In this case it is easy to check that the single requirement $\mathbf{W} \cdot \mathbf{x}_1 = d_1$ is satisfied by a weight vector of the form

$$\mathbf{W} = c_1 \mathbf{x}_1 \tag{2.22}$$

with $c_1 = d_1/(\mathbf{x}_1 \cdot \mathbf{x}_1)$. Moreover, if we add to $\mathbf{W}$ a vector $\mathbf{u}$ such that $\mathbf{u} \cdot \mathbf{x}_1 = 0$, the result is not altered. In conclusion, the general solution to the task is of the form

$$\begin{cases} \mathbf{W} = \dfrac{d_1}{\mathbf{x}_1 \cdot \mathbf{x}_1}\mathbf{x}_1 + \mathbf{u} \\ \\ \mathbf{u} \cdot \mathbf{x}_1 = 0 \end{cases} \tag{2.23}$$

By extending this reasoning, it can be shown that the weight vector that realizes a number μ of input-output mappings can be written as

$$\begin{cases} \mathbf{W} = \sum_{\alpha=1}^{\mu} c_\alpha \mathbf{x}_\alpha + \mathbf{u} \\ \\ \mathbf{u} \cdot \mathbf{x}_\alpha = 0 \qquad \forall \alpha = 1, \ldots, \mu \end{cases} \tag{2.24}$$

The general expression for the c_α's, however, is not as simple as in equation (2.23). We show how to solve this problem in the case of two input-output mappings. This simple task will be used throughout this section to illustrate and compare different weight setting techniques. It may be helpful to think of a discrimination task where $\mathbf{x}_1$ has to be mapped to a high output and $\mathbf{x}_2$ to a low one, although the procedure is the same for all desired outputs.

Following equation (2.24), we guess that $\mathbf{W}$ should be of the form $c_1\mathbf{x}_1 + c_2\mathbf{x}_2$, so our task is to find the values of c_1 and c_2 that satisfy the requirements $\mathbf{W} \cdot \mathbf{x}_\alpha = d_\alpha$. If we substitute $c_1\mathbf{x}_1 + c_2\mathbf{x}_2$ for $\mathbf{W}$, the requirements read

$$\begin{aligned} (c_1\mathbf{x}_1 + c_2\mathbf{x}_2) \cdot \mathbf{x}_1 = c_1\mathbf{x}_1 \cdot \mathbf{x}_1 + c_2\mathbf{x}_2 \cdot \mathbf{x}_1 = d_1 \\ (c_1\mathbf{x}_1 + c_2\mathbf{x}_2) \cdot \mathbf{x}_2 = c_1\mathbf{x}_1 \cdot \mathbf{x}_2 + c_2\mathbf{x}_2 \cdot \mathbf{x}_2 = d_2 \end{aligned} \tag{2.25}$$

For brevity, we write $\mathbf{x}_\alpha \cdot \mathbf{x}_\beta = X_{\alpha\beta}$:

$$\begin{aligned} X_{11}c_1 + X_{12}c_2 = d_1 \\ X_{12}c_1 + X_{22}c_2 = d_2 \end{aligned} \tag{2.26}$$

(note that $X_{12} = X_{21}$). The preceding is a system of two linear equations in the unknowns c_1 and c_2. It can be solved by a variety of methods, yielding

$$\begin{aligned} c_1 &= \frac{d_1X_{22} - d_2X_{12}}{X_{11}X_{22} - X_{12}^2} \\ \\ c_2 &= \frac{d_2X_{11} - d_1X_{12}}{X_{11}X_{22} - X_{12}^2} \end{aligned} \tag{2.27}$$

As an exercise, the reader can check that these expressions are really such that $\mathbf{W} \cdot \mathbf{x}_1 = d_1$ and $\mathbf{W} \cdot \mathbf{x}_2 = d_2$. If the network has to satisfy many input-output mappings, the same techniques apply, but the calculations get boring. Software for numeric or formal mathematics will come in handy, whereas textbooks on linear algebra will provide the theoretical foundation.

Note that equation (2.24) expresses the weight vector $\mathbf{W}$ in terms of the vectors $\mathbf{x}_\alpha$. This gives a concrete interpretation to the intuitive statement that the $\mathbf{x}_\alpha$'s have been "stored in memory." Moreover, it allows one to understand the output to a generic input $\mathbf{x}$ in terms of how $\mathbf{x}$ relates to the $\mathbf{x}_\alpha$'s (and to the vector $\mathbf{u}$). This will be discussed in Chapter 3.

Equation (2.24) also holds if the output node is not linear. Such a network can be written as $r = f(\mathbf{W}\cdot\mathbf{x})$. We mention this case because a nonlinear output node is used often to limit network output between 0 and 1 and interpret it as probability of responding. The proof follows from noting that the requirement $d_\alpha = f(\mathbf{W}\cdot\mathbf{x}_\alpha)$ can be written $\bar{d}_\alpha = \mathbf{W}\cdot\mathbf{x}_\alpha$, where $\bar{d}_\alpha = f^{-1}(d_\alpha)$ and f^{-1} is the inverse function of f. Thus we are back to a linear problem, albeit with different d_α's. This means that different choices of f result in different values of the c_α's but do not alter the structure of the weight vector.

2.3.1.1 *Limits and abilities of linear networks*

We expand here our remarks about the perceptron's limitations (page 40) by asking when the task in equation (2.21) can be solved by a perceptron. We then extend the reasoning to general linear networks. It is easy to determine when the task *cannot* be solved. Note, in fact, that the definition of a linear network $\mathbf{r} = \mathbf{W}\cdot\mathbf{x}$ implies

$$\mathbf{r}(\mathbf{x}_\alpha + \mathbf{x}_\beta) = \mathbf{r}(\mathbf{x}_\alpha) + \mathbf{r}(\mathbf{x}_\beta) \tag{2.28}$$

because $\mathbf{W}\cdot(\mathbf{x}_\alpha + \mathbf{x}_\beta) = \mathbf{W}\cdot\mathbf{x}_\alpha + \mathbf{W}\cdot\mathbf{x}_\beta$ (from equation 2.8). In words, the output to a sum of inputs equals the sum of the outputs to each input (indeed, this is the technical meaning of *linear*). It is then clear that an input-output map where, say, $\mathbf{x}_\gamma = \mathbf{x}_\alpha + \mathbf{x}_\beta$ but $\mathbf{d}_\gamma \neq \mathbf{d}_\alpha + \mathbf{d}_\beta$ cannot be realized by a linear network. In general, if among the inputs $\mathbf{x}_1, \ldots, \mathbf{x}_\mu$ there is one, say, $\mathbf{x}_\beta$, that can be written as a weighted sum of the others, e.g.,

$$\mathbf{x}_\beta = \sum_{\alpha\neq\beta} b_\alpha \mathbf{x}_\alpha \tag{2.29}$$

(for some choice of the b_α's), then a weight vector $\mathbf{W}$ realizing the input-output map in equation (2.21) does not exist unless

$$\mathbf{d}_\beta = \sum_{\alpha\neq\beta} b_\alpha \mathbf{d}_\alpha \tag{2.30}$$

That is, the same relationship must hold among outputs that holds among inputs. Conversely, if it is impossible to write any of the $\mathbf{x}_\alpha$'s in the form of equation (2.29), the input-output map in equation (2.21) can be realized by a linear network for any choice of the $\mathbf{d}_\alpha$'s. Indeed, the reader may have noted that equation (2.27) is meaningless if $X_{11}X_{22} = X_{12}^2$ (fraction denominators vanish). This happens precisely when $\mathbf{x}_1 = b\mathbf{x}_2$, a special case of equation (2.29). In fact, in this case we have

$$\begin{aligned} X_{11} &= \mathbf{x}_1\cdot\mathbf{x}_1 \\ X_{22} &= b^2\mathbf{x}_1\cdot\mathbf{x}_1 \\ X_{12} &= b\mathbf{x}_1\cdot\mathbf{x}_1 \end{aligned} \tag{2.31}$$

so that $X_{11}X_{22} = X_{12}^2 = b^2(\mathbf{x}_1\cdot\mathbf{x}_1)^2$. Further information on this topic can be sought in linear algebra textbooks (sections on systems of linear equations).

2.3.2 Gradient descent methods (δ rule, back-propagation)

In this section and the next we introduce weight-setting methods that require a so-called objective function for their operation. These methods can work with non-linear networks of, in principle, arbitrary complexity. Indeed, the methods are very general and work with any input-output map with adjustable parameters (Mitchell 1996; Widrow & Stearns 1985). An *objective function* is simply a measure of network performance. It may indicate either how much the actual and desired maps differ (*error measure*) or how close they are (*performance measure*). It is a matter of convenience and partly of historical accident what measure is used in each particular case. If we use an error measure, the aim is to find weights that minimize it, whereas performance measures should be maximized. Setting weights in neural networks thus can be viewed as a function minimization or maximization problem (Aubin 1995; Widrow & Stearns 1985).

While many measures of performance are in use (Section 2.3.3), the *squared error* is by far the most popular error measure. Relative to the task (2.21), it is defined as follows. We introduce first the error relative to the input-output pair $(\mathbf{x}_\alpha, \mathbf{d}_\alpha)$ by summing the squared deviations between desired and actual output for each output node:

$$E_\alpha(\mathbf{W}) = \frac{1}{2}\sum_i (d_{\alpha i} - r_{\alpha i}(\mathbf{x}_\alpha))^2 \tag{2.32}$$

The notation $E_\alpha(\mathbf{W})$ emphasizes that the error is a function of the weights. The leading half serves merely to simplify notation later on. The square error relative to the whole task in equation (2.21) is defined as

$$E(\mathbf{W}) = \sum_\alpha E_\alpha(\mathbf{w}) \tag{2.33}$$

Clearly, the smaller the squared error, the closer we are to meeting the requirements of equation (2.21). Only when the squared error vanishes is the desired map established.

We are now ready to illustrate *gradient descent* methods of error minimization. Let us consider a network with only two weights, $\mathbf{W} = (W_1, W_2)$, and let us assume that the squared error $E(W_1, W_2)$ relative to the task is known for every value of W_1 and W_2. The triplet $(W_1, W_2, E(W_1, W_2))$ can be represented as a point in a three-dimensional coordinate space. When the weights vary, so does $E(W_1, W_2)$, and the point describes a surface known as the *error surface* (Figure 2.9). This allows us to visualize the values of the error for different weight choices. The quickest way to decrease the error is to travel on the error surface in the direction where the slope downward is steepest. To find what weight changes correspond to such a direction, one considers the *gradient* vector relative to the surface. This is written $\nabla E(\mathbf{W})$ and is defined as the vector whose components are the (partial) derivatives of $E(\mathbf{W})$ with respect to the weights:

$$\nabla E(\mathbf{W}) = \begin{pmatrix} \dfrac{\partial E(\mathbf{W})}{\partial W_{11}} \\ \vdots \\ \dfrac{\partial E(\mathbf{W})}{\partial W_{NN}} \end{pmatrix} \tag{2.34}$$

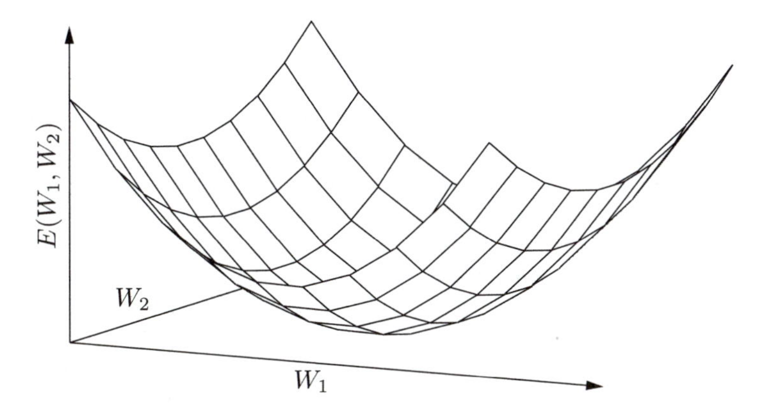

Figure 2.9 An error surface for a network with two weights. For a given task, the height of the surface at (W_1, W_2) is the error $E(W_1, W_2)$ relative to this particular choice of weights.

It can be shown that this vector points in the direction of steepest ascent. To descend the error surface, one thus should travel in the opposite direction. Additionally, the steps should not be too long: since the path to follow generally will be curved, the direction of travel needs to be updated regularly. This reasoning is formalized by writing the update ΔW_i to weight W_i as

$$\Delta W_i = -\eta \nabla_i E(\mathbf{W}) \tag{2.35}$$

where $\nabla_i E(\mathbf{W})$ is the ith component of the gradient vector (i.e., the derivative of $E(\mathbf{W})$ with respect to W_i), and η is a small positive number that calibrates step length. Once equation (2.35) has been applied to all weights, the gradient is calculated anew, then equation (2.35) is applied once more, and so on until error has decreased to a satisfactorily low figure. That is, gradient descent is an *iterative procedure*, and equation (2.35) is applied at each iteration. Note that although we started from an example with only two weights, equation (2.35) can be applied to networks with an any number of weights.

We must consider two important complications that are not apparent in Figure 2.9. In fact, the figure portrays a particularly favorable case where the error surface has a unique minimum, and the gradient always points in a direction that leads to the minimum along a rather short path. In general, neither is guaranteed to hold. The error surface may have many minima, of which only one or a few are satisfactory. Moreover, following the gradient may result in a very long path, which could be avoided with a fuller knowledge of the error surface. In defense of gradient methods, we can say that taking the path of steepest descent is a reasonable thing to do having only *local* information about the error surface (e.g., about how the error would change if the weights were changed slightly from their current values). The basic algorithm presented here can be improved in many ways, as explained, for instance, in Haykin (1999) and Golden (1996).

2.3.2.1 The δ rule

To compute the squared error and its gradient, one needs to know the output to all inputs and how it changes when changing any weight. A real input-output mechanism may not have all this information. For instance, input-output pairs are necessarily experienced in a sequence, and to remember previous inputs and outputs requires a memory external to the network. Without this memory, the best one can do is to compute the error relative to the current input, equation (2.32), rather than the full squared error, equation (2.33). This leads to the best known applications of gradient methods to neural networks (in particular feedforward networks). Applied to networks without hidden nodes, the method is known as the δ *rule*, thus named by McClelland and Rumelhart (1985) but known since Widrow and Hoff (1960) under the less appealing name of *least-mean-squares algorithm*. We begin with a single output node, in which case the instantaneous error is (we drop for simplicity the index α in equation 2.32):

$$E = \frac{1}{2}(d-r)^2 \tag{2.36}$$

where $r = \mathbf{W} \cdot \mathbf{x}$ is the response to $\mathbf{x}$. To calculate the derivative $\partial E/\partial W_i$, note first that the derivative $\partial r/\partial W_i$ is simply x_i. Hence

$$\begin{aligned} \frac{\partial E}{\partial W_i} &= -(d-r)\frac{\partial r}{\partial W_i} \\ &= -(d-r)x_i \end{aligned} \tag{2.37}$$

Thus, following equation (2.35), each weight is incremented by

$$\Delta W_i = \eta(d-r)x_i \tag{2.38}$$

which is the δ rule. It is so called because the difference $d-r$ was written δ by McClelland and Rumelhart (1985), yielding the even simpler expression

$$\Delta W_i = \eta\delta x_i \tag{2.39}$$

(having dropped the index α, we recall that $\mathbf{x}$ represents current input, i.e., one of the $\mathbf{x}_\alpha$'s for which a desired output d_α must be established; likewise, δ is the error relative to the current input).

The δ rule, it can be proved, will find a weight vector that, across the presentation of different inputs, keeps close to the one that minimizes the total squared error, provided the inputs are presented in random order and η is small enough (Widrow & Stearns 1985). Thus, in the case of networks without hidden layers, dealing with only one input at a time is not a serious restriction (because the error surface of these networks has a single minimum; see Widrow & Stearns 1985).

A reason why the δ rule is so popular is that equation (2.38) can be derived and understood intuitively without much mathematics as follows (Blough 1975; extending Rescorla & Wagner 1972). The term $d-r$ is the difference between desired and actual output, which is positive (negative) when actual responding is lower (higher) than desired responding and null when actual behavior matches desired behavior. That is, $d-r$ has the same sign as the desired change in response. Moreover, its absolute value is larger the more actual output departs from desired output. Thus, it

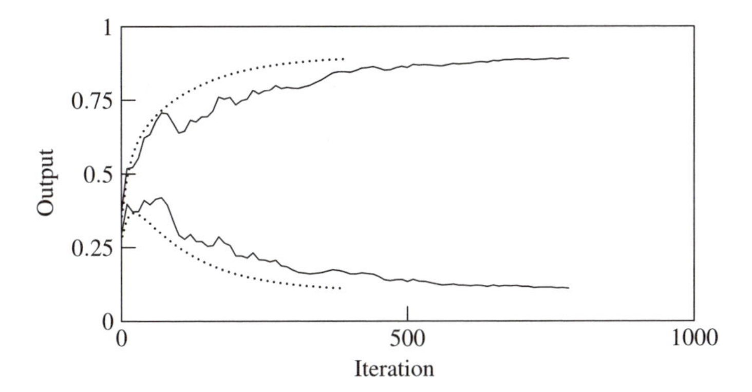

Figure 2.10 Comparison of δ rule learning (solid lines, equation 2.38) and full gradient descent (dotted lines, equation 2.40). A perceptron-like network with 10 input nodes and a linear output node is trained with either method to react with an output of 0.9 to an input $\mathbf{x}_1$ and with an output of 0.1 to $\mathbf{x}_2$.

can be used to command positive or negative weight changes of roughly appropriate magnitude. Multiplying $d - r$ by x_i has the further effect of changing more the weights that are linked to more active input nodes, under the intuitive justification that these weights contribute most to an incorrect output to $\mathbf{x}$.

The δ rule can easily solve the example task introduced earlier. The solid lines in Figure 2.10 are *learning curves* (network output as a function of algorithm iterations) from a simulation that uses the δ rule to solve the task. The dotted lines show that full gradient descent (based on the complete squared error, equation 2.33) solves the same task faster and more smoothly. The formula for full gradient descent is simply the δ rule formula summed over all input-output relationships:

$$\Delta W_i = \eta \sum_\alpha (d_\alpha - r_\alpha) x_{\alpha i} \tag{2.40}$$

The δ rule can also be extended to a nonlinear output node, where $r = f(\mathbf{W} \cdot \mathbf{x})$. Repeating the steps leading to equation (2.38), we get

$$\Delta W_i = \eta (d - r) f'(\mathbf{W} \cdot \mathbf{x}) x_i \tag{2.41}$$

The only addition is the term $f'(\mathbf{W} \cdot \mathbf{x})$. Actually, equation (2.38) can be used even when the output node is nonlinear (Hinton 1989). The δ rule can also be applied to networks with more output nodes. The weight matrix $\mathbf{W}$ is seen as made of weight vectors $\mathbf{W}_1, \ldots, \mathbf{W}_n$, where $\mathbf{W}_j$ connects the input nodes to output node j. Since these weight vectors are independent of each other, one simply applies the δ rule to each one, getting

$$\Delta W_{ij} = \eta (d_j - r_j) x_i \tag{2.42}$$

2.3.2.2 Back-propagation

Back-propagation is a generalization of the δ rule to feedforward networks with hidden layers. Although gradient descent is an old idea, it was not fully exploited in the field of neural networks until publication of the back-propagation algorithm by Rumelhart et al. (1986). For simplicity, we begin with a network with a single hidden layer. We write **V**, the weights from input to hidden nodes; **h**, the activities of hidden nodes; and **W**, the weights from hidden to output nodes. The activity of output node k is thus

$$r_k = f\Big(\sum_k W_{ki} h_i\Big) \tag{2.43}$$

with

$$h_i = f\Big(\sum_j V_{ij} x_j\Big) \tag{2.44}$$

Technically, the back-propagation algorithm can be obtained by calculating the derivative of the preceding equations with respect to the weights, and then using the gradient descent formula, equation (2.35). The algorithm, however, can also be described starting from the δ rule. Note first that the hidden and output layers form a two-layer network that we can train by the δ rule:

$$\Delta W_{ki} = \eta f'(y_k) \delta_k h_i \tag{2.45}$$

where $y_k = \sum_k W_{ki} h_i$ is the input received by output node k, and $\delta_k = d_k - r_k$. What we lack is a formula to change the weights between input and hidden nodes. It seems that we cannot use the δ rule again because for hidden nodes *we have no concept of a desired output*, which means that we cannot calculate a δ term. The solution, with hindsight, is rather simple: the term δ_i, relative to hidden node i, is a weighted sum of the terms δ_k relative to the output nodes, where the weight of δ_k is exactly W_{ki}, i.e., the weight linking hidden node i to output node k (Haykin 1999; O'Reilly & Munakata 2000). Formally,

$$\delta_i = \sum_k \delta_k W_{ki} \tag{2.46}$$

Now that a δ term is defined, we can apply the δ rule as usual:

$$\Delta V_{ij} = \eta f'(y_i) \delta_i x_j \tag{2.47}$$

where $y_i = \sum_j V_{ij} x_j$. Equations (2.45) and (2.47) are the back-propagation algorithm. The algorithm can be generalized easily to networks with more than one hidden layer using δ terms of later layers to define δ terms in earlier layers. This backward construction of δ terms gave the algorithm its name.

2.3.3 Random search

Random search is an increasingly popular way of setting network weights, mostly because of its simplicity and power in many practical cases. Random search methods are also called *genetic algorithms* or *evolutionary algorithms* for reasons that will soon be clear (Holland 1975; Mitchell 1996). The basic procedure is as follows. From an initial network, typically with randomly set weights, some "mutants" are generated. Mutants differ from the initial network and from each other in a few weights, usually picked at random and modified by adding a small random

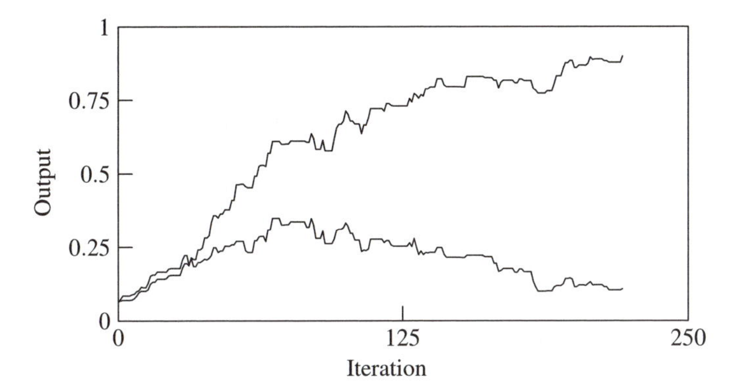

Figure 2.11 Random search simulation. A perceptron-like network with 10 input nodes and a linear output node is trained by random search to react with an output of 0.9 to an input $\mathbf{x}_1$ and with an output of 0.1 to $\mathbf{x}_2$.

number. The mutants and the initial network then are evaluated by means of a performance measure (often called *fitness*). The latter has been constructed so that it is maximized by desired behavior. The network yielding the highest performance is retained, and the process of mutation, evaluation and selection is repeated until a network capable of satisfactory performance is obtained or until one suspects that this will never happen. The basic algorithm can be modified in several respect, e.g., retaining more than one network from one iteration to the next or using different ways of generating mutants (which may include, e.g, changes in network architecture and node properties).

To illustrate random search, we consider our example task, using equation (2.15) with a logistic transfer function (equation 2.7). We need to define a performance function that reaches its maximum value when the task is solved. Since network output is limited between 0 and 1, the quantity $|d_\alpha - r_\alpha|$ (absolute value of the difference between the actual and the desired output for a given input-output mapping) can vary between 0 and 1, with 0 being its desired value. Hence the desired value of the quantity $1 - |d_\alpha - r_\alpha|$ is 1, and this quantity increases as r_α gets closer to d_α. If we need to establish two input-output mappings, we can define a performance function as follows:

$$p(\mathbf{W}) = \sqrt{(1 - |d_1 - \mathbf{r}(\mathbf{x}_1)|)(1 - |d_2 - \mathbf{r}(\mathbf{x}_2)|)} \tag{2.48}$$

(the square root lets $p(\mathbf{W})$ increases at a similar rate over its entire range). The maximum value of $p(\mathbf{W})$ is 1 and is reached only if $r_1 = d_1$ and $r_2 = d_2$. Equation (2.48) can be generalized easily to an arbitrary number μ of input-output mappings:

$$p(\mathbf{W}) = \sqrt[\mu]{\prod_{\alpha=1}^{\mu}(1 - |d_\alpha - \mathbf{r}(\mathbf{x}_\alpha)|)} \tag{2.49}$$

Figure 2.11 shows the results from a simulation using the simplest random search algorithm, whereby a single mutant network is generated at each iteration. Compared with gradient descent (Figure 2.10), progress on the task is more irregular.

Although a shortcoming in this simple case, in more complex tasks this may be a blessing. We have observed already that gradient descent methods may reach unsatisfactory local minima of the error surface. Random search is less prone (though not immune) to this kind of failure because occasional large mutations may lead out of local minima. Random search methods, moreover, can mimic genetic evolution, thereby allowing us to study the evolution of behavior (Chapter 5).

2.3.4 Learning from limited information (reinforcement learning)

The weight-setting methods described so far are based on detailed knowledge of the desired behavior map. Here we describe techniques that require less information. For instance, it may be enough to know whether the output is correct or not. These techniques are often called *reinforcement learning* because they resemble instrumental conditioning experiments in animal psychology (Barto 2003; Sutton & Barto 1998; see Chapter 4). In these experiments, animals do not (obviously) know what behavior the experimenter wants to train but can track what actions are rewarded or punished ("reinforced"). Reinforcement learning in artificial systems usually is achieved by increasing the probability of actions that, in a given situation, are followed by positive consequences and decreasing the probability of actions followed by negative consequences. In a self-contained learning device, this clearly requires a mechanism that identifies an event as positive or negative, as we will discuss in Chapter 4.

In our example task, reinforcement learning can be implemented as follows. First, network output must be interpreted as an action rather than as the probability of performing an action. This is achieved by using threshold transfer functions so that the output node is either active (interpreted as performing the action) or inactive. Moreover, we use a stochastic threshold node (equation 2.18) to generate variable output to the same input. This allows the network to try out different responses to the same stimulus, which is crucial to discover favorable responses. Let $\lambda = 1$ if reinforcement has been received and $\lambda = 0$ if not. Then the rule for changing the weights is

$$\Delta W_i = \begin{cases} \eta(r - \mathbf{W}\cdot\mathbf{x})x_i & \text{if } \lambda = 1 \\ \\ \eta(1 - r - \mathbf{W}\cdot\mathbf{x})x_i & \text{if } \lambda = 0 \end{cases} \tag{2.50}$$

where $r = 1$ if the action has been performed, and $r = 0$ otherwise. To understand equation (2.50), note that both expressions have the same form of the δ rule, with r and $1 - r$ taken as desired responses (Widrow et al. 1973). Thus if performing the action has lead to reinforcement, it is taken as a goal. This works also to suppress responding to a input, in which case not performing the action is reinforced.

Now that network output is binary (react or not react), the d_i's cannot be interpreted as desired outputs. Their interpretation is rather that d_i is the probability that reinforcement will follow if the network responds to $\mathbf{x}_i$. Figure 2.12 shows the result of a simulation of equation (2.50). The outcome is that the network performs the action with probability d_i.

If the reinforcement can take on a continuous range of values, say, between 0 and 1, it is useful to combine the two parts of equation (2.50) into a single equation

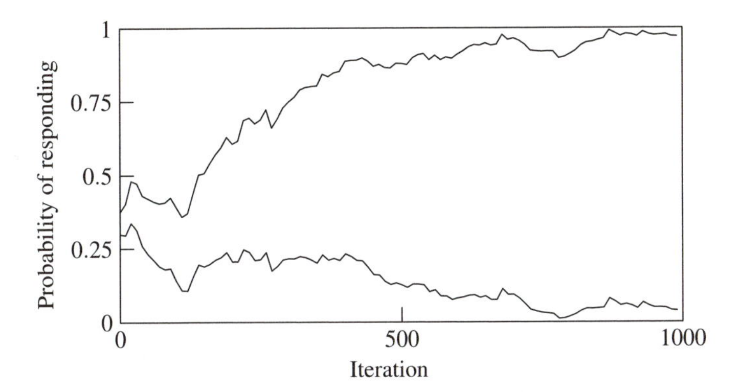

Figure 2.12 Reinforcement learning simulation relative to the same problem in Figures 2.10 and 2.11.

as follows (Barto 1995):

$$\Delta W_i = \eta \big(\lambda(r - \mathbf{W}\cdot\mathbf{x}) + \beta(1-\lambda)(1 - r - \mathbf{W}\cdot\mathbf{x})\big)x_i \qquad (2.51)$$

If $\lambda = 1$ or $\lambda = 0$, this formula reduces to the upper or lower part of equation (2.50), respectively (but the learning rate for the lower part is $\eta\beta$ instead of η). If λ has an intermediate value, ΔW_i is a weighted sum of the two parts in equation (2.50), the weight being the parameter β. It can be shown that equation (2.51), called the *associative reward-penalty rule*, performs best for small values of β, e.g., 0.01 or 0.05. Nevertheless, the second part of equation (2.51) is important because it regulates weight changes when the action taken resulted in poor reinforcement.

2.3.5 Learning about time

So far we have discussed how to realize input-output maps with no time structure. Here we consider algorithms that can set weights so that desired output sequences are produced from given input sequences. We consider discrete time steps (Section 2.2.2), and we write $\mathbf{x}(t)$, the input at time t.

Different network architectures have been used to handle time. One idea is to maintain a feedforward architecture but to replicate the input layer so that it contains copies of the input at different times. This is called a *time-delay network*. Thus, if there are N input nodes receiving the input $\mathbf{x}(t)$, the network would have a second set of N nodes whose input is $\mathbf{x}(t-1)$, and so on (Figure 2.13, top). All techniques introduced to set weights in feedforward networks can be applied. For instance, a perceptron-like network may learn by the δ rule to react the sequence $\{\mathbf{x}_1, \mathbf{x}_2\}$ but not to $\{\mathbf{x}_2, \mathbf{x}_2\}$. This is impossible for a standard feedforward network (Section 2.2.2), but a time-delay network simply sees the two sequences as two different input patterns that thus can elicit different outputs. This simple approach to time, however, has a number of shortcomings. First, it may require a very large network. For instance, if we consider that signals in nervous systems propagate in millisecond times, we would need several hundred copies of the input layer to cover intervals of the order of 1 s. It does not appear that nervous systems manage time

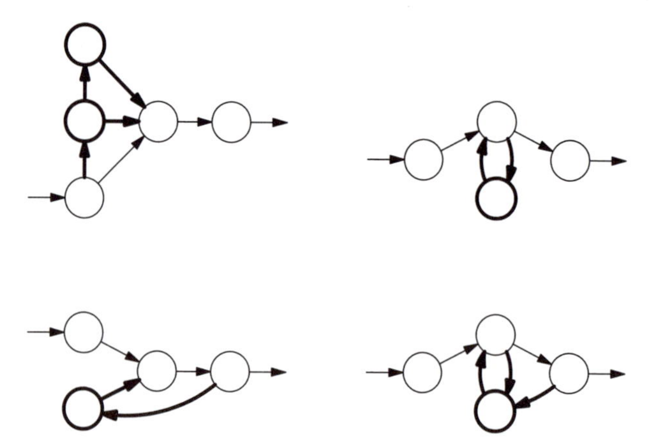

Figure 2.13 Four different modifications (thicker lines) to a feedforward network that add some time-processing abilities. Each circle represents a group of nodes. Top left: A network with no feedback but with multiple feedforward pathways that introduce a delay in the input. In this way the present activity of some nodes represents input received in the past (two delay steps are shown). Top right: The "simple recurrent network" of Elman (1990) that uses feedback between groups of hidden nodes to build internal states that depend on the time structure of the input. Bottom left: Network output is fed back as input. The network thus can combine information about current input and past output (Jordan 1986). Bottom right: A combination of the two previous architectures.

in this way (Buonomano & Karmarkar 2002). Second, everything is remembered with equal weight, even unimportant things.

Recurrent networks provide a superior alternative for temporal processing. These networks can build a more useful representation of input sequences than the simple bookkeeping of the past. Different recurrent architectures have been analyzed, e.g., networks with feedback between internal nodes (Elman 1990) or feedback from output nodes to internal nodes (Jordan 1986). Both features may be helpful, and both are important in nervous systems (Chapter 3). How can we establish a desired input-output map in a recurrent network? The question is difficult from the points of view of technique and biological realism alike. Gradient descent techniques may be applied that are conceptually like to the δ rule and back-propagation, but it is more difficult to calculate the error gradient (Doya 2003; Haykin 1999). The extension of random search techniques, on the other hand, is easier. We only need a performance function that considers *sequences* of inputs and outputs rather than single input-output mappings. Note, however, that the number of possible sequences grows exponentially with the number of time steps considered. Thus a naive extension of random search can be impractical. The very task of prescribing the correct output to every sequence of inputs can be prohibitive. Indeed, it is common that time-dependent tasks are not even phrased as an input-output map. For example, consider a network that must control a model arm so that it can reach toward objects. It would be practically impossible to analyze all possible input sequences to determine what is the network action that is more likely to lead to reaching the object and then to teach this to the network. On the other hand, the task is clearly expressed as "reach the object." In such situations it is often most practical to define a performance function that considers the task at a higher level, e.g., a function that increases as the arm approaches the object (Nolfi & Floreano 2000). Temporal factors important for learning will be considered further in Chapter 4.

2.4 ORGANIZING NETWORKS WITHOUT SPECIFIC GUIDANCE

So far we have considered how to establish a number of given input-output mappings, but not all weight setting procedures are of this kind. Another possibility is to explore how weights develop when the network is exposed to sequences of inputs given a rule of weight change. For instance, experimental data or intuition about neurons and synapses may suggest a particular rule of weight change. We might then investigate whether the emerging weight organization resembles some feature of nervous systems. Alternatively, a rule of weight change may have been designed to produce a particular network organization, e.g., an input-output transformation that reduces noise in input patterns or enhances some of their features. In contrast with the preceding section, in these cases network organization emerges without specific feedback about whether a realistic or useful input-output map is being constructed. This has led to the term *self-organization*. It should be borne in mind, however, that whatever organization emerges depends crucially on the input fed to the network, in addition to properties of the network itself.

It is especially interesting to consider rules of weight change based on information that could be available at the junctions between neurons. These may include, for instance, state variables of both neurons and variables that describe the state of the extracellular environment, such as abundance of neuromodulators. These remarks will be expanded in Section 4.2. We now show a few examples of how interesting organizations of the weights can emerge from the following simple rule:

$$\Delta W_{ij} = \eta (z_i - a)(z_j - b) \tag{2.52}$$

where z_i and z_j are the activities of the nodes connected by W_{ij}, whereas a and b are two constants. Equation (2.52) implies that W_{ij} will increase if $z_i > a$ and $z_j > b$ or if $z_i < a$ and $z_j < b$ and decrease otherwise. We consider the network in Figure 2.14, with one output node that receives input from nodes arranged as a simple model retina. We feed this network with input patterns consisting of diffuse blobs of excitation (Figure 2.15, left) and apply equation (2.52) after each presentation. After repeating these steps a few hundreds times, we visualize the weights as a gray-scale image. Figure 2.15 shows three examples of how an initially random weight matrix can develop different organizations. The latter are often called *receptive fields* because they describe from which part of the retina a node receives input. Thus nodes with the receptive fields pictured in Figure 2.15 would react preferentially to images falling, respectively, on a well-delimited portion of the retina, on the left part of the retina, and anywhere on the retina but a specific portion. In a network with many such nodes, each one could react to stimuli on a different part

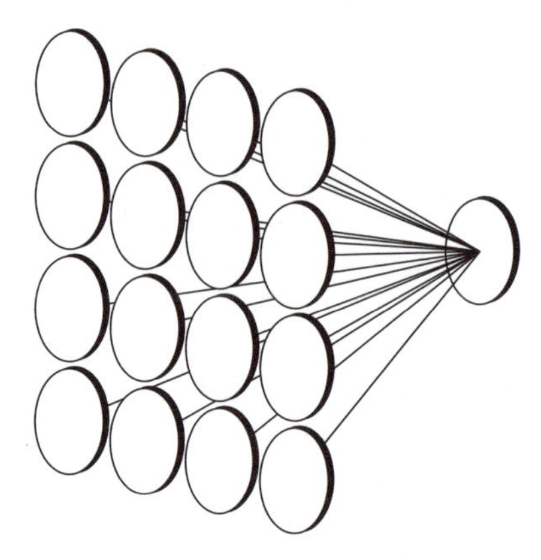

Figure 2.14 Network architecture used for simulations of self-organization. Input nodes are arranged in a two-dimensional grid simulating the physical arrangement of receptors on a retina.

of the retina, and further processing stages could use this information in tasks such as locating objects or orienting in the environment.

How can we produce a network with receptive fields spanning the whole retina? Simply adding output nodes that operate independently of each other is not enough. In fact, Figure 2.16 (left) shows that most nodes develop very similar receptive fields (this means that the sequence of inputs is a strong determinant of weight organization). We obtain a better result if, after presenting an input but before updating the weights, we set to zero the activity of all nodes except the most active one. Biologically, this may correspond to strong mutual inhibition between neurons. Receptive fields obtained in this way are spread more evenly across the retina, as shown in Figure 2.16 (right). To understand why, consider just two nodes and suppose that because of differences in initial weights, node 1 reacts more strongly to input $\mathbf{x}$ than node 2. Then, according to equation (2.52), the sensitivity of node 1 to $\mathbf{x}$ and similar inputs will be increased, and the sensitivity of node 2 (whose activity has been set to zero) to the same inputs will be decreased. Such weight changes make it even more likely that node 1 will be more active than node 2 to inputs similar to $\mathbf{x}$. In summary, we have a self-reinforcing process (positive feedback) that confers on the two nodes sensitivity to different input patterns.

2.5 WORKING WITH YOUR OWN MODELS

Students of animal behavior who wish to work with neural networks can choose among a diversity of software tools, many of which are surveyed by Hayes et al. (2003) and at http://www.faqs.org/faqs/ai-faq/neural-nets/part5. The focus of such software, however, is often not animal behavior. Here we offer a few guidelines for beginners. We broadly divide software in neural network simulators, general-purpose programming languages and software for scientific computation.

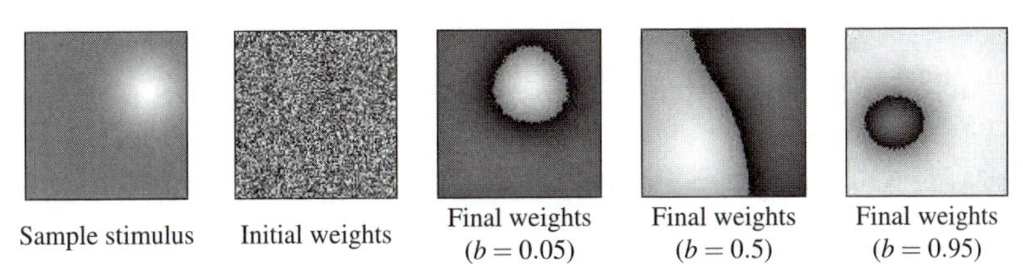

Figure 2.15 Simulations using equation (2.52) with the network in Figure 2.14 (input grid of 100×100 nodes). From left to right: A sample stimulus exciting nodes in a part of the input grid; initial weight configuration; three organizations of the weights, corresponding to three choices of the parameter b in equation (2.52), and $\eta = 5$, $a = 0.05$. From left to right: Excitatory center and inhibitory surround, excitatory and inhibitory hemifields, inhibitory center and excitatory surround. Simulations consisted of 1000 presentations of stimuli like the one shown, positioned randomly on the model retina. After each presentation, equation (2.52) was applied to all weights. A lighter shade of gray represents a stronger weight; negative weights are darker than the darkest gray in the sample stimulus.

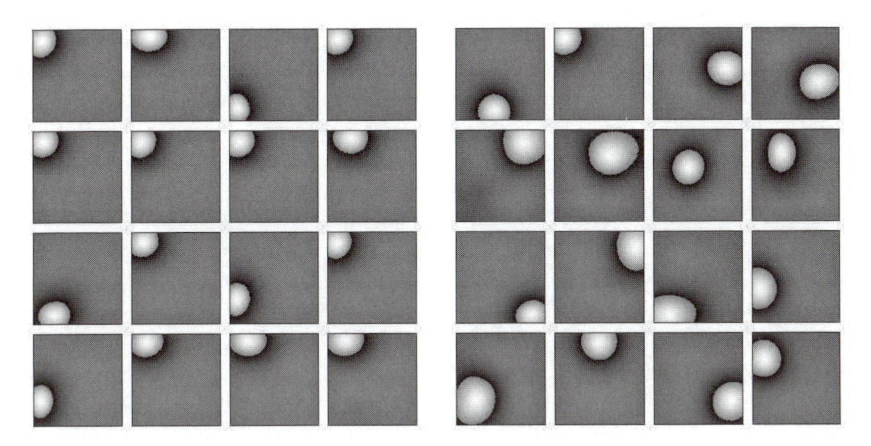

Figure 2.16 Weight organizations induced by equation (2.52) (with $a = 0.5$ and $b = 0.05$) in two different networks. Left: Receptive fields developing in a network like the one in Figure 2.14 but with 16 independent output nodes (all weights are updated independently at each input presentation). Right: Receptive fields developing in a similar network, but with strong inhibition between the 16 output nodes (only the weights feeding to the most active node are updated). In the latter case, receptive fields are spread more evenly.

```
1  #include <stdio.h> /* these two lines are needed to use the */
2  #include <math.h>  /* functions printf() and exp(), see below*/
3
4  /* this function calculates node input, see equation (2.3) */
5  /* W = weight vector, x = input vector, N = size of W and x */
6  float weighted_sum( float *W, float *x, int N ) {
7    int i;
8    float y = 0.;
9    for( i=0; i<N; i++ ) y += W[i]*x[i];
10   return y;
11 }
12
13 /* logistic node transfer function, see equation (2.7) */
14 float transfer_function( float y ) {
15   return 1/(1+exp(-y));
16 }
17
18 /* a C program starts executing from the "main" function */
19 int main( void ) {
20   float x[] = { 1, 1, 1, 1 };     /* input vector */
21   float W[] = { 1, 0, 1, 0 };     /* weight vector */
22   float y;                        /* input to output node */
23   float z;                        /* activity of output node */
24   y = weighted_sum( W, x, 4 );    /* calculate y */
25   z = transfer_function( y );     /* apply the transfer function */
26   printf( "Output is %f\n", z ); /* print result */
27   return 0;                       /* end of the program */
28 }
```

Listing 2.1 A C program that simulates a two-layer network with $N = 4$ input nodes and 1 output node. Numbers are represented as variables of type `float` and vectors as arrays of such variables. Variables of type `int` represent integers (e.g., vector indices). The expression `x[i]` is the *i*th element of array `x` (C indices start from 0). The functions `weighted_sum` and `transfer_function` translate equation (2.3) and equation (2.7), respectively, and provide the basic structure to simulate two-layer networks. Explanatory comments are enclosed between `/*` and `*/`.

Neural network simulators are ready-to-use programs, often with graphical user interfaces, explicitly developed to study neural networks. Some of the most developed simulators are intended for neurophysiologically realistic simulations (Hayes et al. 2003 and references therein). They tend to consider many details of neurons, resulting in node models with many state variables and parameters that are, presently, of uncertain value to behavioral modeling (see the end of Section 2.1.2). Neural network simulators have also been developed within artificial intelligence, computer science and engineering. The drawback of such software is that it often requires some technical background and tends to focus on different questions than animal behavior. Two fairly general neural network simulators that, while requiring some study, are not too difficult for beginners and do not require programming are PDP++ (http://psych.colorado.edu/~oreilly/PDP++/PDP++.html) and JavaNNS (http://www-ra.informatik.uni-tuebingen.de/software/JavaNNS/welcome_e.html). The former focuses on (human) cognition, including behavior-level phenomena (O'Reilly & Munakata 2000), whereas the latter is more neutral regarding the interpretation of the network models.

It is also possible to do research on neural networks without specific software using general-purpose programming languages such as Pascal, C, C++, etc. These have no knowledge of neural networks as such, yet simple neural network simulations can be implemented easily. For instance, to simulate a two-layer network with one output node, we need simply two arrays of numbers to store the input and weight vectors, one number for the response, as well as the ability to compute weighted sums and define transfer functions. As an example, Listing 2.1 shows a complete C program that calculates the output of a two-layer network with a single output node and weights $\mathbf{W} = (1,0,1,0)$ to the input $\mathbf{x} = (1,1,1,1)$. The program would not look too different in most other programming languages. People have also developed additions to the languages, called *libraries*, that allow programming of more complex simulations with little effort. Typically, these libraries provide data structures to represent various network architectures, as well as the ability to perform common tasks such as calculating network responses or gradient descent learning. Exploring such language extensions probably would be rewarding for readers with some programming experience or those who wish to start programming. Many neural network libraries are listed at http://www.faqs.org/faqs/ai-faq/neural-nets/part5.

Software for scientific computation is well suited for beginners wishing to simulate simple or moderately complex networks using a rather straightforward programming language. The key insight is that the operation of neural networks can be expressed in terms of vector and matrix algebra, for which extensive support exists. In the following we present a few examples in a syntax common to the programs Octave (http://www.octave.org, available free of charge) and Matlab (http://www.mathworks.com). Similar software, although with different programming syntax, are Scilab (http://www.scilab.org, available free of charge) and Mathematica (http://www.wolfram.com). Many of the models in later chapters require only a little more programming than the following examples, and we encourage the reader to try to reproduce some of our results. We do not expect readers without programming experience to understand each line of the code that follows but rather to get a flavor of neural network coding with scientific software.

2.5.1 Two-layer feedforward networks

Simple neural networks are coded rather compactly in scientific software. For instance, the following code is equivalent to Listing 2.1:

```
1  function r = f( y )
2    r = 1 ./ (1+exp(-y)); % cf.equation (2.7)
3  end
4
5  x = [ 0 1 0 1]';           % input vector
6  W = [ 1 1 1 1];            % weight vector
7  r = f( W*x )               % cf. equation (2.17)
```

Note how the last line translates equation (2.17) almost literally, with `W*x` as equivalent to the notation $\mathbf{W} \cdot \mathbf{x}$ for a weighted sum. By convention, the result of a computation is printed out if the corresponding line does *not* end with a semicolon; hence the code above will display the value of `r`, i.e., `r = 0.88080`.

```
N = 10;             % number of input nodes
d = 0.9;            % desired output
t = 0.0001;         % target error
eta = 0.01;         % rate of weight change

x = rand(N,1);  % random input vector (uniform distribution)
W = randn(1,N); % random weight vector (normal distribution)

e = 2*t;                      % does not matter so long as e>t
while( e>t )
  r = W*x;                    % calculate output
  W = W + eta*(d-r) * x'; % change weights
  e = .5*(d-r)^2;             % calculate error
end
```

Listing 2.2 Code to establish a single input-output mapping by the δ rule in a two-layer network with $N = 10$ input nodes and 1 output node.

A technical comment: the prime (`'`) following the input vector `x` denotes a *column* vector (a matrix with one column and many rows), whereas `W` is a *row* vector (a matrix with one row and many columns). The distinction, while not crucial in this book, is important in matrix algebra and thus enforced in scientific software. In short, the matrix multiplication `W*x` is the weighted sum $\sum_i W_i x_i$ only if `W` is a row vector and `x` a column vector. Further explanations can be found in textbooks about linear algebra (Hsiung & Mao 1998).

It is easy to code networks with many output nodes. In fact, the transfer function defined earlier will return a vector of transformed values if we supply a vector of input values (this requires writing `./` rather than / on line 2 to perform the correct vector operation). In practice, this means that a network with more output nodes is coded simply by replacing the weight vector above with an appropriate weight matrix. A network with two output nodes, for instance, is coded like this:

```
x = [ 0 1 0 1]';
W = [ 1 1 1 1     % weights of 1st output node
      2 2 2 2 ]; % weights of 2nd output node
r = f( W*x )
```

This will display `r = 0.88080 0.98201`, showing the output of both nodes. The extension to multilayer networks is also trivial. The following example calculates the output of a three-layer network with four input nodes, three hidden nodes and two output nodes:

```
x = [ 0 1 0 1]'; % input vector
W = [ 1 1 1 1    % weights from input to hidden nodes
      2 2 2 2    % (3x4 matrix)
      3 3 3 3 ];
V = [ 1 1 1      % weights from hidden to output nodes
      2 2 2 ];   % (2x3 matrix)
r = f( V*f( W*x ) )
```

2.5.2 The δ rule

An algorithm to establish a number of input-output mappings by the δ rule is as follows:

```
1   N = 10;                 % number of input nodes
2   M = 4;                  % number of input-output mappings
3   t = 0.0001;             % target error
4   eta = 0.01;             % rate of weight change
5
6   rand("seed",time()); % initialize random number generator
7   d = rand(1,M);          % M random outputs
8   x = rand(N,M);          % M random inputs arranged in a matrix
9   W = randn(1,N);         % random weight vector
10
11  e = 2*t;
12  while( e>t )                        % while error is too big
13    e = 0;                            % reset error
14    for i = randperm(M)               % loop through inputs in random order
15      r = W * x(:,i);                 % calculate output to x(:,i)
16      W = W + eta*(d(i)-r)*x(:,i); % change weights
17      e = e + .5*(d(i)-r)^2;          % add to error
18      end
19  end
```

Listing 2.3 Generalization of the code in Listing 2.2 to many input-output mappings. The expression `randperm(M)` on line 14 gives a random arrangement (permutation) of numbers from 1 to M. The expression `x(:,i)` on lines 15 and 16 is the ith column of the matrix `x`, i.e., the ith input vector.

1. Create input-output pairs.
2. Set weights at small random numbers.
3. Set total error to zero.
4. Arrange inputs in random order.
5. Loop through all inputs:
    a. Calculate output.
    b. Change the weights using the δ rule (Section 2.3.2.1).
    c. Calculate the error for this input, add to total error.
6. If the total error is low enough, end; else go to step 3.

Listing 2.2 is an implementation of this algorithm for a single input-output mapping, whereas Listing 2.3 is a generalization to an arbitrary number of input-output mappings (4 in the example). Lastly, Listing 2.4 shows how to implement full gradient descent by computing the full gradient in an auxiliary vector `G`, and then using equation (2.35) to update the weights.

2.5.3 Random search

Here is a description of the simplest random search algorithm:

1. Define a performance function.
2. Create two weight vectors. $\mathbf{W}_1$ represents the best vector found so far; $\mathbf{W}_2$ is an attempt to improve.
3. $\mathbf{W}_1$ is initialized with small random numbers.
4. $\mathbf{W}_2$ is made equal to $\mathbf{W}_1$.

```
1   M = 4;       % number of input-output mappings
2   N = 10       % number of input nodes
3   t = 0.0001;  % target error
4   eta = 0.01;  % rate of weight change
5
6   d = rand(1,M);          % M random outputs
7   x = rand(N,M);          % N random inputs arranged in a matrix
8   W = randn(1,10);        % random weight vector
9
10  e = 2*t;
11  while( e>t )                      % while error is too big
12    e = 0;                          % reset error
13    G = zeros(size(W));             % gradient vector, same size as W
14    for i = randperm(M)             % loop through inputs in random order
15      r = W * x(:,i);               % calculate output to input i
16      G = G - (d(i)-r)*x(:,i);      % update gradient
17      e = e + .5*(d(i)-r)^2;        % update error
18    end
19    W = W - eta*G;                  % change weights, equation (2.40)
20  end
```

Listing 2.4 Modification of Listing 2.3 that uses full gradient descent instead of the δ rule (see Section 2.3.2.1). The auxiliary vector `G` stores the error gradient.

```
1   function p = perf( W, x, d ) % performance function: reaches
2     p = 1 - abs( d - W*x );    % a maximum of 1 when W*x=d
3   end;
4
5   N = 10;          % number of input nodes
6   d = 0.9;         % desired output to x
7   t = .9999;       % target performance
8
9   x = rand(N,1);   % create a random input vector
10  W = randn(1,N);  % create a random weight vector
11
12  p1 = perf( W1, x, d );  % initial performance
13
14  while ( p1<t )                       % while performance too low
15    W2 = W1;                           % let W2 equal W1
16    i = ceil( N*rand() );              % choose a weight to mutate
17    W2(i) = W2(i) + .01*randn();       % add small random number
18
19    p2 = perf( W2, x, d );   % calculate performance of W2
20    if( p2>p1 )              % accept W2 if it performs better
21      W1 = W2;
22      p1 = p2;
23    end
24  end
```

Listing 2.5 Code for the simplest random search algorithm, establishing a single input-output mapping in a two-layer network. The expression `i=ceil(N*rand())` selects a random integer between 1 and N by selecting a random number in $[0, 1]$, multiplying by N and finally rounding up to the next integer.

5. Add a small random number (equally likely to be positive or negative) to a randomly selected weight of $\mathbf{W}_2$.
6. Evaluate the performance of $\mathbf{W}_2$, and compare it with the performance of $\mathbf{W}_1$. If $\mathbf{W}_2$ performs better, it replaces $\mathbf{W}_1$ as the best vector so far.
7. If performance is satisfactory, end; otherwise return to step 4.

Listing 2.5 implements this algorithm for one input-output mapping.

CHAPTER SUMMARY

- Neural networks are mathematical models made of interconnected nodes that attempt to capture the most important features of neurons and synapses.
- Even simple networks can solve tasks that appear relevant for behavioral modeling.
- General feedforward networks can realize arbitrary input-output maps, provided the output to each input is fixed. A number of powerful weight-setting techniques exist to actually find the appropriate weights.
- Other weight setting techniques can be used to organize the network in response to incoming input (e.g., to encode statistical regularities) rather than to achieve a desired input-output mapping.
- Recurrent networks add the ability to process complex input sequences and generate time-structured output sequences. Recurrent networks often are more difficult to train, even though many techniques exist.
- If mathematical analysis of a neural network model is too difficult, we can simulate it on a computer using either ready-made software or programming our own.

FURTHER READING

Arbib MA, 2003. *The Handbook of Brain Theory and Neural Networks*. Cambridge, MA: MIT Press, 2 edition. A collection of about 300 articles on neural networks from different perspectives.

Dayan P, Abbott LF, 2001. *Theoretical Neuroscience: Computational and Mathematical Modeling of Neural Systems*. Cambridge, MA: MIT Press. Many topics treated formally but accessibly, from single neurons to large-scale network models.

Haykin S, 1999. *Neural Networks: A Comprehensive Foundation*. New York: Macmillan, 2 edition. A thorough introduction to almost anything technically important about neural networks, with good cross-field connections.

Mitchell M, 1996. *An Introduction to Genetic Algorithms*. Cambridge, MA: MIT Press. A general introduction to random search methods.

Chapter Three

Mechanisms of Behavior

This chapter presents neural network models of short term and reversible changes in behavior (traditionally referred to as *motivation*). The topics covered include reactions to stimuli, making decisions and the control of movement. In our terminology, these are properties of a given behavior map (Chapter 1). *Learning* and *ontogeny* (development), in contrast, correspond to changes in a behavior map and are considered in Chapter 4.

3.1 ANALYSIS OF BEHAVIOR SYSTEMS

Before considering neural network models, we discuss in some detail the nature of motivational variables, how they cause behavior and how they enter models of behavior systems. The aim is to sketch a conceptual frame in which the different material causes of behavior are conveniently analyzed, building on the discussion in Section 1.1.

Motivational variables commonly are divided into external and internal (McFarland 1999). The former usually are referred to as *stimuli*. For instance, courtship behavior in a peacock can be caused or motivated by visual stimulation provided by a peahen. Stimulation is picked up by sense organs, in this case the peacock's eyes, and then is processed by recognition mechanisms that can determine whether a peahen is present or not. Finally, motor mechanisms produce observable courtship behavior. Such a division in sensory processing, decision making and motor control, while not of universal validity, often is convenient and is used in this chapter.

Whether the peacock will court or not also depends on internal motivational variables. For instance, a hungry peacock may look for food rather than court. Internal variables can be further classified as state variables of the body and those of the nervous system (see SENSE ORGANS in McFarland 1987). The former will be referred to as *body states*. Examples are body water volume, blood sugar level, body temperature and so on. States of the nervous system can be further divided into *memory states* and *regulatory states*. By *memory*, we mean long-term storage, encoded in structural properties such as the pattern and strength of connections between neurons. By *regulatory states*, we mean states that are not memory in this strict sense but that nevertheless contribute to structuring behavior in time. Instances are the activity state of neurons or the concentration of neurotransmitters and neuromodulators (note that the expression *short-term memory* is used for many phenomena and may imply both nervous activity and changes in synaptic strengths; Gibbs 1991; Gibbs & Ng 1977).

Table 3.1 Principal sensory modalities of humans. Adapted from Ganong (2001). Asterisks indicate senses directed to the external world (note though that pain arises from both internal and external stimuli).

Sensory modality	Receptor	Body part
Vision*	Rods and cones	Eye
Hearing*	Hair cells	Ear (organ of Corti)
Smell*	Olfactory neurons	Olfactory mucous membrane
Taste*	Taste receptor cells	Taste buds on tongue
Rotational acceleration*	Hair cells	Ear (semicircular canals)
Linear acceleration*	Hair cells	Ear (utricle and saccule)
Touch-pressure*	Nerve endings	Various
Temperature*	Nerve endings	Various
Pain*	Naked nerve endings	Various
Joint position and movement	Nerve endings	Various
Muscle length	Nerve endings	Muscle spindles
Muscle tension	Nerve endings	Golgi tendon organs
Arterial blood pressure	Stretch receptors	Carotid sinus and aortic arc
Central venous pressure	Stretch receptors	Walls of great veins, atria
Inflation of lungs	Stretch receptors	Lung parenchyma
Blood temperature (head)	Neurons	Hypothalamus
Arterial O_2 pressure	Neurons?	Carotid and aortic bodies
pH of CSF	Neurons	Medulla oblungata
Osmotic pressure of plasma	Neurons	OVLT (inside the brain)
Glucose level	Neurons	Hypothalamus, liver

The reason for distinguishing between states of the nervous system and body states is that the latter can be treated similarly to external stimuli. Both external stimuli and body states, in fact, influence behavior via specialized receptors that transform various physical realities into neural activity. Internal senses, indeed, are as many as or more than external ones. Table 3.1, for instance, lists the senses known to exists in humans: more than half are internal. Modeling external stimuli and body states amounts thus to knowing how external and internal conditions stimulate the relevant sense organs. Just as not all sense are external, the nervous system does not produce only overt behavior. Internal responses include the control of internal organs and the secretion of hormones by specialized nerve cells. The conceptual components of behavior systems introduced so far are summarized in Figure 3.1.

Memory and regulatory states differ from body states in one crucial aspect: their function is solely to regulate behavior. A body state such as water volume does influence behavior, but first of all water is there because of the physiology of the organism. States of the nervous system are important because they allow storage of information and structuring of behavior in time. Without them, behavior would be totally controlled by current input from the senses. For instance, responding to

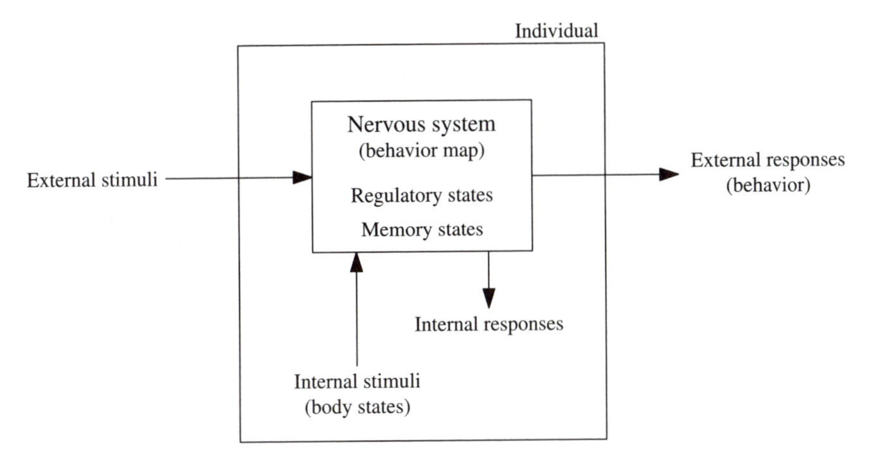

Figure 3.1 Main conceptual components of behavior systems.

a stimulus may be delayed by including a state variable that changes slowly as a consequence of repeated stimulation (see below).

To understand how behavior sequences unfold, we need to know not only what response is produced at each instant of time but also how state variables change. These interactions lie at the heart of both biological reality and the state space modeling approach (Section 1.2). Consider the following example: when a male pigeon courts a female, she may respond behaviorally by displaying her interest. In addition, male courtship causes a series of internal changes, eventually resulting in the female producing and laying eggs (Lehrman et al. 1965). The female is not ready to reproduce when the male starts courting. Rather, visual stimulation from courtship leads to brain activity, including neurosecretions that ultimately cause the secretion of follicle-stimulating hormone (FSH) into the blood. FSH stimulates follicle growth and is accompanied by secretion of estrogen from the same tissue. Estrogen, in turn, affects the brain and makes the female prepared to copulate and take part in nest building, given appropriate external stimuli. Further interactions between external stimuli, the body and the brain lead to egg productions, egg laying and incubation. Experiments clearly have shown the importance of a courting male and that the female requires some amount of stimulation to start breeding. This arrangement makes functional sense because it safeguards the female from males that would court only briefly (perhaps because they are already paired with another female). It also shows how external and internal state variables together bring about functional behavior. Hormone levels in the brain and in the blood can be viewed here as regulatory states that build up with time. In general, it is important to view behavior and body physiology as a whole, keeping in mind that the brain causes physiological changes as well as overt behavior.

In the following we apply the concepts introduced here to build neural network models of behavior. We focus on how external stimuli, body states and regulatory states control behavior, leaving the study of memory states to the next chapter. In this chapter, memory (the pattern and strength of network connections) is considered as fixed parameters that do not change (Section 1.2).

3.2 BUILDING NEURAL NETWORK MODELS

The aim of this chapter is to show how neural networks allow us to model all the various kinds of state variables considered above. In brief, external stimuli and body states are treated as identical and influence the network via input nodes that model external and internal sense organs. Regulatory states can arise from feedback loops within the network that can build persistent activity in some nodes. Additionally, special variables may be included to model regulatory states other than nervous activity (e.g., hormones). Memory corresponds naturally to connection weights (we recall that in motivational models memory is considered as fixed parameters, whereas in models of learning and ontogeny memory is a state variable that can change, see Section 1.1).

3.2.1 An introductory example

In practice, building a neural network model includes choosing a network architecture, setting network weights and testing the model. To connect the model with reality, we also need an interpretation of input and output nodes. That is, we must know how to translate physical stimulus situations (including body states) into network inputs, and we must be able to decide whether network output agrees with observed behavior. To illustrate these topics, we start with a simple example that shows how a neural network can model a specific behavior: the discrimination between two visual stimuli. The latter are a cross and a rectangle, and the network should respond only to the cross. This can be accomplished by a feedforward network, as shown in Figure 3.2. The first part of the network is a model retina of 5×5 nodes on which the cross and the rectangle can be projected. The activity of input nodes is fed to a layer of 10 hidden nodes (the figure shows only 3 for simplicity), each connected with all input nodes. The network is completed by an output node connected with all hidden nodes. The activity of the output node is interpreted as the probability that the network responds to the stimulus.

Although there are only two stimuli, the problem is not trivial because the same stimulus can yield many different input patterns. Figure 3.3 shows some retinal images obtained by projecting the cross in different positions on the retina. In most cases, the cross shape is blurred. Despite these complications, a network that responds to cross images but not to rectangle images can be obtained rather easily with the weight-setting algorithms introduced in Section 2.3, e.g., back-propagation or random search. We use the latter exactly as described in Sections 2.3.3 and 2.5.3. The result is a behavior map that can respond properly to two kinds of stimulation, as shown in Figure 3.4.

This simple example already hints at several important features of neural networks. One feature, already stressed a number of times, is the natural interpretation of the network as a model nervous system, comprising all steps from stimulus

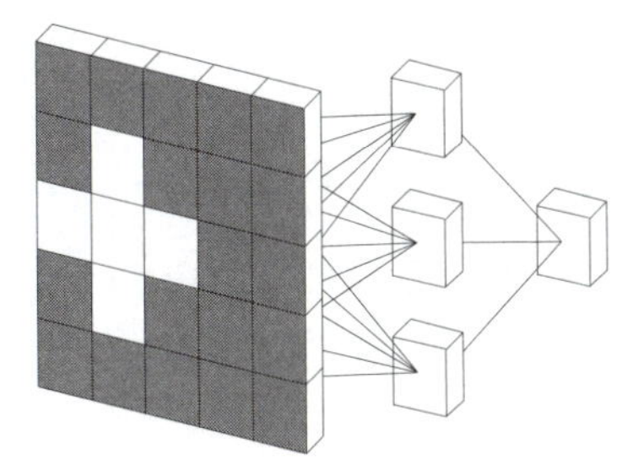

Figure 3.2 The model retina of a feedforward network is stimulated by a cross-shaped stimulus, resulting in a pattern of activity of the input nodes. The latter is fed through the network to generate an output response, interpreted as the probability of reacting to the stimulus.

to response. A second feature is power and flexibility. Although the recognition problem was fairly complex, we did not have to consider it in detail: the weight-setting algorithm could find suitable weights by itself. We easily could have introduced more stimuli and more responses, resulting in increasingly complex behavior maps. That the same basic structure can realize many behavior maps is important because it allows for both individual flexibility (learning) and evolutionary flexibility (Chapter 5). If each problem had to be solved by a dedicated mechanisms, behavior systems would be very complex to build.

A third feature concerns responding to novel stimuli. Considering our example, one may object that we simplified the problem greatly because we tested the network with only some of the infinitely many retinal images potentially projected by each stimulus. How does the network react to images not included in the weight-setting procedure? Actually, we already did this test. The network was trained only with *half* the cross images in Figure 3.3, but the test in Figure 3.4 comprises *all* images. Testing with still more images would produce similar results. We will see below that neural networks reproduce very well how animals react to novel stimuli.

3.2.2 More complex models

In general, a neural network model of behavior includes one or several groups of input nodes modeling particular sense organs, an internal architecture with connections between receptors and effectors (direct or mediated by hidden nodes) and finally effectors modeling how the nervous system controls muscles and produces other physiological responses. There are many potentially useful choices of internal architecture. For instance, some weights might be fixed, whereas others could change with experience or as a consequence of maturation processes. It is often useful to have one or a few layers of fixed processing of the input, followed by a network with adjustable weights that can be molded to respond flexibly to the processed patterns (Section 3.4). Another possible architectural element is an internal clock that allows the network to vary its sensibility to different inputs with time of day. The clock either could feed its activity to the network in the same manner as input nodes or could be built out of network nodes.

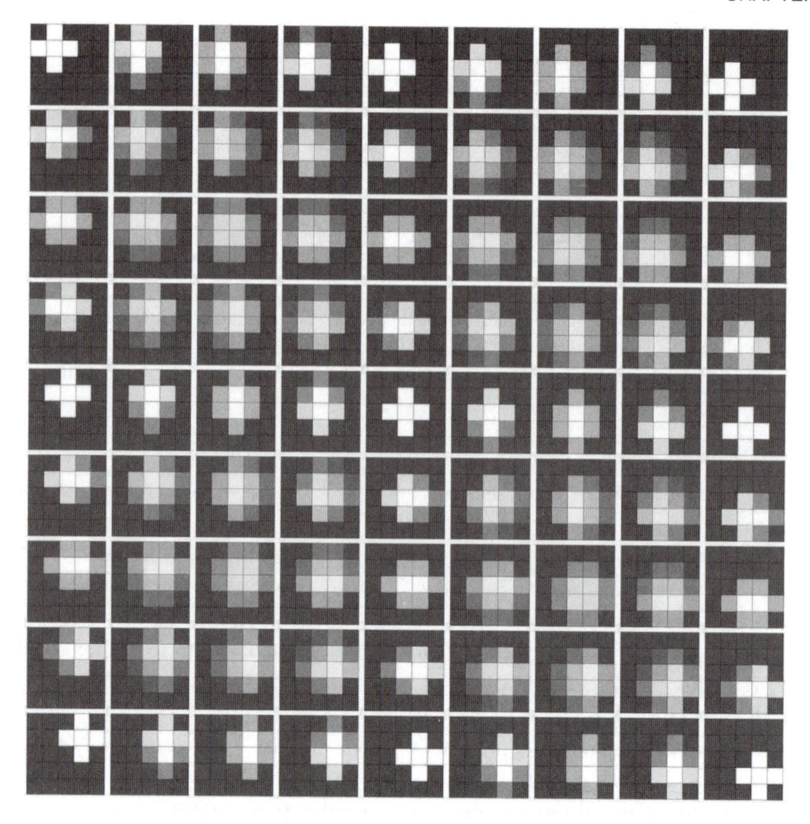

Figure 3.3 Reception of a cross-shaped visual stimulus by a 5×5 model retina showing that the same stimulus can produce quite different patterns of activity depending on how it falls on the retina. Each retina node is active in proportion to the part of the stimulus that covers it (a lighter shade of gray corresponds to higher activity). A node that is not covered at all has a background activity of 0.05; a node that is fully covered has an activity of 1.

How complex different parts should be depends on our aims and on the complexity of the behavior we want to model. For instance, if we are interested in reactions to stimuli or decision making we usually avoid modeling the control of muscles and represent the animal's response with the activity of just one output node. A model of motor control, on the other hand, must include more details in the output part of the network. Figure 3.5 presents a few examples of how neural network components can be combined into motivational models.

In Chapter 1 we stressed that to account for sequences of behavior, models of behavior must predict how the animal motivational state changes, in addition to predicting current behavior. Here we face several different situations. Regulatory states are internal to the system itself and thus should get updated as a part of the system's own dynamics. For instance, nodes that receive recurrent connections are updated based on the activity of other nodes in the network. Other regulatory states,

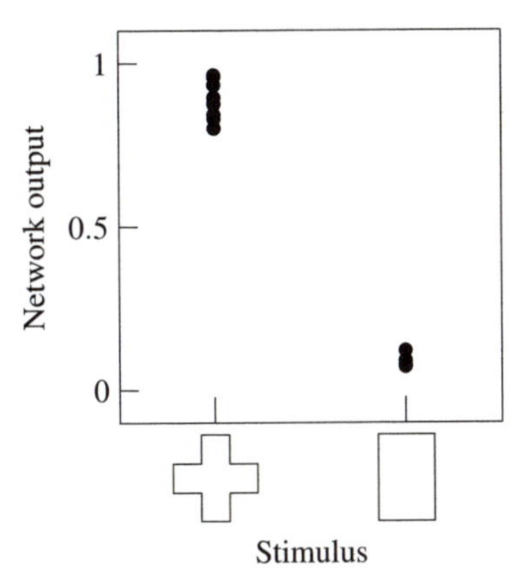

Figure 3.4 Reactions of a feedforward network (with 10 hidden nodes) to retinal images produced by a cross and a rectangle pasted at different positions on the retina (Figure 3.3). The cross was translated horizontally and vertically in steps of length 0.25 (given that an input node is a square of side 1), resulting in the 81 images in Figure 3.3. The same procedure, applied to the 2×3 rectangle, results in 117 images. Network weights have been obtained by random search as described in Sections 2.3.3 and 2.5.3.

such as those modeling hormone levels, may need special update equations. State variables modeling body physiology are instead external to the network and change only indirectly as a consequence of responding. For instance, drinking affects water volume only as a consequence of ingestion of water. Behavior may also change the relation with the environment (e.g., position and orientation), resulting in different input at later times. More subtle changes in later input are possible if behavior can alter the physical or social environment. We will see below some instances of these different cases.

3.2.3 Setting network weights

Once we have decided on a network architecture, we need to adjust network weights so that the model operates according to some requirements. Requirements on network behavior may come either from observations of animals or from functional considerations such as being able to survive in a given environment. In this chapter we are not concerned with how suitable weights are obtained: we consider them as given (in practice, we find suitable weights using the methods described in Chapter 2). Chapter 4 considers weight changes from the point of view of understanding learning and ontogeny of behavior.

3.2.4 Testing models

How can neural network models and models of behavior in general be tested? First, a model should be able to reproduce the observations that were specifically used to develop the model or solve the functional task it was given. While this is a fun-

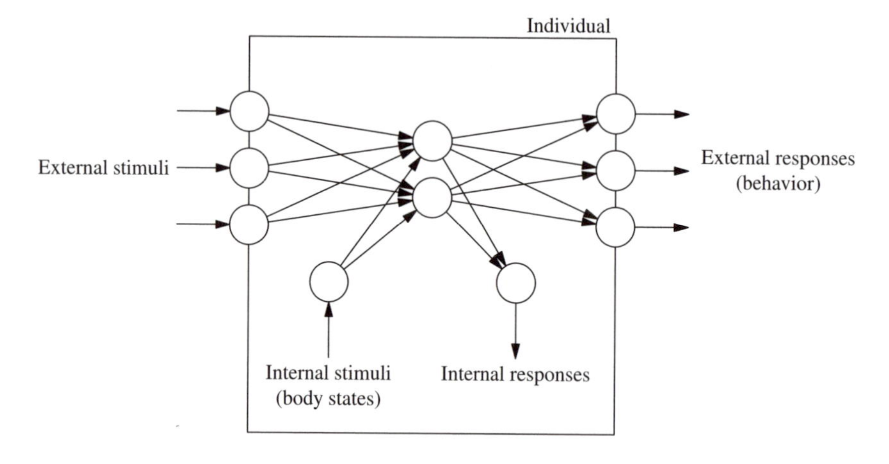

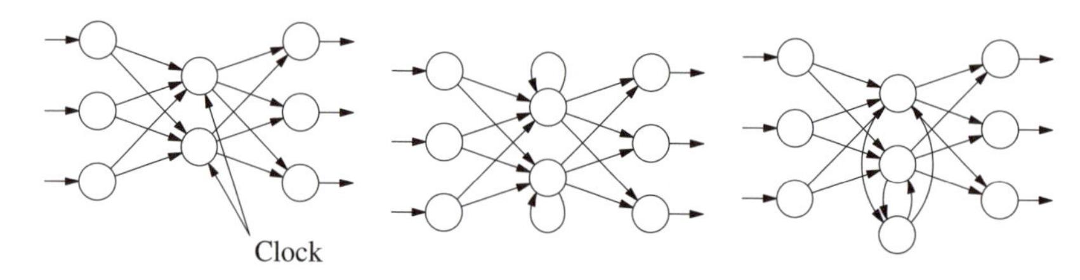

Figure 3.5 Various examples of how neural network components may be combined into a model of behavior. Top: A feedforward network fitted to the conceptual scheme in Figure 3.1. Bottom: Three alternative architectures that could replace the former one. From left to right: A "clock" internal input allows one to model rhythms in behavior; recurrent connections on the hidden nodes allow one, for example, to delay responses to stimuli; adding hidden nodes allows one to build internal states that retain information about previous stimulation. These and other examples will be analyzed more carefully in the following.

damental requirement, it may be easy to achieve with systems as flexible as neural networks. An important additional test is to evaluate the model in situations that were not included in the fitting procedure. We will employ both these conditions frequently. If we aim to develop physiological models, the model should agree with known physiological data about the functioning of nervous systems. Below and in the next chapter we will describe several neural network models with support from neurophysiology.

3.2.5 The receptor space

Based on our subjective experience, it is tempting to assume that animals simply know what stimuli are there and have direct access to characteristics such as color, size and distance. Some models of behavior actually make similar assumptions (see Section 3.3.3). However, perception is often more complex than it seems.

For instance, yellow appears to us as a color that cannot be split into components but actually is perceived by integrating signals from three kinds of photoreceptors. Figure 3.3 is a further example of how difficult it is to make sense of signals from receptors.

The analogy between neural networks and nervous systems encourages us to consider perception in more detail. We will use the term *receptor space* to mean a network input space built with consideration of how sense organs receive stimulation. Most of the results below depend on a degree of realism in the construction of network inputs, but building a receptor space that models a given physical situation adequately is not always trivial. A basic fact is that only those aspects of physical stimuli that can be received can influence behavior. For instance, some animal species can perceive ultraviolet light, whereas others cannot. If we consider only the physical properties of light, we cannot understand why just some species can discriminate between light that differs only in its ultraviolet component. In general, we need to consider the sensitivity of individual receptors to physical characteristics of stimuli, as well as the arrangement of receptors in sense organs. Note also that thinking about receptors includes naturally the indexcontext*context* in which a stimulus is presented as part of the stimulation received by the animal: a change in context is modeled simply as a change in stimulation.

Studies of sensory physiology reveal both similarities and differences between the senses of different animal species (Coren et al. 1999; Gegenfurter & Sharpe 1999; Goldsmith 1990; Warren 1999; Zeigler & Bischof 1993). It is possible to highlight some basic similarities that can be used to build simple yet general models of how stimuli activate sense organs (see also Figure 2.2):

- A receptor is usually maximally sensitive to a particular value along a stimulus dimension (e.g. light wavelength or sound frequency); receptor activity declines in an orderly fashion as physical stimulation departs from the value of maximum sensitivity.
- One important exception is intensity dimensions: when a physical stimulus increases (decreases) in intensity, the activity of receptors sensitive to the stimulus also increases (decreases).
- Since receptors have different characteristics (even within one sense organ), a given stimulus usually elicits a distributed activity pattern, with some receptors activated strongly, some mildly and some negligibly.
- Even when stimulation is seemingly lacking, receptors usually are slightly active rather than being entirely silent (e.g., sound receptors in a quiet room).

3.3 REACTIONS TO STIMULI

To predict how animals react to external stimuli is a very common problem, and students of behavior within ethology and psychology have dealt with it extensively (Baerends & Drent 1982; Hinde 1970; Lorenz 1981; Mackintosh 1974, 1994; Pearce 1997; Tinbergen 1951). Ethologists often speak of "stimulus selection" and experimental psychologists of "stimulus control." In practice, we usually know how

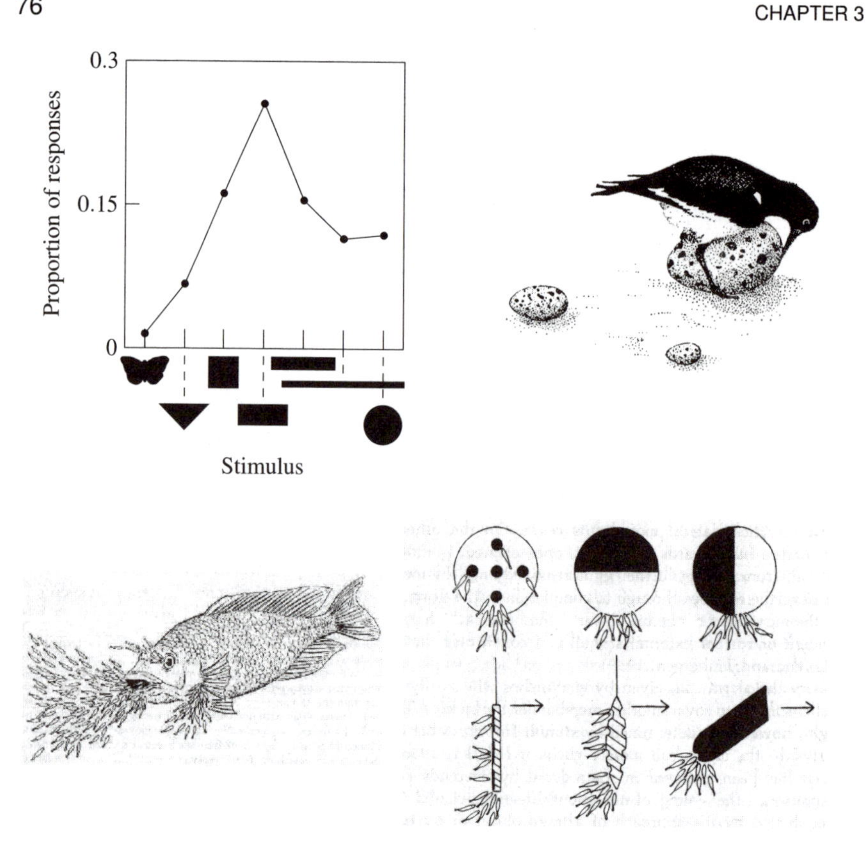

Figure 3.6 Ethological studies of reactions to stimuli. Top left: Magnus (1958) placed in a field different cardboard shapes of the same color as *Argynnis paphia* female butterflies, and then observed how often the different shapes were approached by males of the same species. The shape most closely resembling the natural stimulus was not the preferred one. Top right: Tinbergen (1948) let oystercatchers (*Haematopus ostralegus*) choose among eggs of different sizes. Surprisingly, the birds preferred giant-sized eggs to eggs of natural size (the smallest one in the figure). Reprinted from Tinbergen (1951) by permission of Oxford University Press. Bottom: Young of the mouthbreeding fish *Tilapia mosambica* swim into the mother's mouth when large objects approach or the water is disturbed (left). Baerends & Baerends-van Roon (1950) studied what features of visual stimuli direct this response (right). The young are attracted by dark spots and the bottom part of receding objects, and they find hollows by pushing against the surface. Reprinted by permission of the McGraw-Hill Companies from Hinde (1970). *Animal Behavior*. McGraw-Hill Kogakusha.

an animal reacts to a set of *familiar stimuli*, and we want to predict how it will react to one or more *test stimuli* (Figure 3.6). In this section we test two- and three-layer neural network on tasks modeled on actual experiments, and we show that the networks react very similarly to animals. To understand the basic biological problem and how neural networks can be applied to it, consider an animal that reacts to a

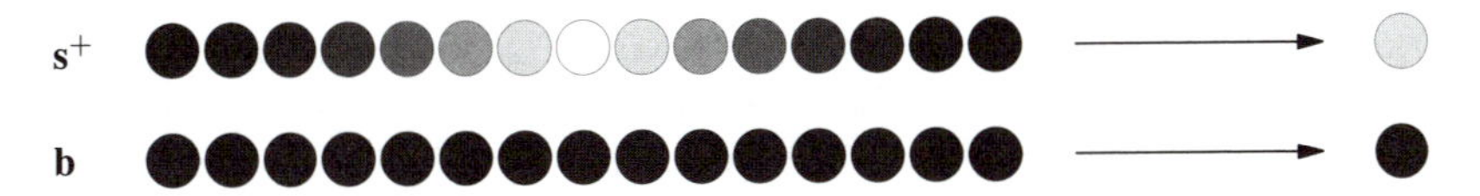

Figure 3.7 Input-output map for our first model of reaction to stimuli. The input $\mathbf{s}^+$ must be associated with a high output and input **b** with a low one. Input **b** can be interpreted as a background stimulus (e.g., low ambient noise), whereas $\mathbf{s}^+$ may correspond to a stimulus such as a tone.

stimulus S^+ and ignores background stimulation B (the notation S^+ is traditional in experimental psychology for a stimulus to which a response has been established, e.g., through classical or instrumental conditioning). The question is: how will the animal react to other stimuli? Our model is a simple network consisting of an array of input nodes connected directly to one output node (Section 2.2.1.1; see Figure 3.9 for technical details). We require the network to reproduce the input-output map in Figure 3.7. The input pattern $\mathbf{s}^+$ models S^+ and should produce a high activity in the network's output node, interpreted as a reaction to S^+. As introduced in Chapter 2, $\mathbf{s}^+$ is a *vector* whose element s_i is the activity of input node i. The input pattern **b** models a low-intensity background and should produce low activity in the output node. The difference between **b** and $\mathbf{s}^+$ is increasing activity toward the center of the node array. This models the fact that a stimulus typically activates many receptors to different extents, there being cells that are most active, some that are mildly active and some that are unaffected. The activity of sound receptors in the ear of many species, for example, would be rather similar to the figure, when S^+ is a single sound frequency. In this interpretation, **b** would model silence or ambient noise, where all receptors are active at low levels.

The network can discriminate easily between $\mathbf{s}^+$ and **b**, and weight vectors that solve the task can be obtained by the techniques presented in Section 2.3. In the following we assume that the discrimination has been established and explore how the network reacts to novel input patterns resembling those used in actual experiments. Before proceeding, we introduce the concept of *overlap* between patterns (also referred to as *scalar product* or *inner product*). Overlaps are easier to work with than network mathematics and allow us to understand how simple networks (and, to varying extents, more complex ones) operate. To the extent that networks are good models of behavior, overlaps also allow us to predict animal reactions to stimuli. Formally, the overlap between patterns **a** and **b** is written $\mathbf{a} \cdot \mathbf{b}$ and is defined as

$$\mathbf{a} \cdot \mathbf{b} = \sum_i a_i b_i \tag{3.1}$$

where the index i runs over all nodes in a network layer (usually we consider the input layer, so **a** and **b** are input patterns). What does this mean in practice? A first aspect corresponds to the everyday use of the word *overlap*: similar input patterns tend to overlap more than dissimilar ones. A second aspect relates to the intensity of activation: the overlap of **a** with **b** increases if we increase the values of elements of **a**. In summary, based on equation (3.1), we can say that the overlap $\mathbf{a} \cdot \mathbf{b}$ is large if **a** and **b** have *high activity on the same nodes*. Equation (3.1) applies also to two-dimensional patterns (the index i merely identifies nodes and does not imply that they are arranged in any particular way). For instance, patterns that are close to each other in Figure 3.3 overlap more than patterns that are far away. In the following we will gain more familiarity with overlaps and see how they are used in practice.

The reason why overlaps are useful can be summarized as follows. The reader may recall from Chapter 2 that a network node essentially computes the overlap between the input and weight vectors (equation 2.15 on page 40). In particular, the response of a two-layer network with a single output node is completely described in terms of the overlap between the input pattern and the weight vector. In turn, the weight vector can be expressed as a weighted sum of familiar patterns (Section 2.3.1). Thus the response to an input pattern can be predicted based on its overlap with familiar patterns without explicitly considering the network.

3.3.1 Generalization gradients

How animals react to stimuli often is explored by modifying the familiar stimuli along specific physical dimensions, such as sound frequency or light wavelength. Such tests are referred to as *generalization tests* and the resulting response curves over the dimension as *generalization gradients*. Generalization has been studied extensively in psychology (Kalish 1969a,b,c,d; Mackintosh 1974), ethology (Baerends & Drent 1982; Baerends & Kruijt 1973) and behavioral ecology (Enquist & Arak 1998; Ryan 1990). These empirical studies show that generalization is a fundamental phenomenon. The main features of generalization are independent of what species we study (e.g., human, rat, chicken, pigeon, goldfish or butterfly), what behavior system is engaged by the stimuli (e.g., hunger, thirst or sex), what sensory modality is probed (e.g., vision, sound or touch) and whether the response is learned or innate (Ghirlanda & Enquist 2003).

Below we analyze our model along two important kinds of stimulus dimensions called *rearrangement* and *intensity* dimensions (details follow). These are the most empirically investigated dimensions and can be given a simple explanation in a receptor space. These two kinds of dimensions give very different results in animal experiments and therefore are a good test for models of stimulus control. The testing procedure is conceptually identical to studies of generalization in animals: a behavior map such as the one in Figure 3.7 is established, and then network responses to a set of test stimuli, modeling a particular empirical test, are observed.

3.3.1.1 Rearrangement dimensions

Varying stimulation along a "rearrangement dimension" means changing the *pattern* of stimulation while keeping total *intensity* unchanged (at least approximately). An example is variation in sound frequency. When frequency is either increased or decreased, the receptors in the ear most sensitive to the original frequency are less stimulated, whereas some other receptors get more active (Fay 1970; Galizio 1985; Jenkins & Harrison 1960). For visual stimuli, the spatial distribution of stimulated

Figure 3.8 Stimulus dimensions obtained by variation of one parameter in the formula $x_i = l + h\exp(-(i-m)^2/2\sigma^2)$, where x_i is the activity of the node at position i in the linear input array. The most active node is at position $i = m$ and has an activity of $l + h$. Left: A rearrangement dimension obtained by varying m. Right: An intensity dimension obtained by varying h. Values of parameters, when not varied, are $l = 0.2$ (background activity), $h = 0.5$ (maximum level beyond background activity), $m = 8$ (position of maximally active node), $\sigma = 3$ (regulates how quickly activity drops when departing from the maximally active node).

receptors may change, whereas the total amount of stimulation is unchanged, such as when changing the position or orientation of a stimulus (see also Section 3.4) or when changing object shape and keeping its size constant. Changing the wavelength of a monochromatic light (at least in a restricted range) is another example (Hanson 1959; Marsh 1972; Ohinata 1978). In reality, there are almost endless ways of constructing rearrangement dimensions for all senses.

Empirical gradients of generalization along rearrangement dimensions are almost invariably bell-shaped, with maximum responding observed at the S^+ or close to it (Ghirlanda & Enquist 2003). Figure 3.9 compares empirical findings with our model. As can be seen, the network model generalizes very similarly to real animals. We constructed our rearrangement dimension by translating the activity pattern in $\mathbf{s}^+$ over the input layer, obtaining the set stimuli in Figure 3.8 (left). We could have employed other rearrangement dimensions while still obtaining a bell-shaped gradient. This can be understood either by visualizing the network's weight vector or reasoning about overlaps. We consider both explanations. Weight values corresponding to the behavior map in Figure 3.7 are depicted in Figure 3.10. We see both positive and negative weights. The $\mathbf{s}^+$ pattern stimulates mostly input nodes linked to positive weights, and this explains why the network responds strongly to $\mathbf{s}^+$. In contrast, $\mathbf{b}$ stimulates equally nodes linked to positive and negative weights. As a result, output to $\mathbf{b}$ is low. Let us now turn to generalization. In a test pattern obtained by rearranging $\mathbf{s}^+$, number of high-valued elements have been removed from the center and exchanged with lower-valued elements (see Figure 3.8, left).

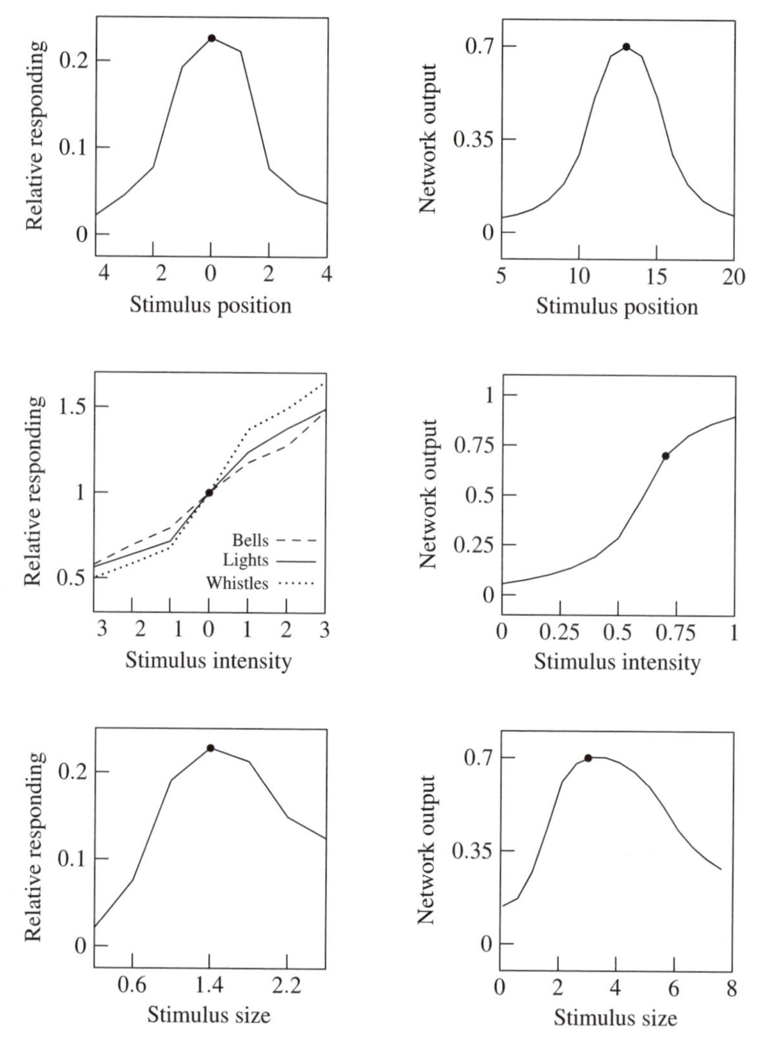

Figure 3.9 Left column: Empirical generalization gradients. The closed circle indicates the training stimulus. From top to bottom: Responding in pigeons to stimuli in different positions (a rearrangement dimension) after training on a particular position (data from Cheng et al. 1997); responding in dogs along three kinds of intensity dimensions (data from Razran 1949; summarizing data from Pavlov's laboratory); size generalization in rats (data from Brush et al. 1952). Right column: Generalization in a two-layer feedforward network trained on the input-output map in Figure 3.7 and tested along a rearrangement dimension (top), an intensity dimension (middle) and a size dimension (bottom). The first two dimensions correspond to the stimuli in Figure 3.8; the size dimension is obtained by varying the standard deviation of the Gaussian activity profile in similar stimuli (the σ parameter in the stimulus formula in Figure 3.8). Formally, network response r to the input pattern $\mathbf{x}$ is given by $r = f(\sum_i w_i x_i)$, where f is a sigmoid function of maximum slope equal to 1 and such that $f(0) = 0.1$. See Chapter 2 for further details.

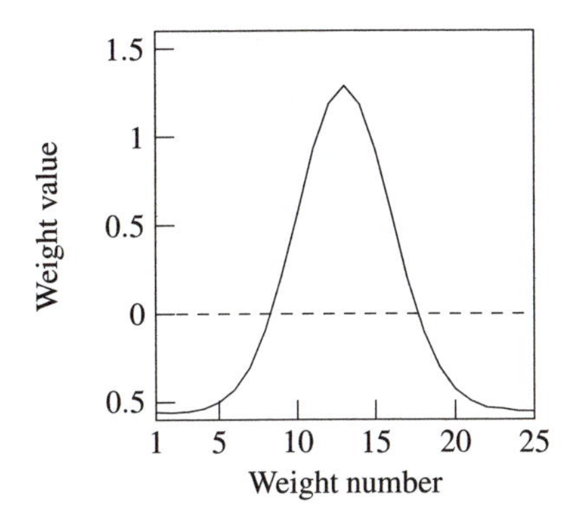

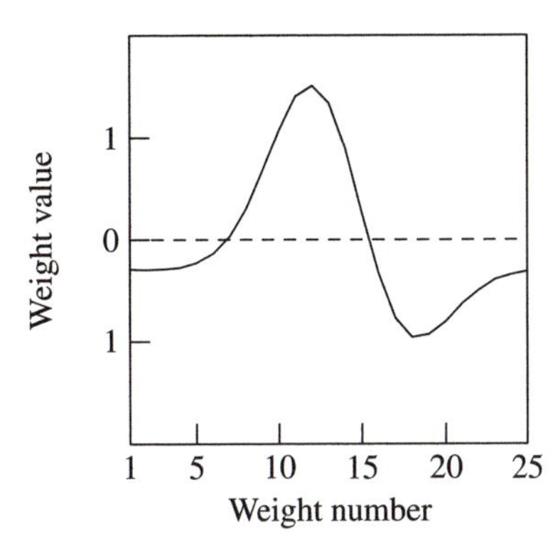

Figure 3.10 Weight values resulting from training a two-layer network to solve the discriminations in Figure 3.7 (left) and Figure 3.11 (right).

Nodes linked to positive weights thus will receive less input and nodes linked to negative weights more input. The more rearranged a stimulus is relative to $\mathbf{s}^+$, the less of its high-valued elements will correspond to high-valued weights. Thus we see that network output drops with greater rearrangement, and maximum output corresponds exactly to $\mathbf{s}^+$ (no rearrangement).

We reach the same conclusion considering overlaps. We showed in Section 2.3.1 that after the behavior map in Figure 3.7 is established, responding to any test stimulus $\mathbf{x}$ is (nearly) equal to

$$r(\mathbf{x}) = f\left(c_1 \mathbf{s}^+ \cdot \mathbf{x} + c_2 \mathbf{b} \cdot \mathbf{x}\right) \tag{3.2}$$

where the numbers c_1 and c_2 are such that $r(\mathbf{s}^+)$ and $r(\mathbf{b})$ have the desired values (Section 2.3.1). Thus responding to $\mathbf{x}$ depends on its overlap with $\mathbf{s}$ and $\mathbf{b}$. The latter is of little interest in the present case: since $\mathbf{b}$ is uniform and all test stimuli are rearrangements of $\mathbf{s}^+$, the overlap $\mathbf{b} \cdot \mathbf{x}$ is constant and does not contribute to differences in output. Variation in overlap with $\mathbf{s}^+$ is what produces different responding to the test stimuli. It is easy to see that the more rearranged $\mathbf{x}$ is relative to $\mathbf{s}^+$, the less it overlaps with it.

3.3.1.2 Intensity dimensions

In contrast to rearrangement, changing a stimulus along an intensity dimension does not alter which receptors are stimulated; rather, the activity of all receptors is increased or decreased simultaneously. Empirical generalization gradients along intensity dimensions are not bell-shaped: the gradient grows monotonically, and stimuli of higher intensity than the S^+ evoke stronger responses. This finding has been replicated along many dimensions, including sound intensity (Razran 1949; Zielinski & Jakubowska 1977), light intensity (Razran 1949; Tinbergen et al. 1942) and chemical concentration (Tapper & Halpern 1968). Figure 3.9 shows some examples.

To test how our network model generalizes along an intensity dimension, we used a different set of stimuli, in which the height of the activity peak in $\mathbf{s}^+$ varies

(Figure 3.8, right). This models the increase in receptor activity observed in receptors (or other neurons) when the intensity of stimulation increases. Figure 3.9 shows that the network generates monotonic gradients along intensity dimensions. This result is explained as above: if we increase the intensity of the test stimulus, its overlap with $\mathbf{s}^+$ increases more than its overlap with $\mathbf{b}$, resulting in increased output.

3.3.1.3 Other dimensions

Most empirical studies of generalization have dealt with either rearrangement or intensity dimensions. One exception is size generalization (Dougherty & Lewis 1991; Grice & Saltz 1950; Magnus 1958; Mednick & Freedman 1960). An example is a series of circles that vary in size. Size generalization is peculiar to visual stimuli, although one could imagine similar dimensions for other senses. Size generalization gradients often are asymmetrical, with stronger responses to stimuli that are larger than S^+ (Ghirlanda & Enquist 2003). Our network model generalizes similarly, as shown in Figure 3.9.

Sometimes stimuli are used that do not lie on a simple stimulus dimension. In ethology, for instance, animals often are presented with modifications of natural stimuli that vary along many dimensions (Figure 3.6). A strength of the receptor space approach is that predictions about responding can be generated for any stimulus, given that we can determine how it stimulates receptors. For instance, we can easily generate predictions about responses to stimuli that vary in both intensity and rearrangement relative to $\mathbf{s}^+$. Another example of this flexibility is compound stimuli composed by adding or removing discrete stimuli. For instance, presenting S^+ together with a novel stimulus A often produces a weaker response than presenting S^+ alone (*external inhibition*; Pavlov 1927; Pearce 1987). This can be explained by noting that an input pattern $\mathbf{a}$ (modeling stimulus A), to the extent that it differs from $\mathbf{s}^+$, will activate mostly nodes linked to negative weights. Hence the vector that models the simultaneous presence of A and S^+ will also provide more intense input to negative weights.

3.3.2 Finer discriminations

So far we have analyzed discriminations between a stimulus S^+ and its absence, or background B. Here we analyze generalization after introducing a "negative stimulus" S^- that is similar to S^+. This means that animals respond to S^+ but not to S^- or B. The addition of a negative stimulus introduces new phenomena along both rearrangement and intensity dimensions. Let us first consider rearrangement. Extensive empirical data show that if S^+ and S^- are close enough, generalization along the stimulus dimension is no longer symmetrical around S^+. The gradient remains bell-shaped but is shifted away from S^-. Maximum responding does not occur at S^+ but at a stimulus more distant from S^- (Hanson 1959). Psychologists have called this phenomenon *peak shift* (Hanson 1959; Mackintosh 1974). Ethologists too have noted that novel variants of familiar stimuli, referred to as *supernormal stimuli*, may elicit a stronger response than the familiar stimuli themselves (Tinbergen 1951). The oystercatcher in Figure 3.6 is a classical example. In general, we may refer to such phenomena as *response biases*. We already saw in Section 3.3.1.2 that biases exist favoring responding to more intense stimuli.

We can model discriminations between similar stimuli by adding an input pattern $\mathbf{s}^-$, similar to $\mathbf{s}^+$, to which the network should not react. One instance is the input-output map in Figure 3.11, where $\mathbf{s}^-$ is a rearrangement of $\mathbf{s}^+$. This is a good model of fine discriminations along rearrangement dimensions such as sound frequency or light wavelength. Figure 3.10 shows the weights that result from training a network on this discrimination. Positive weights correspond to nodes where $\mathbf{s}^+$ is more intense and negative weights to nodes where $\mathbf{s}^-$ is more intense. However, the largest positive weights do not match $\mathbf{s}^+$ exactly: they are shifted away from $\mathbf{s}^-$. This means that maximum responding will not be at $\mathbf{s}^+$, as shown in Figure 3.12 (right). In terms of overlaps, responding to a test stimulus $\mathbf{x}$ should now be explained by taking into account its overlap with both $\mathbf{s}^+$ and $\mathbf{s}^-$. The latter tends to decrease responding to $\mathbf{x}$. Peak shift arises because some stimuli along the dimension retain most of their overlap with $\mathbf{s}^+$ but overlap significantly less with $\mathbf{s}^-$. Peak shift can also be viewed as arising from interaction among memory parameters, since the same weights encode the reaction to both $\mathbf{s}^+$ and $\mathbf{s}^-$. In Chapter 5 we will discuss the evolutionary relevance of this phenomenon.

Response biases are also observed along intensity dimensions, when animals are trained to discriminate between two similar intensities. When S^+ has a stronger intensity than S^-, the generalization gradients is very similar to that obtained when S^+ is discriminated from the background (indeed, the latter can be considered an S^- with very low intensity). However, if S^- is more intense than S^+, the result is somewhat surprising. The generalization gradient now decreases rather than increases with intensity. This shows that it not intensity per se that control responding but the interference between memories along the intensity dimensions.

Discriminations between stimuli of similar intensity can be modeled by picking $\mathbf{s}^+$ and $\mathbf{s}^-$ from the dimension in Figure 3.8 (right). We meet here one limitation of our model, namely, the two-layer network cannot realize the input-output map where $\mathbf{s}^+$ is less intense than $\mathbf{s}^-$. Technically, this happens because the three stimuli $\mathbf{b}$, $\mathbf{s}^+$ and $\mathbf{s}^-$ lie on a line in the network's input space (i.e., one of them can be written as a weighted sum of the other two; Section 2.3.1.1). Along such a line, the output of a two-layer network (or any linear network) is either ever-increasing, ever-decreasing or constant. Adding a layer of nonlinear nodes removes this limitation. The three-layer network solves the task easily and generalizes realistically, as shown in Figure 3.12. For completeness, we note that the three-layer network also reproduces the findings considered earlier (Ghirlanda & Enquist 1998).

3.3.3 Comparison with other models

Ethologists and psychologists have researched extensively how animals react to stimuli (Baerends & Drent 1982; Hinde 1970; Kalish 1969a,b,c,d; Lorenz 1981; Mackintosh 1974). We can distinguish two closely related issues. The first deals with how stimuli are modeled; i.e., what is the input to the behavior mechanism? The second is how the behavior mechanism operates based on such input.

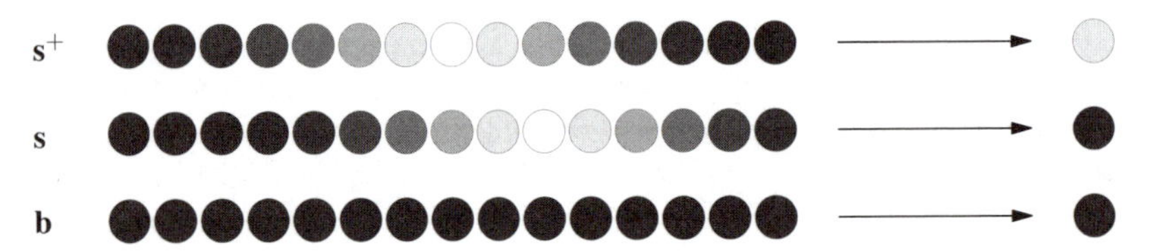

Figure 3.11 Input-output map modeling discrimination between two similar stimuli $\mathbf{s}^+$ and $\mathbf{s}^-$ (see Figure 3.7). The network must react to $\mathbf{s}^+$ and ignore both $\mathbf{s}^-$ and a background stimulus **b**.

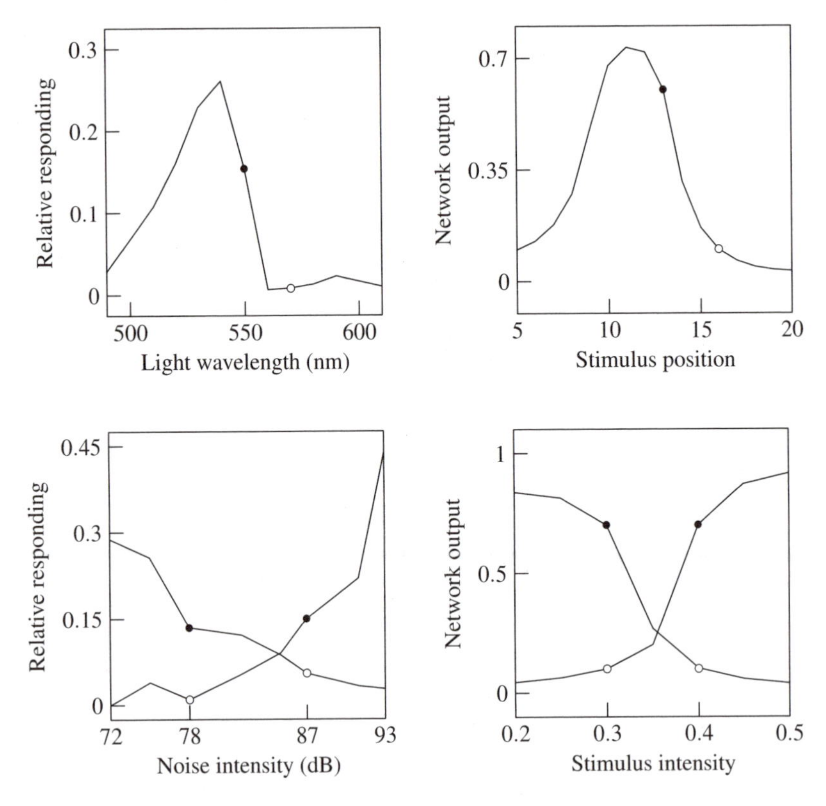

Figure 3.12 Response biases along rearrangement and intensity dimensions in animals and networks. Top left: Light-wavelength generalization in pigeons trained to react to a 550-nm light but not to a 570-nm one (data from Marsh 1972). Top right: Network output to the stimuli in Figure 3.8 (left) after training on the behavior map in Figure 3.11. Bottom left: Noise intensity generalization in rats trained to respond to one noise intensity (full circles) but not to another (empty circles) (data from Huff et al. 1975). Bottom right: Output of a three-layer network along the intensity dimension in Figure 3.8 (right). The network was trained by random search (Section 2.3.3) to react to an intense stimulus and not to a less intense one (ascending gradient), or vice versa (descending gradient). The relative intensity of the model stimuli was approximately matched to the relative intensity of the physical stimuli used by Huff et al. (1975).

A common approach is to assume that animals "know" what stimuli are present or absent; i.e., this information is directly available to the behavior mechanism. Based on this, it is often assumed that responding to a compound stimulus (e.g., a light and a sound simultaneously present) arises as the sum of independent contributions from component stimuli (*summation model*). Thus responding to the simultaneous presence of three stimuli may be written as

$$r(1,2,3) = W_1 + W_2 + W_3 \tag{3.3}$$

where W_i is the contribution of element i. Responding when only stimuli 1 and 2 are present would be given by $r(1,2) = W_1 + W_2$. This approach is followed in many psychological models of associative learning (Atkinson & Estes 1963; Rescorla & Wagner 1972); in this context, W_i is called the *associative strength* between stimulus i and the response. Summing contributions from different stimuli is also a prominent ingredient of several "cognitive" models of memory, such as *instance theory* (Chapter 3 in Shanks 1995). Ethologists used equation (3.3) as well, calling it the *law of heterogeneous summation* (Baerends & Kruijt 1973; Lorenz 1981; Seitz 1940–1941, 1943). In neural network terms, the model is equivalent to assuming that each input node only reacts to a given stimulus. Such a "detector" node would have an activity of 1 when the stimulus is present and of 0 when the stimulus is absent (Section 1.3.3). Thus equation (3.3) can be written as a two-layer network:

$$r = \sum_i W_i x_i \tag{3.4}$$

where x_i is 0 when stimulus i is absent and 1 when it is present.

A major problem of summation models is that they can handle only stimuli that can be broken up into discrete components. Generalization along continuous dimensions such as object size, sound frequency and light wavelength cannot be dealt with. The *gradient-interaction theory* of Spence (1937) and Hull (1943) instead deals with just such issues. Spence hypothesized that when a response is established to a stimulus having a given value along a continuous dimension, a "gradient of excitation" builds around it. Stimuli with different values on the dimension would thereby elicit a response proportional to the height of the gradient at that point. Nowadays this seems just an acknowledgment that generalization occurs, but generalization gradients had not yet been observed when Spence wrote! Spence used gradients as theoretical constructs to show that the interaction between an excitatory gradient around S^+ and an inhibitory one around S^- could result in a shifted gradient, inducing animals to prefer to S^+ a stimulus more removed from S^- than S^+ itself (*transposition*). This predated by more than 20 years the observation of actual shifted gradients by Hanson (1959). Despite these successes, the main shortcoming of Spence's theory, as well as its refinements proposed by Hull (1943, 1949), is still that generalization is assumed rather than explained. The theory cannot predict why a given shape of the gradient is observed, e.g., why intensity generalization gradients are not bell shaped (Ghirlanda 2002). Moreover, the theory has a drawback complementary to those of summation models: it cannot deal with discrete stimuli!

The model by Blough (1975) combines features from the summation tradition and the gradient-interaction one. Blough assumed that every stimulus along a given dimension activates an array of "stimulus elements." In contrast with earlier summation models, each stimulus activates many elements, and different stimuli induce partially overlapping activity patterns. Responding to a pattern of activity is still given by equation (3.4) but the x_i's are continuous values corresponding to the degree to which element i is active. Blough further assumed that each element has a "preferred" value along the dimension, to which activity is maximal. Element activity decreases in a smooth fashion as the actual stimulus value departs from the "preferred" one. The patterns of activity generated by this scheme along a continuous dimension match the stimuli in Figure 3.8 (left).

Formally, Blough's model is the same as the two-layer neural network considered here. A key difference, however, lies in the interpretation of the "stimulus elements." Blough regarded them as a somewhat arbitrary abstract device, a price to pay to describe behavior accurately. This has lead to considerable criticism (Pearce 1987, 1994). We have instead interpreted the "elements" as models of receptors, which gives them a concrete interpretation and establishes a general method to build input patterns. A first consequence is that the difference between discrete stimuli and continuous dimensions disappears. Every stimulus situation, not only continuous dimensions, is represented as an input pattern, just as it happens with real sense organs (Blough proposed instead to merge his model with Rescorla & Wagner's summation model, resulting in one set of associative strengths for "within-stimulus" generalization and one set for "between-stimuli" generalization). Moreover, the correspondence between network inputs and sense organ activity patterns allows the model to account for phenomena that require additional assumptions in formally identical models such as Blough's and Rescorla & Wagner's. Two examples are intensity generalization (Section 3.3.1.2) and external inhibition (p. 82). The former is often explained by introducing ad hoc explanations that favor responding to intense stimuli, with only partial success (Ghirlanda 2002). The latter has sometimes been explained as an effect of attention. A third example is *summation.* If r_1 and r_2 are the responses to stimuli 1 and 2, responding to the compound $(1,2)$ is usually less than $r_1 + r_2$ (Hull 1943). A simple way to achieve this is to assume a nonlinear transfer function for the output node, appealing to nonlinearities in neuron operation. However, we may also note that the physical situation of 1 and 2 being present together does not induce in the sense organs an activity pattern that is the sum of those induced by 1 and 2 presented alone. Usually, total activity is less (e.g., owing to nonlinearities in receptors), thus naturally yielding a response less than $r_1 + r_2$.

We have seen in Chapter 1 that a number of neural network-like models of stimulus control have been developed within ethology (Sections 1.3 and 1.4). In particular, we discussed the models by Baerends and Sutherland shown in Figure 1.11 on page 26. Here we consider a further example, resulting from Baerends and coworkers' impressive work on egg retrieval in the herring gull (*Larus argentatus*). In short, if an egg accidently falls out the nest, gulls retrieve it back into the nest (as do many other birds; see, e.g., Figure 3.34). Observations that odd objects such as cans and stones could sometimes be found in the nests of gulls prompted research

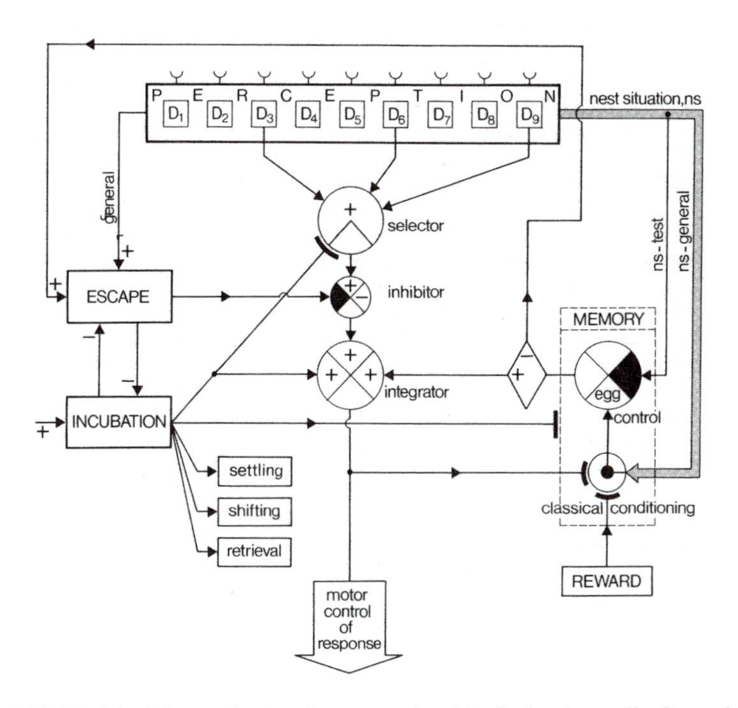

Figure 3.13 Model of the mechanism for egg retrieval in the herring gull. General visual perception (top) is represented as a series of detectors (D_1 to D_9) that respond to particular features of visual stimuli. Some detectors feed on to a specific detector for egg recognition, which in turn feeds on to motor control for egg retrieval. Egg retrieval is performed during the period of incubation (positive contribution from the "incubation" box) but may be overridden by other factors such as the need to escape (inhibitory influences from the "escape" box). Whether a particular object will be retrieved or not depends also on the bird's memory based on experience with real eggs (right). Reprinted from Baerends (1982) by permission of Brill Academic Publishers.

on what determines a gull to retrieve an object into the nest. Based on countless tests with objects placed on the nest rim, Baerends and coworkers were able create a detailed model of how various features (color, speckles, size and shape) and other motivational factors determined the efficiency of a stimulus in eliciting the retrieval behavior. The resulting model is shown in Figure 3.13. Egg detection occurs in two steps. First, a stimulus is processed by a number of "feature detectors" serving general-purpose vision. Some of these detectors (corresponding to, say, "elliptic shape," "speckled object" and so on) feed onto an "egg detector" with the specific purpose of determining whether an egg is present. The output signal is not a "yes" or "no" but a continuous value based on what egg features are present and on the effectiveness of each feature. Different features interact according to a summation model such as equation (3.3). This can explain why stimuli vary in their efficiency to elicit egg retrieval. Finally, the egg detector determines behavior in combination with other motivational factors such as fear or the bird's experience with real eggs. As we argued in Section 1.4.3, it is not easy to see why models of this kind did not lead to more extensive use of neural network models.

A rather different approach to stimulus control is taken by some models from cognitive psychology. Here we only have space to discuss the very popular notion that generalization is determined by the distances between the representations of stimuli in a "psychological space" (Shanks 1995; Shepard 1987; see also Section 1.3.5). Thus the closer two stimuli are represented, the greater generalization will occur between them. Distance in the psychological space has been related to the physical similarity between stimuli (i.e., more similar stimuli lie closer in the space), but modifications of distances owing to learning have been considered (Shanks 1995). One drawback of these theories is that at present they cannot predict how the psychological space looks (Staddon & Reid 1990). Rather, distances between representative points are fit to reproduce observed generalization (Nosofsky 1986; Shanks 1995; Shin & Nosofsky 1992). However, this does not always work. Intensity generalization, for instance, cannot be described satisfactorily in this way (Ghirlanda 2002). In a nutshell, the problem is that knowing whether stimulus S_1 is at a given distance from S_2 does not tell which one is more intense. Additional assumptions may be introduced to deal with intensity (such as appealing to "biases," "attention" or "prominence" of different stimuli; Johannesson 1997; Nosofsky 1987, 1991), but the resulting model is no longer based on distance alone. Neural networks, in contrast, deal naturally with stimulus intensity, provided it is modeled realistically as larger activity of input nodes. The reader interested in this topic is referred to Ghirlanda (2002).

In summary, ethological and psychological thinking contains many seeds for neural network models of behavior, whereas some other approaches are markedly different. The advantages that neural networks enjoy over similar or even formally equivalent models stem from their interpretation as models of nervous systems rather than black-box models or models of abstract cognitive processes. This interpretation pinpoints the importance of knowing exactly how a stimulus enters an organism and how it is processed. What a stimulus is and what its parts are have often been arbitrary, generating confusion about the power and limits of models such as equation (3.4) (Ghirlanda 2005). At the same time, many of the phenomena considered earlier arise from basic properties of network architectures, such as the distributed nature of processing. Consequently, bringing in just a little physiology can offer great rewards. For instance, to predict whether generalization along a stimulus dimension will be bell-shaped or monotonic, it is enough to know whether the dimension is perceived as a rearrangement or intensity dimension. We do not need to know in detail, say, the biochemistry of receptors. Such an origin from basic properties agrees with the observed generality of generalization phenomena across species and taxa (Ghirlanda & Enquist 2003). On the other hand, to explain why light wavelength generalization gradients in humans and pigeons show systematic differences, we would need to appeal to finer details of their visual apparatus, such as differences in photoreceptors and subsequent neural pathways.

3.4 SENSORY PROCESSING

The simple models analyzed so far do not show any clear separation between sensory processing and decision making. In two-layer feedforward networks, sensory processing and decision making come "as one package," whereas adding a third layer introduces some potential for separation. In nervous systems, except perhaps the simplest ones, neural layers immediately after receptors mold sensory information, whereas decision making occurs further away from the sense organs (Ewert 1980, 1985; Kandel et al. 2000; Simmons & Young 1999). In other words, a part of processing is independent of how the animal will ultimately react to the stimulus. Figure 1.8 illustrates a case with minimal sensory processing, whereas the processing of visual stimuli in humans is an extreme case in the other direction, built on many layers and with many parallel processing pathways (Kandel et al. 2000).

Why aren't decisions made directly based on information from sense organs? At first, this may sound paradoxical. All sensory information that the organism has available is already present at the level of receptors. Indeed, sensory processing leads to a decrease in information in the technical sense. The trick is that the activity patterns of receptors are transformed into patterns of activity in subsequent layers that are more useful for decision making (Haykin 1999). In fact, many properties of the real world are poorly represented by receptor activities. The same object can give rise to very different patterns of activity, wrongly suggesting that such patterns arise from different objects (Figure 3.3). Complex discriminations, such as the XOR problem encountered in Section 2.2.1.1, are also cases that benefit from processing of input patterns.

In neurophysiological studies, many kinds of sensory processing can be found. Lateral inhibition is one example, whereby information about changes in stimulation is made available for decision making or further sensory processing. An important concept is that of *receptive fields* (Kandel et al. 2000; Simmons & Young 1999). This term refers to the part of a sense organ (or any other neural layer) where stimulation can excite a cell in a subsequent layer. For instance, a neuron might react only to stimulation falling on a restricted part of the retina, which thereby is called the neuron's receptive field. In a more abstract sense, receptive field may also refer to those input patterns able to excite a neuron. Many kinds of receptive fields have been found in nervous systems (Hubel 1988; Kandel et al. 2000). A famous example is orientation-sensitive cells in the cat's visual cortex, discovered by Hubel and Wiesel (1962). Each of these cells is sensitive to the orientation of a line projected on a small portion of the retina, being maximally responsive to lines in a particular orientation.

3.4.1 Processing by lateral connections

Lateral connections are a particular form of recurrence that takes place between nodes in the same layer (Figure 3.14). Lateral connections, as well as feedforward and feedback ones, are clearly identifiable in nervous systems. Indeed, connections internal to an anatomically identifiable group of neurons are often more abundant than connection between different groups of neurons, e.g., in the cortex (Kandel

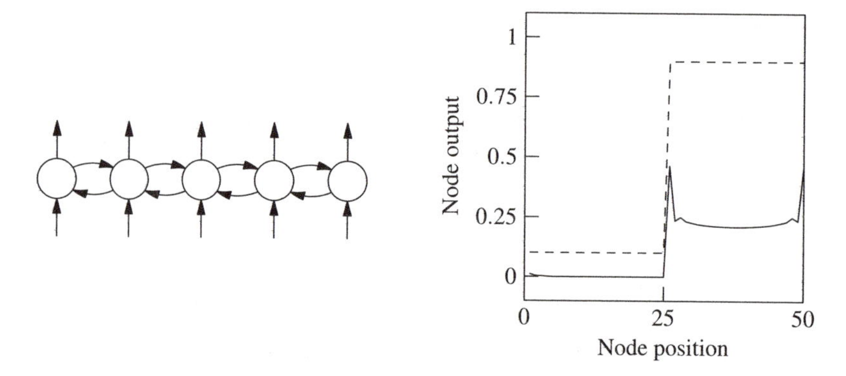

Figure 3.14 Left: Lateral feedback connections between neighboring nodes. Right: Input transformation by inhibitory lateral connections (equation 3.5). The node array receives a sharp edge as input (dotted line) and responds with enhanced activity near the edges (solid line). Network nodes were updated using the asynchronous procedure described in Section 2.2.2.

et al. 2000). The term *lateral*, like *feedforward* and *feedback*, refers to the orientation of connections relative to an identified input-output direction. We give below two examples of the possible roles of lateral connections in sensory processing (their role in decision making is discussed in Section 3.7).

Our first example is *lateral inhibition*, meaning simply that each node inhibits its neighbors. Inhibition may be limited to nearest neighbors or extend over a longer range. Lateral inhibition can highlight changes in input, e.g., in visual processing (Coren et al. 1999). This effect can be demonstrated using a one-dimensional array of nodes with lateral connections given by

$$L_{ij} = \begin{cases} 0 & \text{if } i = j \\ -\dfrac{1}{2|i-j|} & \text{otherwise} \end{cases} \tag{3.5}$$

That is, node j inhibits node i with a strength inversely proportional to their distance, $|i-j|$. Besides inhibition from other nodes, we assume that an excitatory input x_i comes to node i. Figure 3.14 shows the output of this network when stimulated by an input vector $\mathbf{x}$ presenting an "edge," i.e., a sharp transition from low to high values. The profile of node activity reveals clearly the location of the edge. The mechanism is rather simple. Because a node is inhibited by activity in neighboring nodes, nodes near the edge receive less inhibition owing to the lower activity of nodes left of the edge.

Figure 3.15 illustrates graphically another possible structure of lateral connections, whereby nodes separated by a distance of about 25 (shaded region) excite each other, whereas nodes at different distances inhibit each other. As a result, this network is maximally sensitive to inputs having separate peaks at a distance of 25 (in other words, the network is particularly sensitive to a given spatial frequency). Figure 3.15 shows the total output of the network (sum of all node activities) as

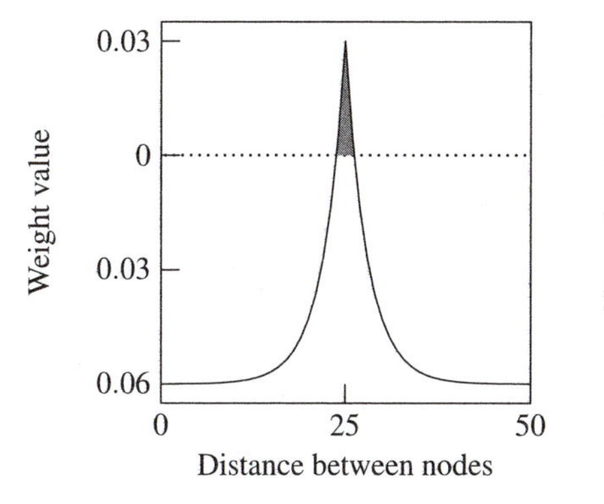

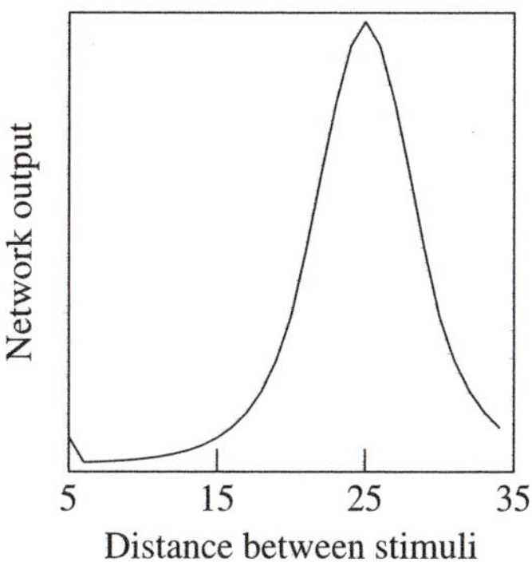

Figure 3.15 Left: Value of lateral connection L_{ij} as a function of the distance between nodes i and j in a network with the architecture in Figure 3.14. Excitatory weights come from nodes at a distance of about 25 (shaded region), whereas weights from other nodes are inhibitory. The exact form of the function is $L_{ij} = \frac{1}{100}\left(9\exp\left(-\frac{1}{3}\big||i-j|-25\big|\right)-6\right)$. Right: Output (summed activity of all nodes) of a recurrent network to inputs consisting of two segments that activate 5 nodes each as a function of the distance between the segments. The network consisted of a one-dimensional array of 50 nodes linked by lateral connections L_{ij} as portrayed in the left panel. Other details as in Figure 3.14.

a function of the distance between two segments of length 5; output is maximum when the segments are 25 nodes apart.

3.4.2 Generalization along "difficult" dimensions

Edges and lines are one example of important structures in the real world that are poorly represented at the level of receptors. Properties of a line such as its length, thickness and orientation can be derived only by combining information from many receptors. Consider, for instance, the image projected by a vertical line on a retina (suppose for simplicity that the line always falls at the same position on the retina). If we tilt the line even slightly, it will influence a whole new set of receptors and very few of the receptors previously affected. Thus a small change in the real world (or its projection onto the retina) has given rise to a big leap at the level of receptors. But this seems hardly to matter to animals. Animals generalize to different line orientations as smoothly as to changes in light wavelength or sound frequency (closed circles in Figure 3.16; note that line orientation is a rearrangement dimension).

The networks employed so far generalize very poorly in this and similar situations (dotted line in Figure 3.16). The reason is that responding is based directly on the overlap of input activity patterns. To address this problem, we focus on a piece of sensory processing that is widespread in animals, i.e., spatially organized receptive fields. This means that neurons in a given layer do not receive input from all neurons in the preceding layer, nor from a random subset of them. Rather, each neuron often gathers input from *spatially neighboring* neurons. This kind of processing occurs, for instance, in the retina of vertebrates, where ganglion cells

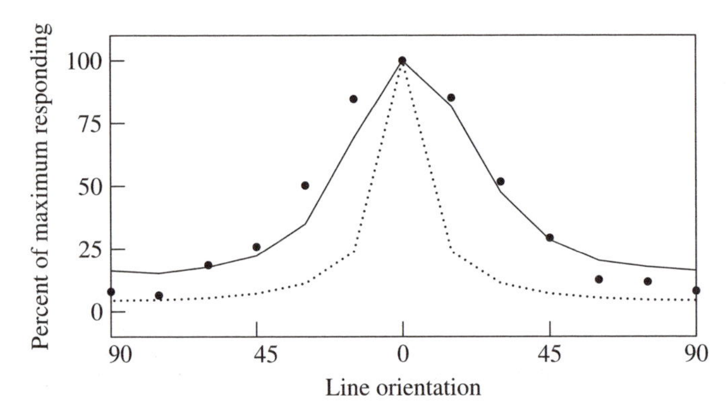

Figure 3.16 Line-tilt generalization in animals and networks. Closed circles reproduce data from an experiment in which pigeons where trained to peck an illuminated key when a vertical line was shown on it and not to peck in the absence of the line (Bloomfield 1967). The lines show generalization in two different network models trained on a similar task. Both networks have a 100×100 artificial retina on which a 60×4 rectangle could be projected in different orientations. The dotted line represents the poor generalization of a two-layer network in which input nodes connect directly to an output node. The continuous line shows the more realistic generalization of a three-layer network with spatially organized receptive fields (Figure 3.17). We used 1000 receptive fields, randomly scattered across the retina and of circular shape with a radius of 7 nodes.

forward to the brain signals that integrate the output of many neighboring photoreceptors. A simple example is shown in Figure 3.17. We use this organization in a three-layer feedforward network, in which hidden nodes are connected to a subset of neighboring nodes in the input layer. For simplicity we assume that these weights are fixed, while the weights between the hidden and output nodes are set to discriminate between a vertical line and absence of the line. Such a network generalizes realistically to lines in different orientations, as shown by the continuous line in Figure 3.16. In terms of overlaps, the effect of receptive fields is to transform input patterns that overlap little into patterns of activity in the hidden nodes that overlap more and whose overlap changes more smoothly as line orientation is changed. This allows smooth generalization by the two-layer network made by the hidden nodes and the output node.

It is instructive to note that the same behavior of our three-layer network could be achieved by a two-layer network with an appropriate weight matrix (because the model is still linear; see Section 2.3.1.1). On the other hand, simply training a two-layer network to recognize a vertical line from absence of a line does not produce such a matrix, as we have seen. Adding features such as more layers and nonlinear nodes would not help. The reason is that the training procedure contains no requirements about generalization. On the other hand, endowing the network with receptive fields somehow anticipates that the same object may appear in a different orientation. We may say that such a network has a *predisposition* for generalizing over different orientations (see Section 4.2.3). This suggests that receptive fields

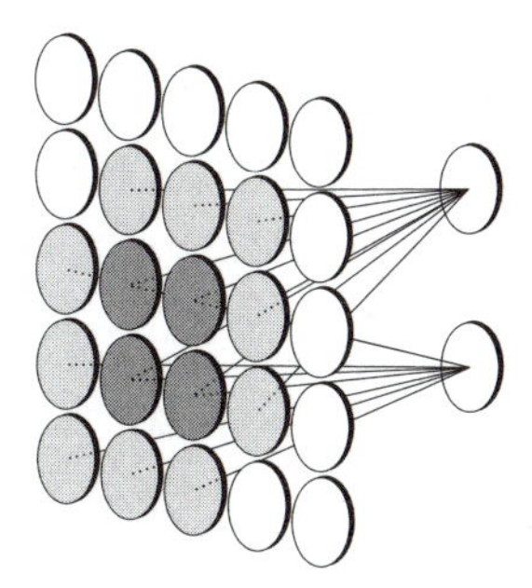

Figure 3.17 Spatially organized overlapping receptive fields in a small network with a 5×5 retina. Two output nodes are shown, each gathering input from a subset of the retina (light gray). Dark gray indicates input nodes in the receptive field of both output nodes.

are, at least to some extent, an adaptation for generalization because they provide the network with abilities that do not "come for free" with a network structure. A downside of increasing the overlap between input patterns is that discriminations become more difficult. Sections 2.4 and 4.7.2 contain some information on the ontogeny of receptive fields.

3.4.3 Perceptual constancy

Perceptual constancy refers to the ability of animals to respond similarly to the same physical stimulus despite the many patterns of receptor activity that the stimulus may produce (Figure 3.3). Constancy has been observed for many species and sensory modalities (Walsh & Kulikowski 1998). It is possible to imagine two different explanations. We consider vision for definiteness. The first possibility is that animals have had prior experiences of many different circumstances and have learned (or been selected) to respond with the same behavior to each of them. The second possibility is that sensory processing does some "magic" that compensates for changes in retinal images. The evidence suggests that constancy may emerge from a combination of these two possibilities (Walsh & Kulikowski 1998). Animals are often very good at recognizing transformed images of familiar objects, much better than a simple two-layer network. On the other hand, recognition is not always perfect. This is hardly surprising, since, in general, it is not possible to tell whether two retinal images come from the same object, without knowledge of the object's structure. Recognition improves with experiences of transformed images, and some familiarity with the stimuli seems to be required. Even very familiar stimuli are recognized faster from typical viewpoints than from less typical ones (Walsh & Kulikowski 1998). In other words, although we are often amazed by constancy phenomena (especially if we have tried to program similar abilities in a computer), we must also recognize that animals do not always react in the same way to a familiar object in a new situation. For instance, the empirical data in Figure 3.16 show that pigeons trained to react to a vertical line do not react equally to all orientations of the line. This strengthens the idea that experience with relevant stimuli is a pow-

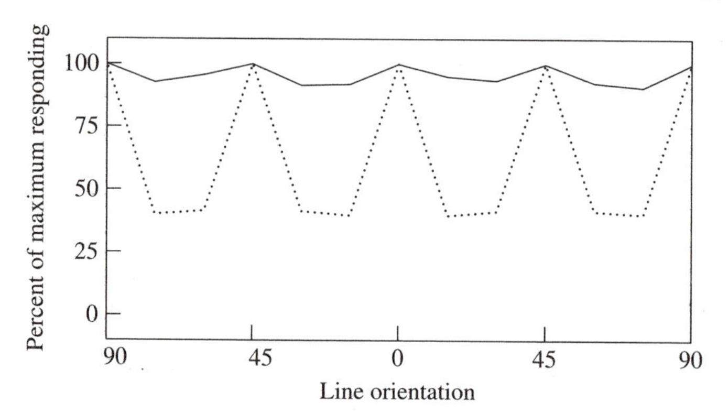

Figure 3.18 Perceptual constancy can result from the interaction of network architecture and specific experiences. Continuous line: When the network in Figure 3.17 is trained to react to a line presented in four different orientation, 45 degrees apart, it reacts with nearly the same output to any orientation. Dashed line: A two-layer network subject to the same procedure does not achieve constancy.

erful determinant of constancy phenomena. An example comes from experiment with the fruit fly (*Drosophila*), which seems to lack sensory mechanisms for achieving constancy, at least for nonmoving visual stimuli. In these experiments, learning about stimuli projected on a part of the fly's compound eye did not transfer to other parts of the eye (Dill et al. 1993). Intriguingly, similar results have been obtained in humans (Dill & Edelman 2001; Dill & Fahle 1998; Kahn & Foster 1985).

We consider now a simple example of how neural networks can achieve constancy with respect to object orientation. This is perhaps the simplest example of how specific experiences can interact with preexisting network structure to produce perceptual constancy. We have seen that a network with spatially organized receptive fields (Figure 3.17) generalizes realistically to lines in different orientations after having been trained to respond to a single orientation (continuous line in Figure 3.16). If the network is further trained to respond to just a few more orientations, it reacts in the same way to lines in any orientation, as shown in Figure 3.18, effectively achieving constancy. The figure also shows that a two-layer network trained in the same way is far from this result. The model thus shows how constancy may arise from "good" generalization plus a relatively limited number of experiences. This is consistent with information from animal behavior. Animals do not seem to show constancy after very little experience with particular stimuli or kinds of stimuli, but constancy is achieved very quickly with increasing experience. We do not doubt, however, that nervous systems are smarter than our simple model (see Arbib 2003 and Walsh & Kulikowski 1998 for models and data).

3.4.4 Stimulus "filtering" and "feature detectors"

A traditional description of sensory processing is that it *filters out* information that is irrelevant to the animal (McFarland 1999; Simmons & Young 1999). This account is valid but incomplete. A more general description is that sensory processing transforms input patterns into patterns of activity in successive neural layers. Such transformations do involve "filtering out" (i.e., making neurons relatively insensitive to some stimuli or stimulus dimensions) but also have similarities with, for instance, principal component analysis and other signal analysis techniques (Field 1995; Haykin 1999). This includes emphasizing some stimulus dimensions at the expense of others, as well as building new dimensions that are not apparent in the raw input.

Stimulus filtering is often thought to be accomplished by "feature detectors," i.e., neurons that respond solely if a stimulus has a particular feature (Figure 3.13; Simmons & Young 1999). In ethology, the concept was linked to the idea that animals respond based on "sign stimuli" identified by a particular feature or combination of features (Lorenz 1981; Tinbergen 1951). Figure 3.13 is a classical example. The idea that recognition in general is based on feature detectors is still a matter of lively debate in psychology and cognitive science (Field 1995; Schyns et al. 1998). Rolls and Deco (2002) show several network models based on feature detectors. It is clear from previous sections, however, that neural networks are very capable even without any feature detectors. This does not exclude the possibility that, in networks with three or more layers, patterns of processing may emerge (from training or evolution) that can be described as "feature detectors" (e.g., Kohonen 2001). However, these do not fully resemble the classical idea of, say, line and edge detectors. One example (borrowed from Arbib et al. 1998) comes from studies of the frog retina. Lettvin et al. (1959) classified retinal ganglion cells (whose axons form the optic nerve) into four types. Type 3 cells were described as responding to small moving objects, and it is tempting to call them "prey detectors." A better description, however, is that these cells respond to a moving edge entering their receptive field, not to the presence of a prey within the receptive field. Further processing in the brain determines whether the frog snaps at the moving object or not. Similarly, cells of type 4 are better described as measuring dimming of the visual scene than as "predator detectors" (the link here is that predators coming from above often cast a shadow over the frog).

In conclusion, we think that a better description of sensory processing is that it makes relevant information available to decision making and control of behavior. The receptive fields that have been found in network models and nervous systems seem to differ from the classical idea of feature detectors in several respects. First, they tend to be more general purpose. Type 3 cells in the frog retina, for instance, are important in the processing of all kinds of visual stimuli, not prey only. Second, they are not as sharply tuned. For instance, a "line detector" will react to various extents to many kinds of objects, although lines might be those stimuli to which the cell responds the most. Lastly, perceptual learning and other findings suggest that the nervous system may be more dynamic than implied by the idea of a fixed set of feature detectors (Chapter 4; Schyns et al. 1998 and following commentaries).

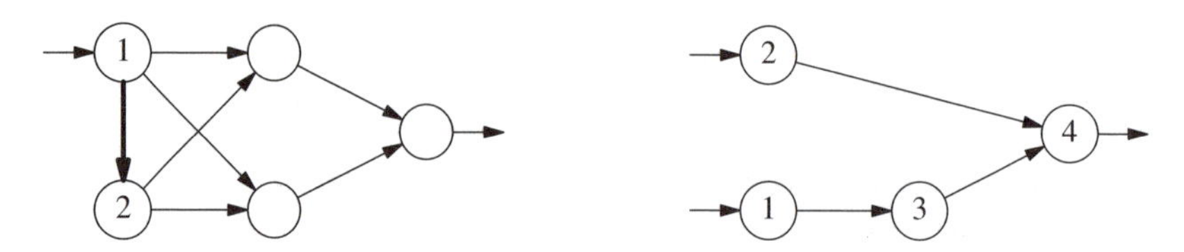

Figure 3.19 Simple networks with elementary time-processing abilities. Left: A network that reacts to changes in input. The network is based on the XOR network in Figure 2.8 (left), with one added connection (thicker arrow) and input to only one node. Right: An elementary mechanism with an ability to detect motion (Hassenstein & Reichardt 1959). The output of node 1 is delayed through node 3 and then combined with the output of node 2. Node 4 reacts only if nodes 2 and 3 simultaneously provide input. The mechanism is a more selective motion detector if nodes 1 and 2 react only to changes in stimulation (otherwise a constant input to both nodes would also elicit a response; see Section 2.1.3 for a suitable node model).

3.5 TEMPORAL PATTERNS

Temporal patterns are a very relevant feature of the world. Duration, rhythm and other temporal structures are crucial, for example, in many visual and auditory displays used in animal communication. Different species of woodpeckers are an interesting example. They can be identified on the basis of their drumming, a rhythmic series of sounds caused by hammering against a resonant object such as a tree trunk. Drumming varies among species in duration, frequency and amplitude. Temporal patterns are also crucial for detecting movement (Simmons & Young 1999), which can have a number of uses, such as detecting prey or predators (Ewert 1985; Lettvin et al. 1959). How visual images move over the retina is important also when moving in the environment, e.g., for maintaining a straight course (Section 3.5.1).

Obviously, responding to temporal patterns cannot be based on sensory information at a single instant of time. A single hammering does not tell woodpecker species apart, nor does a single visual image tell whether an object is approaching or receding. Consequently, to make decisions based on temporal patterns, sensory processing beyond the level of receptors is needed. Somehow information about prior stimulation must be retained and made available to the decision mechanism. Perhaps the simples example of time structure is a change in input. Imagine that an input signal $x(t)$ is mostly constant but sometimes changes value from 0 to 1 or vice versa. The network in Figure 3.19 reacts only to change and ignores periods of constant input. To see how this is achieved, note that this network is derived from the network in Figure 2.8 on page 44. The modification is simply that now node 2 receives input from node 1 rather than from external stimuli. In this way, nodes 1 and 2 hold the values $x(t)$ and $x(t-1)$, respectively. Thus, reacting to a change is the same as reacting when exactly one of nodes 1 and 2 is active, and this is what the network in Figure 2.8 was designed to do. By a small change in network organization, a temporal pattern has been transformed into a pattern of activity of nodes *at the same time*, which can be used for further processing and decision making.

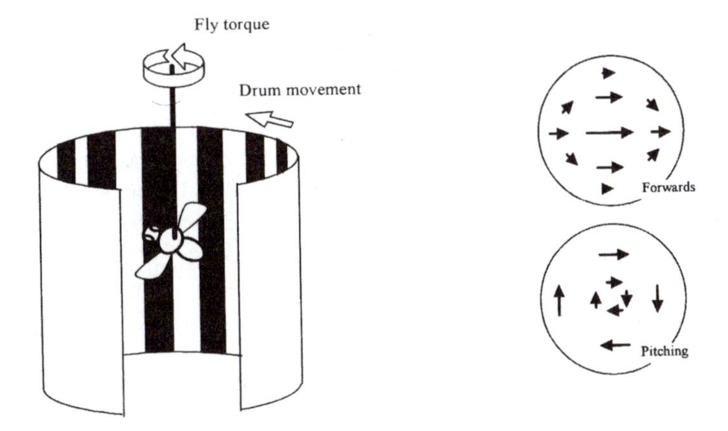

Figure 3.20 Left: Apparatus for the study of optomotor responses in flies. Right: Examples of flow fields (see text). Reprinted by permission from Simmons and Young (1999). *Nerve Cells and Animal Behavior*. Cambridge University Press.

Figure 3.19 shows a further example: an elementary network that provides some information about motion (Hassenstein & Reichardt 1959). The output of node 1 is delayed (passing through node 3) and then combined with the output of node 2; node 4 has a threshold such that it is activated only if nodes 2 and 3 provide input at the same time. Owing to the delay, this means that node 1 had been stimulated some time before node 2, as it happens when a moving object passes over the retina. The circuit is also sensitive to the direction of motion because stimulating node 2 before node 1 does not produce a response. A functionally similar mechanism has been identified in flies, although its neural architecture is still unknown (Simmons & Young 1999).

3.5.1 Optomotor responses in flies

Biologists often admire the ability with which animals perform complex tasks. Although some abilities may seem to require very specific engineering, we can also realize that general principles may be used and need not be reinvented. One such principle is that by tapping information from preceding neural layers we can often build a more useful representation that makes the task a little simpler. Consider an animal such as a fly, which needs to maintain a particular flight course relying on continuous feedback from its surroundings. We can study how the fly uses information from the environment by placing a tethered fly within a cylinder with vertical stripes (Figure 3.20). If the cylinder is rotated, the fly perceives a deviation from the intended course and responds to maintain course relative to the movement of the cylinder (Simmons & Young 1999). Such a transformation between visual input and flight control movements is called an *optomotor response*.

How are optomotor responses achieved? To rely directly on motion detectors

(Figure 3.19) is not enough because the change in input produced by deviating from the intended course is different for different parts of the eye. However, particular courses (flying straight, turning, approaching the ground, etc.) create global movement patterns over the whole visual field, referred to as *flow fields* (Simmons & Young 1999; Figure 3.20). For instance, in an animal with laterally placed eyes, moving straight produces the same flow field on both eyes, whereas turning produces opposite flow fields. Thus flow fields are an interesting example of a transformation from sensory information scattered over many receptors into simpler dimensions on which decisions can be taken more easily.

Neurons in the fly brain have been identified that respond as if they were monitoring flow fields, and empirical studies suggest that this is achieved by integrating signals from motion detectors functionally equivalent to the one in Figure 3.19. Some simple models of how this may be accomplished have been proposed, building on the known structure and functional properties of the fly visual system. Harrison and Koch (2000) built an electric circuit consisting of one artificial neuron that simply sums the output of a linear array of motion detectors that receive input from an array of light sensors. Half the detectors were tuned to motion in one direction and half in the opposite direction. The sum of the motion detector signals thus estimates the direction and speed of rotation. The circuit was placed in an apparatus like the one in Figure 3.20 and subject to the same stimulation used in experiments with flies in that apparatus. The circuit output, interpreted as a motor command, was able to cancel the apparent rotation produced by the apparatus, behaving similarly to flies. Planta et al. (2002) developed a neural network of similar architecture but with four distinct arrays of motion detectors tuned to different directions and receiving input from two cameras. These fed into four nodes that could thus respond to flow fields produced by left/right rotation and up/down translation. The output of these nodes controlled the four propellers of a flying robot (a Zeppelin over 1 m long), allowing forward motion at different speed, left and right turning, and changing altitudes. The task of the robot was to fly straight without changing altitude. This was achieved with considerable accuracy when the flight control mechanism was turned on and rather badly when it was switched off. The network could thus correct for the rather poor aerodynamics of the robot.

3.5.2 Call recognition in the Túngara frog

A valuable application of neural networks to behavior in response to temporal structures has been developed by Phelps and Ryan (1998), studying the vocal communication of the Túngara frog, *Physalaemus pustulosus*, and related species (see also Phelps 2001; Phelps & Ryan 2000). The aim of this work was to gain insight into how female mechanisms for recognition may have influenced the evolution of calling in males (Chapter 5). The technical challenge is that important aspects of the Túngara call unfold in time. Phelps and Ryan employed a recurrent network that had been shown previously to be capable of detecting structures in time by Elman (1990). The network is depicted in Figure 2.13 on page 56. It consists of a three-layer feedforward network with the addition of a second group of hidden nodes linked with the first group by recurrent connections. The network input nodes model the frog's ear, with each node being sensitive to a range of sound frequencies. Actual male calls where sampled at small intervals, obtaining a number of time slices for each call. These slices are fed one at a time to the network (an approximation to the continuous time processing that takes place in nervous systems). The activity of the model ear at each time is sent to the first group of hidden nodes. This connects to an output node and to the second group of hidden nodes, which feeds back its activity to the first group. Through this feedback the input to the first group of inner nodes becomes a function of past as well as current input. This allows the network to set up internal states over time and become sensitive to temporal structure across a series of call slices. The network was trained by a random search ("evolutionary") algorithm (Section 2.3.3) to respond to the Túngara call but not to noise in the same frequency range. After training, the network was able to accurately discriminate between calls of different species, and it was also able to predict responding in real females, to both familiar and novel stimuli (Figure 5.14 on page 202; Phelps 2001; Phelps & Ryan 1998).

3.6 MANY SOURCES OF INFORMATION AND MESSY INFORMATION

The information relevant to a task often comes from many different sources. For instance, cues for navigation can come from magnetic fields, terrestrial landmarks and the position of celestial bodies such as the sun and stars. Moreover, the cues are not always present and sometimes are unreliable (e.g., the sky can be overcast). It has been suggested that animals may cope with such situations using "rules of thumb" or "simple heuristics" that consider only one or a few cues and discard information about others (Stephens & Krebs 1986). Our intuition suggests that rules of thumb are easier to realize than more complete strategies. For a neural network, however, it is not necessarily easier to discard information than to take it into account. Indeed, an important aspect of neural networks is that they make it easy to integrate all kinds of information. By means of different receptors, all sorts of external and internal events can be converted in the "universal currency" of neuron activity (Table 3.1). Moreover, recurrent connections allow the network to build up regulatory states that provide sensibility to time patterns. Whether incoming information is precise or messy does not matter much. Networks are able to decide what is important or not, e.g., by lowering weights that convey input from unreliable sources.

How these properties of neural networks can help us to understand behavior is illustrated by navigation in migratory birds. Suggestions of what information guides this extraordinary ability have been growing steadily. The list includes the above-mentioned cues, as well as smells, infrasounds, inertial, gravitational and Coriolis forces, and others (Alerstam 1993). At the same time, there has been a tendency to ask which one is "actually" used. But why not use all sources? A neural network can naturally integrate many kinds of information, given that it can be received. This also provides a more robust solution that would work even when some sources of information are unavailable. The mechanism could also be sensitive to the current reliability of sources. That biological navigation systems embody these principles has been shown eventually in pigeons. They can find the way home on sunny days when the magnetic filed is distorted, and they can use magnetic fields when the sky is overcast. As more and more data became available on bird orientation and navigation, a multiple-source hypothesis gained support. Today it is accepted that pigeon use all sorts of information, including three compasses and other supplementary information (Alerstam 1993).

3.7 CENTRAL MECHANISMS OF DECISION MAKING

Decision making refers here simply to the fact that an animal takes one behavioral action rather than another without implying consciousness or intent (McFarland 1999). A simple view is that a behavior is performed whenever relevant causal factors are present. But a second thought, as well as empirical observations, reveals that decision making is a more difficult issue (Hinde 1970; McFarland 1999; McFarland & Sibly 1975). A basic consideration is that animals can usually do only one thing at a time, whereas causal factors for incompatible behaviors may be present at the same time. A hungry and thirsty animal might have both food and water available, but it cannot eat and drink at the same time. In general, many activities are necessary for survival and reproduction, but only one or a few can be performed at any given time. Decision making is thus about structuring behavior in time in a sensible way, taking into account both the animal's internal state and external conditions (McFarland 1999). In this section we aim to show that neural networks have natural abilities for animal-like decision making and can embody a number of ideas that have been developed within ethology, psychology and control theory (for reviews, see Baerends 1982; McFarland 1999; Shettleworth 1998; Toates 2001). We begin by considering how a single decision is influenced by the animal's internal state, as well as by external stimuli, and then we consider decisions among activities.

3.7.1 Behavior and body states (homeostasis)

The previous sections considered very simple decisions. The only causal factors were external stimuli, and the decision was whether to respond or not. Our models of such decisions were networks with a single output node, whose activity was taken as the probability of responding. Here we continue to discuss such simple decisions, but we introduce body states (internal stimuli) as additional causal factors. Decisions between different activities are considered later.

Both body states and external stimuli are sensed by means of specialized receptors (Table 3.1). Thus to model body states we do not need new concepts but rather some knowledge of how they are sensed. One crucial difference, however, is that body physiology requires that variables such as water volume, body temperature, blood glucose level, etc. be maintained within strict bounds for life to continue. This is referred to as achieving *homeostasis*, and often behavior plays a crucial role. For instance, moving between sunlight and shade is a powerful way to regulate body temperature, on which animals such as reptiles rely extensively.

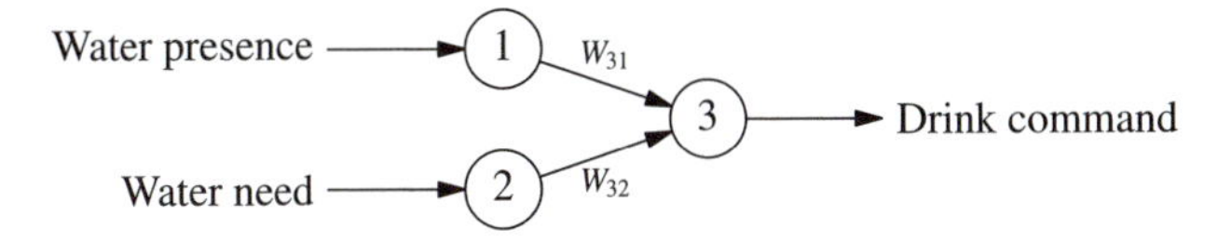

Figure 3.21 A simple model of deciding whether to drink or not based on external (availability of water) and internal (need of water) stimuli.

We begin our discussion of neural network models of homeostasis with a simple model of drinking. Organisms constantly lose water, which must be replenished by drinking. Thus a network model of homeostatic drinking should explain how body water volume is maintained close to what is required by body physiology. The model is comprised of three nodes, as shown in Figure 3.21. Nodes 1 and 2 convey information about body water volume and availability of water to node 3. Similar to models presented earlier in this chapter, the activity of node 3 is assumed to correspond to the probability of drinking:

$$\Pr(\text{drinking at time } t) = z_3(t) = W_{31}z_1(t) + W_{32}z_2(t) \tag{3.6}$$

To turn this generic equation into a model of drinking, we need to be more specific about (1) how water volume changes with time, both when an organism drinks and when it does not drink, and (2) what are the internal and external stimuli that form the raw material for the decision to drink. Note that assumptions of this kind are needed in all models including body states as well as external stimuli.

Let $v(t)$ denote body water volume at time t. We assume that water loss at any time is $-av(t)$ (proportional to water volume) and that drinking causes a gain in water at a constant rate b:

$$v(t+1) = \begin{cases} -av(t) & \text{if not drinking} \\ -av(t) + b & \text{if drinking} \end{cases} \tag{3.7}$$

Let us now consider network inputs. The model has two input nodes. Node 1 models a cell able to sense water deficits (Table 3.1). The node is assumed to operate as follows:

$$z_1(t) = \begin{cases} V - v(t) & \text{if } v(t) < V \\ 0 & \text{otherwise} \end{cases} \tag{3.8}$$

The parameter V may be thought of as a *set point* for water level regulation, i.e., the ideal water level that should be maintained. We shall soon return to this concept. According to equation (3.8), node 1 is inactive if the organism has enough or more than enough water; otherwise, it is active in proportion to water deficit. Neurons with similar properties are thought to exist, for instance, in the central nervous system of mammals (Toates 2001; Table 3.1). Node 2 is a "water detector"; i.e., it is active if and only if water is accessible:

$$z_2(t) = \begin{cases} 1 & \text{water present} \\ 0 & \text{water absent} \end{cases} \tag{3.9}$$

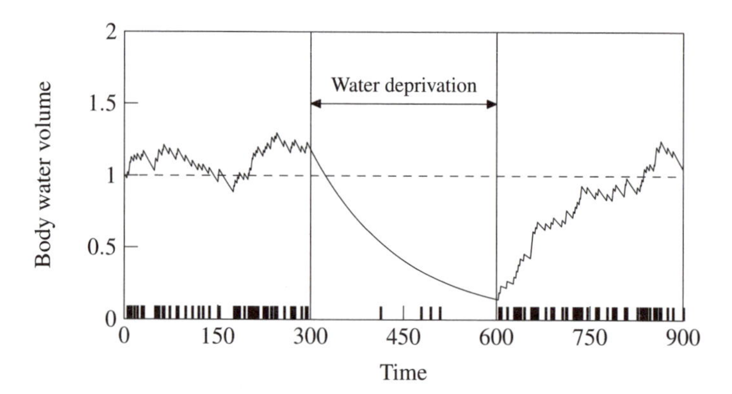

Figure 3.22 Sample behavior of the model of drinking in Figure 3.21. Dashed line: Body water volume to be maintained (value of V in equation (3.8)). Continuous line: Actual water volume. Marks on the horizontal axis: Times of occurrence of drinking. Water was unavailable between times 300 and 600.

This rough model of recognition is used here for simplicity and can be replaced by more realistic ones, as seen earlier.

With appropriate weights, the model can maintain water balance close to the set point V, as shown in Figure 3.22. The figure also shows what happens if water is made unavailable for some time: drinking does not occur despite continuing water loss. The network can thus integrate information from external and internal stimuli, refraining from performing an internally motivated behavior if the appropriate external conditions are not present. When water is reintroduced, drinking resumes at a high rate until the water deficit is eliminated. Note that under conditions of severe deficit, occasional drinking occurs even when water is absent (Figure 3.21). If water volume is particularly low, in fact, node 2 contributes a larger than usual input to node 3, which can result in a sizable probability of drinking even in the absence of input from node 1. Animals may indeed perform a behavior under inappropriate conditions if motivation is very high. Ethologists call this a *vacuum activity*. Water-deprived pigeons, for instance, may attempt to drink from smooth and shiny surfaces (see REDIRECTED BEHAVIOR in McFarland 1987) or often visit places where water is usually available (our model is too crude to distinguish between different behaviors motivated by thirst). Another classical example is a dove lacking a mate for a long time that may court a stuffed dove or even a piece of cloth (Hinde 1970).

Control of drinking in animals, of course, is more complex. In addition to water levels sensed by osmoreceptors, drinking is controlled by factors such as presence of water in the stomach and mouth stimulation resulting from water ingestion. The functional reason for these additional control factors is that it takes time to move water from the stomach to other tissues. Before a change in water volume could be sensed by osmoreceptors, an excessive quantity of water would have been drunk already. This complication is ignored in our model.

The existence of multiple control factors is at odds with simple examples of set

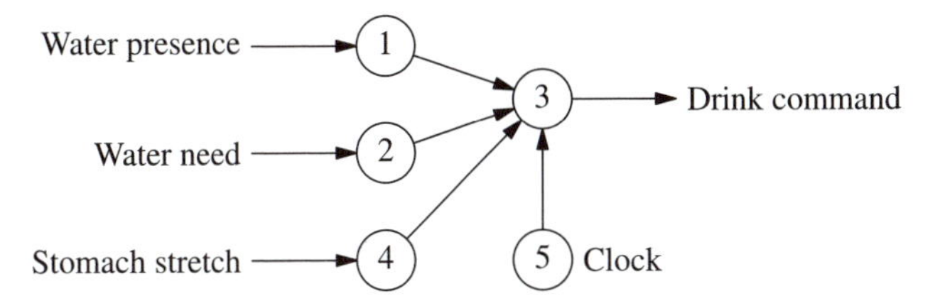

Figure 3.23 A model of drinking that takes many sources of information into account (see Figure 3.21).

points in human-made devices such as a thermostat. For this reason, the set-point analogy has been judged misleading (Toates 1980, 1986). A further critique concerns the fact that the model operates solely based on of *negative feedback. Feedback* refers here to the fact that the system, via specialized receptors, continuously compares the actual value of the controlled variable with a set point or ideal value. *Negative* refers to the fact that when a departure from the set point is sensed, an action is taken leading to a change that is opposite in sign. Negative feedback control is an important aspect of how animals achieve homeostasis, but not the only one (Toates 1986). For instance, chickens eat more during the hour before dusk (Hogan 1998). This does not reflect increased need but rather the anticipation that during the night the bird will not eat. Figure 3.23 shows a refined neural network model of drinking that can take into account information from many sources. We have included a temporary water store, the stomach, from which water is transferred to the rest of a body, as well as receptors (stomach stretch receptors) that can sense water in the stomach store and thereby inhibit drinking. Still other factors could be included if necessary. The model also has a "clock" component (i.e., a time-varying input with a period of 24 h). This provides time-varying input to the decision system and allows for circadian influences (see also Section 3.7.5.2).

3.7.2 Choice among alternatives

As already mentioned, the starting point for research into animal decision making is the basic observation that incompatible activities may be motivated simultaneously. A water hole may motivate both drinking and hunting in a lion, given that the animal is both hungry and thirsty, but hunting requires that the lion hides near the water, whereas drinking requires visiting the water. In ethology this is called a *conflict situation* (Hinde 1970; McFarland 1999; Tinbergen 1952), meaning conflict between tendencies to do incompatible things. In some species, a few fixed rules govern the resolution of such conflicts (Figure 3.24), but in most cases the animal's internal state and the external conditions influence what decision is taken.

Most theories of decision making appeal to the core notion that mutually incompatible activities inhibit each other (Hinde 1970; Tinbergen 1952). For instance, Baerend's model in Figure 3.13 features mutual inhibition between incubation and escape behaviors. Despite its popularity, this mechanism seems to have been studied concretely only a few times (Amari & Arbib 1977; Didday 1976; Ludlow 1980).

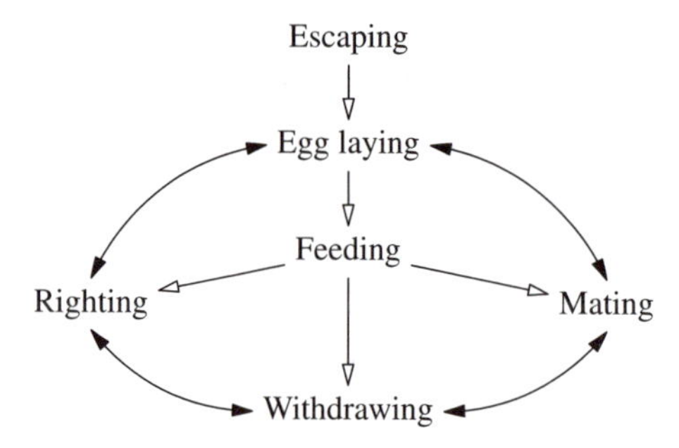

Figure 3.24 Resolution of motivational conflicts in carnivorous snails (Gasteropoda: *Pleurobranchea*). An empty arrow from an activity to another indicates that the former inhibits the latter; bidirectional filled arrows indicate lack of a strict hierarchy. *Righting* refers to the animal turning upright from a reversed position. Turning the animal upside down and offering food causes a conflict between feeding and righting. The animal feeds and can wait up to 1 hour after feeding to turn upright. Interestingly, food offered to a satiated animal also inhibits righting. Adapted from Davis (1979) and Toates (1998).

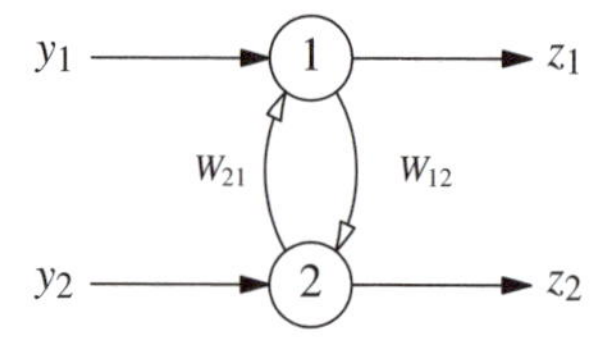

Figure 3.25 A simple "decision maker" based on Ludlow (1980). The weights between nodes 1 and 2 are inhibitory.

The simplest decision making mechanism of this kind consists of just two nodes inhibiting each other (Figure 3.25) and can be formalized as

$$\begin{cases} z_1(t+1) = f\big(\, y_1(t) - W_{12} z_2(t) \,\big) \\ z_2(t+1) = f\big(\, y_2(t) - W_{21} z_1(t) \,\big) \end{cases} \tag{3.10}$$

where $y_i(t)$ is the input that node i receives from other parts of the network, and the positive numbers W_{12} and W_{21} give the strength of inhibition. The transfer function f limits node activity between 0 and 1 but is otherwise linear:

$$f(y) = \begin{cases} 0 & y \le 0 \\ y & 0 < y < 1 \\ 1 & y \ge 1 \end{cases} \tag{3.11}$$

The characteristics of this circuit are as follows. If the weights W_{12} and W_{21} are both smaller than 1, nodes 1 and 2 can be active simultaneously; i.e., a decision is not guaranteed. If both weights are equal to or larger than 1, however, the network settles into a state where only one of z_1 and z_2 is different from zero; i.e., a decision is taken. The larger the weights, the shorter time the decision takes (i.e., smaller number of iterations of equation 3.10 before one node is shut off). Figure 3.26

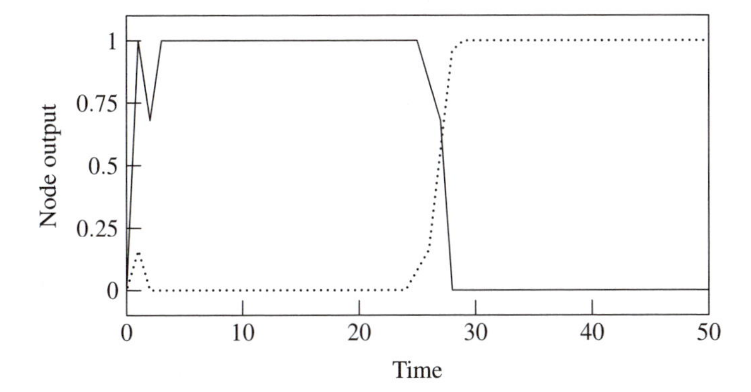

Figure 3.26 Example of operation of the decision maker in Figure 3.25. Inhibition between the two nodes was set to -2 (i.e., $W_{21} = W_{12} = 2$ in equation 3.10). Node 1 receives a constant input of 1 throughout the simulation. The input to node 2, instead, increases linearly from 0 to 4. The simulation is started with an activity of 0.5 in both nodes. Initially, node 1 prevails, but when the input to node 2 overcomes the inhibition exerted by node 1, a transition occurs: node 1 is shut off, and node 2 is activated.

shows one example of a decision using $W_{12} = W_{21} = 2$. Node 1 receives a constant input $y_1(t) = 1$ throughout the simulation, but the input to node 2 increases from 0 to 4. It is important to note that it is not enough that input to node 2 be larger than input to node 1, $y_2(t) > y_1(t)$, for a switch to occur. Rather, node 2 must receive an input greater than the inhibition it receives from node 1, i.e., $y_2(t) > W_{21}z_1(t)$. In the simulation, y_2 overcomes y_1 at $t = 13$, when $y_2(13) = 1.04$ and $y_1(13) = 1$. However, at this point node 2 is receiving from node 1 an amount of inhibition equal to $W_{21}z_1 = -2$, so its total input is still negative. Although y_2 keeps increasing, there is no transition until $t = 26$, when $y_2 = 2.08$. At this point, node 2 begins to have a positive output of 0.08. This inhibits somewhat node 1, which, in turn, lowers the inhibition exerted on node 2. Thus the output of node 2 will increase further, and it's easy to see that this process will continue until node 1 is shut off completely. This is an example of positive feedback; i.e., the reaction to a change is a further change in the same direction rather than in the opposite direction, as in negative feedback.

Ludlow (1980) has shown that this simple mechanism can coordinate a number of systems such as that in Figure 3.21 so that homeostasis is achieved with respect to more than one variable. Didday (1976) used a similar mechanism to model prey selection in frogs (the decision concerned which of two potential preys to attack). To consider decisions between more than two activities, we simply include more output nodes. If all nodes strongly inhibit each other, transitions between activities will be fast, and only one activity will be performed at any given time (sometimes referred to as a *winner-take-all* system). However, if some connections are missing, or if some nodes excite others, more complex (and interesting) outcomes are possible. The inhibitory connections need not be all equal, of course, in which

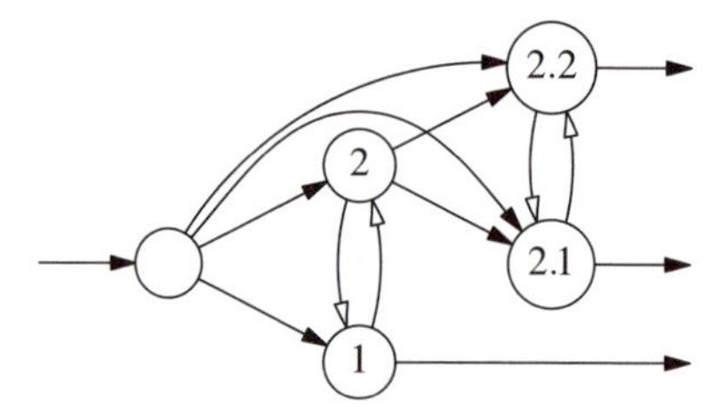

Figure 3.27 An example of a decision hierarchy with two levels. A decision between activities controlled by nodes 2.1 and 2.2 is taken only if node 2 is active, i.e., only if the decision between nodes 1 and 2 has favored the latter. The figure also shows that the same input can potentially affect decision making at multiple levels, as observed in studies of animal behavior (Baerends 1976; Gallistel 1980; Hogan 2001; Tinbergen 1951). We have depicted only one input node for simplicity, but decision making nodes can receive input from multiple sources. See also Hogan (1994b, 2001).

case it may be easier to switch from one activity to another rather than vice versa. We may arrange decisions in hierarchies, as suggested by ethological theory of behavior (Tinbergen 1951), by replicating the basic decision making network several times, as illustrated in Figure 3.27. Figure 3.28 shows an example from classical ethology. Further possibilities are gained by letting each activity be represented by more nodes, which is also more realistic. If nodes relative to one activity project connections of different strengths to nodes relative to other activities, the effective inhibition exerted will depend on what specific nodes are active. This means, for instance, that particular stimuli may be more effective than others (e.g., one kind of food could persuade an animal to switch to feeding, whereas other kinds could not). We may also add random input to some or all nodes to model variability in decisions. Most of these possibilities are largely unexplored.

3.7.3 Stability of behavior

Decisions, besides being taken, should also be reasonably stable. Imagine that our lion, which is both thirsty and hungry, decides to go to the water and drink. If the effect of drinking is simply to reduce thirst, as we may naively guess, thirst would quickly diminish below hunger. The animal would thus stop drinking and go hide, as required for hunting. However, after a short time hiding, thirst would have increased again because the animal drank very little, and the cycle would repeat. Another example is a lizard regulating body temperature by moving between shade and sun. If the lizard is too cold, it will go to the sun. But if it moves in the shade as soon as a comfortable temperature is reached, it will become cold again very soon and will need to go to back to the sun. The lizard would oscillate rapidly between sun and shade. Such instabilities, called *dithering* in engineering, are rarely observed in animals.

The decision making mechanism in Figure 3.25 offers some safeguards against dithering, as shown in Figure 3.29. This simulation is identical to that in Figure 3.26

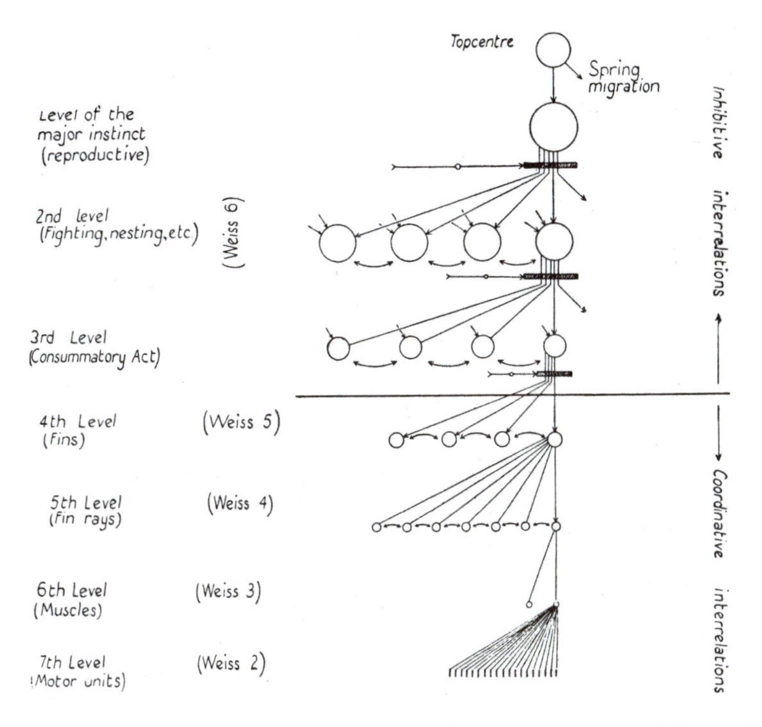

Figure 3.28 Tinbergen's hierarchical model of a behavior system, exemplified by the reproductive behavior of the stickleback (*Gasterosteus aculeatus*). The top level consists of *major instincts*. At the second level are *subsystems* and at the third level are individual activities (e.g., gathering of nest material and gluing, belonging to the nest-building subsystem). At these levels there is competition between systems, and reciprocal inhibition secures that only one has control over behavior at any given time. At lower levels there are "coordinative" interactions that produce coherent motor patterns (Section 3.8). The "Weiss" labels relate this model to the levels of the neurophysiological model of motor coordination of Weiss (1941). Reprinted from Tinbergen (1951) by permission of Oxford University Press.

up to time 28, i.e., when the transition from node 1 to node 2 is under way. At this point, the input to node 2 starts to decrease linearly with time, reaching 0 at time 50. The input to node 1 is constant at 1. This arrangement may simulate the decrease in causal factors for the activity controlled by node 2 as this activity is performed, such as the decrease in causal factors for drinking after the onset of drinking. Contrary to what one might expect, such a decrease does not immediately cause a return to activity 1. Rather, the causal factors for activity 2 must decrease considerably before a switch takes place. As seen earlier, the condition for a switch from 2 to 1 is $y_1 > W_{12}z_2$. This means that if W_{12} is large enough, the network can persist in activity 2 even if the input to node 2 is smaller than the input to node 1 (formally, $y_1 < W_{12}z_2$ is compatible with $y_1 > y_2$, given that $W_{12} > 1$).

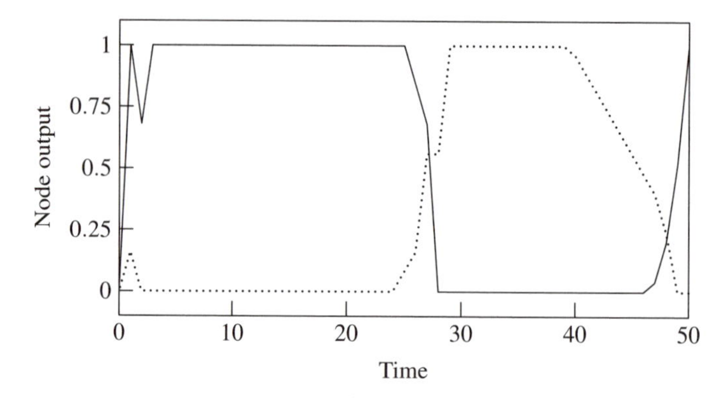

Figure 3.29 Stability of decisions in the decision making circuit in Figure 3.25. This simulation is identical to that in Figure 3.26 up to time 28, i.e., just after the transition from node 1 to node 2 has started. At this point the input to node 2 starts to decrease linearly, reaching 0 at time 50. However, node 2 is not shut off immediately but keeps active for some time.

3.7.4 Indecision

Despite the overall stability of behavior, indecision is observed occasionally and has been studied extensively by ethologists (Hinde 1970). For instance, an animal may be uncertain whether to approach or avoid a familiar food in a novel environment. A few different outcomes are possible in similar conditions:

1. The animal may oscillate between behaviors, e.g., first approaching and then retreating (a familiar thing to everyone who has attempted to feed half-tame animals).
2. Components common to both behaviors may appear (e.g., walking would be appropriate to both approach and avoidance).
3. A mix of components from either behavior can appear.
4. A third behavior may surface that apparently has little to do with those in conflict (*displacement activity*).
5. A few attempts may be necessary to switch between activities (e.g., a bird that has been resting for a while may make several attempts before actually taking off).

These issues are complex and have not been studied with neural network models. We may speculate that at least some of these phenomena may occur in decision making networks of the kind shown in Figure 3.25. For instance, long transition times between activities can arise from inhibitory weights that are not much larger than 1. Similarly, if inhibition between nodes is weak, several nodes could be active simultaneously, at least for some time. This could explain the appearance of hybrid behavior patterns or of components of behavior common to more than one motivated activity. The classical ethological explanation of displacement activities may also be implemented in neural networks. The idea is that the two conflicting behaviors cancel each other by mutual inhibition, thus allowing the expression of a third behavior (see Hinde 1970 for detailed discussion). This would require the nodes that represent the conflicting activities to inhibit each other strongly while exerting a weaker inhibition on the node representing the surfacing activity. The latter would continue to be influenced by the usual causal factors, consistent with ethological findings. For instance, van Iersel and Bol (1958) and Rowell (1961) found that in terns and chaffinches, wetting the plumage increases the rate of both normal and displacement preening in an approach-avoidance conflict situation.

3.7.5 Organization of behavior in time

Often behavior must have a certain organization in time to be functional (Section 3.1). Animals have more specific needs than we have considered so far. For instance, the problem of feeding does not end with the decision to eat: a number of different activities (e.g., searching and handling food and swallowing) must be coordinated in a functional behavior sequence. In this section we show how a number of requirements on behavior sequences can be implemented by appropriately structuring the decision making mechanism.

To understand behavior sequences, we must keep in mind that a behavioral response may partly change the motivational state that caused the response (Sections 1.2.2 and 3.2.2). This may happen via two routes. First, behavior may cause a change in the external or internal environment that the animal perceives at a later time. For instance, drinking changes body water volume, which is one of the causal factors for drinking itself. Second, a state *within* the decision making mechanism can change. This is required, for instance, when a given behavior sequence must be performed in the absence of significant or reliable changes in stimulation. We have called such states *regulatory states* (Section 1.2). The simple decision maker examined above (Figure 3.25) already has regulatory states that are crucial to its abilities. The network, in fact, is not simply input-driven. For instance, in Figure 3.29, the input at $t = 21$ and $t = 35$ is identical, but the pattern of activity is reversed. The recurrent inhibitory connections are able to build a "regulatory state" concretely implemented in the activities of nodes 1 and 2 that enabled the network to overcome dithering. The two nodes thus fulfill a double function: an output function that directly controls behavior and a regulatory function. In the following we consider further examples of the role of regulatory states in structuring behavior. Further discussion of how animals structure behavior in time can be found in Hinde (1970) and McFarland (1999).

3.7.5.1 Internal states that build up with time

As our first example, consider a response that is produced only if a stimulus is present for a given time. For instance, females in monogamous species tend not to respond immediately to male courtship. This makes functional sense because males may court without being willing to bond with the female and because males willing to bond tend to be more persistent (Wachtmeister & Enquist 2000; see also page 69). Another example of delayed responding occurs in the sexual behavior of male rats. The rat does not attempt to copulate immediately when a receptive

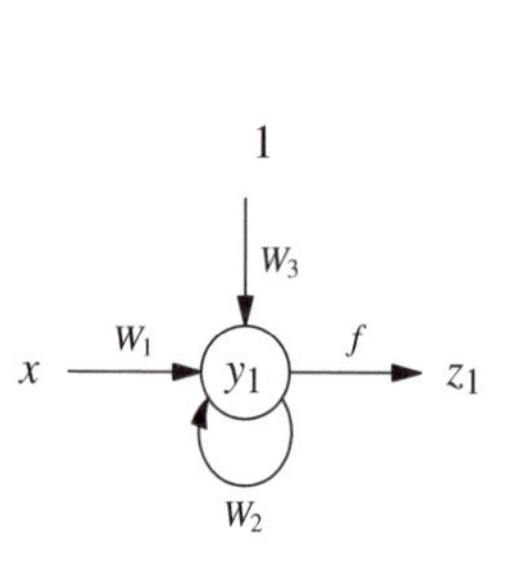

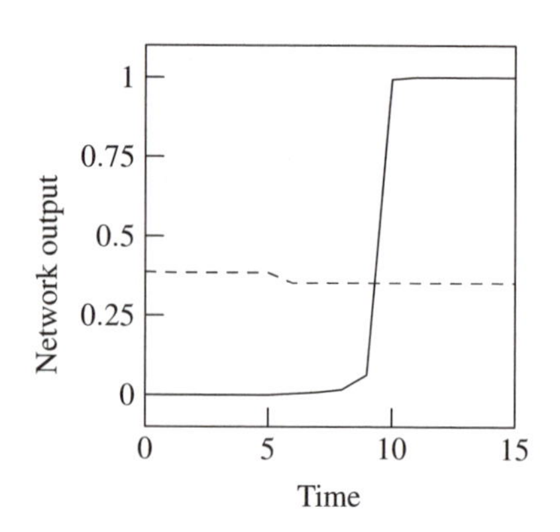

Figure 3.30 Left: Simple recurrent network capable of delayed responding (equation 3.12). Right: Sample response to an input that is low before time step 5 and high afterwards. Dashed line: Network with randomly set weights. Continuous line: Network with properly set weights (the basic random search algorithm was used; Section 2.5.3).

female is present but rather goes through a series of behaviors that ultimately result in copulation.

The simple network in Figure 3.30 is capable of delayed response to stimuli. Formally, the network response is given by

$$r(t+1) = f\big(W_1 x(t) + W_2 r(t) - W_3\big) \tag{3.12}$$

Note that the response at time t sums to the input to jointly determine responding at time $t+1$. The constant negative input $-W_3$ ensures that weak stimuli cannot elicit a strong response. The transfer function f is a sigmoid function whose slope regulates how fast the system can go from a small to a large output. Figure 3.30 (right) demonstrates the abilities of this network.

An internal state that under suitable conditions accrues with time potentially has many uses. The basic mechanism just described can be included in larger networks to provide a diversity of behaviors. For instance, we can include a more capable recognition mechanism by replacing the single input node with a layer of N input nodes. The resulting network is formalized as

$$r(t+1) = f\Big(\sum_{i=1}^{N} W_i x_i(t) + W_{N+1} r(t) - W_{N+2}\Big) \tag{3.13}$$

This network can be more selective as to which stimuli are able to accrue in the internal state. For instance, males of a female's own species could be discriminated from males of other species.

Another interesting example is obtained considering the network in Figure 3.19. As we have seen, this network reacts to changes in input, but not to constant input. If a delay mechanism is added to the output node, the network reacts only when the input presents a series of consecutive changes. Such a mechanism could be

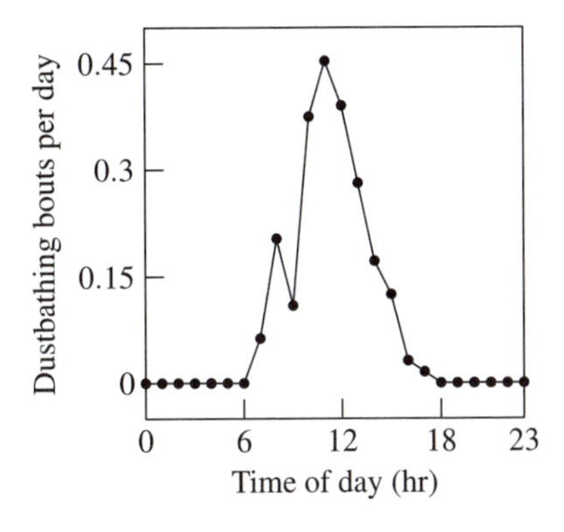

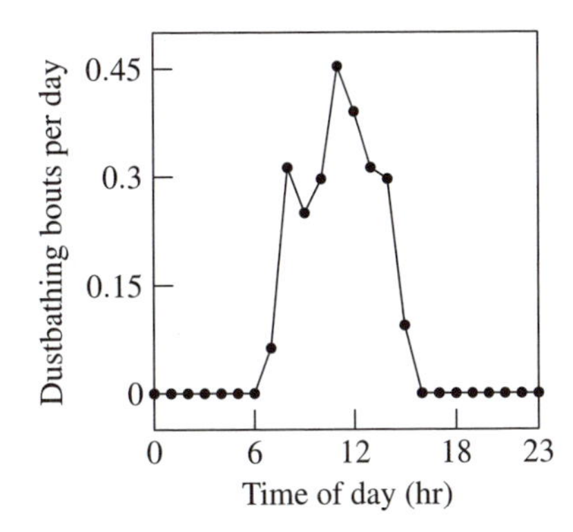

Figure 3.31 Distribution of dustbathing during the day. Left: Empirical data from 2-week-old red Burmese junglefowl, *Gallus gallus spadiceus*. A group of 16 birds was observed in the laboratory for 4 consecutive days, with light provided between 6:00 and 18:00 each day. The birds dustbathed an average of 2.2 times a day, more frequently around midday and never in the dark. (Unpublished data from J. A. Hogan's laboratory; see also Hogan 1997; Hogan & van Boxel 1993.) Right: Simulation using the network in Figure 3.32. A single network was run for $16 \times 4 = 64$ simulated days, yielding a pattern of dustbathing similar to the one observed in birds, with an average of 2.4 dustbathing bouts per day. See the text for further details.

employed, for instance, by an animal such as a glow worm or a firefly to react to a signal consisting of a series of light flashes, but not to a single light flash.

3.7.5.2 Circadian rhythms and "action-specific energy"

In this section we illustrate two additional factors that can organize behavior in time. As a concrete example, we consider dustbathing, a characteristic behavior of many birds that serves to spread dust into the feathers and keeps them clean (see Section 4.7 for further details). The first factor is circadian control, meaning that a behavior is performed more often at some time of day than at others (see, e.g., Toates 1998 for an introduction to circadian rhythms). For example, chickens dustbathe more often around midday, as shown in Figure 3.31. Circadian control may have a number of functions. One is to anticipate events (e.g., to go to a safe place before dusk). Another is to provide stability against environmental noise (e.g., not believing it is dusk if the sky is cloudy). Consideration of dustbathing and feeding in chickens shows that circadian rhythms can also decrease conflict between activities. Eating peaks at dawn and dusk and is stable during the day (about 5 min/h). This makes functional sense, considering that the animals do not eat at night. Dustbathing, as we have seen, shows a reversed pattern: it peaks around midday and is performed rarely at other times. Dustbathing is a time consuming activity: a bout takes almost 30 minutes in adults and is performed every other day. There seems to be no functional reason for dustbathing to be performed at a particular time, but the observed timing reduces conflict with feeding.

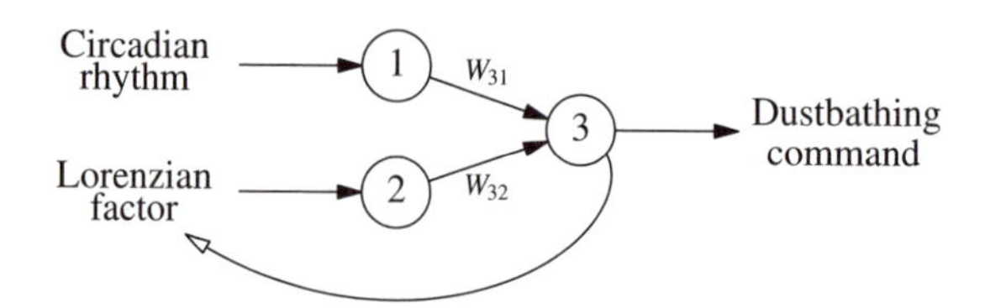

Figure 3.32 A minimal neural network model of a behavior (dustbathing) controlled jointly by a circadian rhythm and a Lorenzian factor. The latter increases whenever the behavior is not performed and decreases during performance of the behavior (indicated by the inhibitory feedback arrow). With suitable choices of weight values and causal factors (explained in the text), this network can reproduce the timing pattern of dustbathing observed in the fowl (see Figure 3.31).

A second kind of causal factor has been called *action-specific energy* by Lorenz (1937, 1981). The concept originates from observations of vacuum activity (see page 102) that suggest that causal factors for a behavior may build up spontaneously with time when the behavior is not performed and decrease only during performance of behavior. While many details of Lorenz's original model can be criticized (reviewed in Hinde 1970), a spontaneous buildup of motivation appears important in the control of many activities (Hogan 1997). Dustbathing is one of the clearest examples. Birds that have been prevented from dustbathing for some time will dustbathe copiously as soon as they are allowed to do so, even at times when dustbathing hardly ever occurs normally. Moreover, external manipulations can influence the timing of dustbathing but not its amount, as if a limited budget of the causal factor was available (Hogan 1997; Hogan & van Boxel 1993).

In summary, while circadian control ensures that dustbathing occurs primarily in the middle of the day, a Lorenzian factor ensures that dustbathing will occur eventually. It may be conjectured that without the Lorenzian factor, dustbathing would be dominated too easily by more urgent activities such as feeding and drinking, with the risk of never being performed (McFarland 1999).

A neural network model of dustbathing can be set up by including a neural "clock" and a source of motivation for dustbathing. A minimal model with only three nodes is shown in Figure 3.32. Although the causal factors are different, the basic architecture of the model is the same as our simple model of drinking (Figure 3.21). The only difference is that the activity of node 3 is interpreted as the probability of dustbathing rather than drinking:

$$\Pr(\text{dustbathing at time } t) = z_3(t) = W_{31}z_1(t) + W_{32}z_2(t) \tag{3.14}$$

(see equation 3.6). An additional similarity is that both the Lorenzian factor and the water deficit increase with time if the relevant behavior (dustbathing or drinking) is not performed, and they decrease when it is performed. Note, however, an important conceptual difference between these two causal factors. Water deficit is a physiological state of the body that exists independently of the drinking control mechanism. The Lorenzian factor, in contrast, is a regulatory state that exists solely to participate in the control of dustbathing (see Section 3.1).

The simple network model formalized by equation (3.14) can reproduce, with a suitable choice of weight values and causal factors, the observed pattern of dustbathing, as shown in Figure 3.31. The remainder of this section explains in more detail our choice of causal factors and network parameters. A circadian rhythm can be included simply by assuming that the activity of node 1 varies according to a sine wave:

$$z_1(t) = \sin\left(\frac{2\pi}{24}(t-5)\right) \tag{3.15}$$

This function oscillates between -1 and 1, with a period of 24 hours appropriate for a circadian rhythm. Using $t-5$ rather than simply t in the function argument ensures that $z_1(t)$ is maximum for $t = 11$, when dustbathing is more likely to occur. Around this time z_1 will provide strong excitatory input to node 3, while at night z_1 will be negative and thus will inhibit activity in node 3.

To model a Lorenzian factor, we assume that the time course of z_2 is described by the equation

$$T\Delta z_2(t) = 1 - z_2(t) \tag{3.16}$$

where $\Delta z_2(t)$ is the variation in z_2 at time step t (a time step corresponds to 1 minute in our simulation). The meaning of equation (3.16) is that z_2 approaches a maximum value of 1 at a rate regulated by the time constant T ($T = 180$ hours in our simulation). This is akin to a node with slow dynamics that receives a constant input of 1 (see Section 2.1.3). During a dustbathing bout, however, the activity of z_2 is assumed to decrease by an amount of 0.12. Lastly, the two weights in the network had values $W_{31} = 0.01$ and $W_{32} = 0.003$. The model is not very sensitive to the precise values of most parameters, and we could find suitable values simply by trial and error guided by a few heuristic considerations (e.g., that the circadian rhythm must provide maximum input around 11:00).

3.7.6 Comparison with control theory

The analysis of decision making in animals owes much to *control theory*. In a typical control theory model, behavior is seen as part of a *control system* with a well-defined task such as maintaining homeostasis. Important ideas such as negative feedback or the need for stability of behavior arise very clearly as control problems (McFarland 1999). To build our neural network models, we have borrowed significantly from applications of control theory to animal behavior (McFarland 1974a, 1999; McFarland & Sibly 1975). This inspiration is not difficult to see. Figure 3.33 (top) shows a simple control model of drinking based on negative feedback (see page 103). The model prescribes drinking rate $d(t)$ as a function of the difference between systemic water volume $v(t)$ and a set point V as follows:

$$d(t) = k(V - v(t)) \tag{3.17}$$

where k is a positive constant. This is very similar to the simple model in Figure 3.21, where the probability of drinking when water is present is $W_1 + W_2(V - v(t))$ (see equation 3.6). Thus in both cases there is a linear relationship between drinking rate (or probability of drinking) and water deficit. The network model thus can be seen as one implementation of the idea that homeostasis can be achieved via

(a)

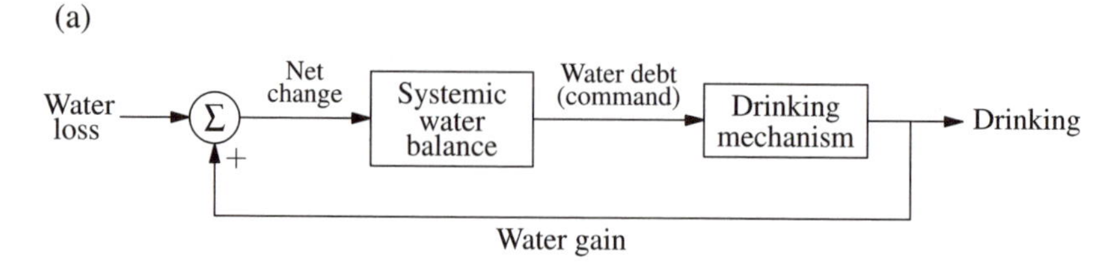

(b)

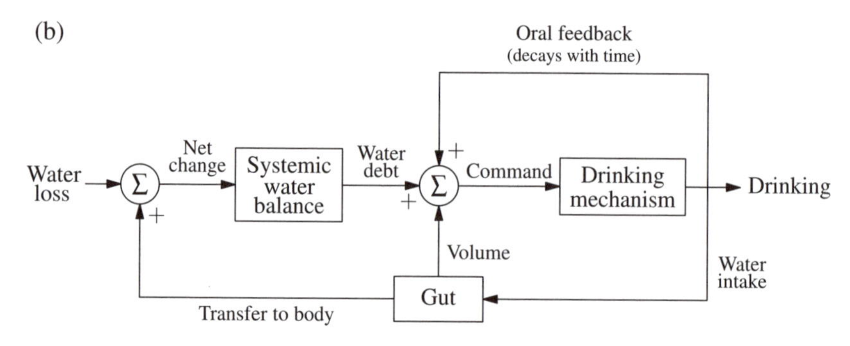

Figure 3.33 Diagrams of two control models of drinking in doves illustrating feedback mechanisms. (Adapted from McFarland 1971.) Both models use systemic water deficit as a control factor. In model (a) this is the only causal factor for drinking, whereas model (b) adds two further factors, the volume of the gut and oral stimulation arising from the ingestion of water. When drinking fills the gut, stretch receptors in the gut are stimulated and inhibit drinking (negative feedback). The oral feedback is positive and increases motivation temporarily once the animal has started drinking. This positive feedback decays with time. Model (b) can describe drinking behavior on a finer time scale than model (a) because it takes into account more details of animal physiology.

negative-feedback control. Our more complex model of drinking (Figure 3.23) likewise is similar to the more refined control schemed in Figure 3.33 (bottom).

Control theory is an extensively developed framework with both theoretical and practical sides. While the former addresses the general problem of how to control a system, employing concepts such as feedback and feedforward control, the latter covers the analysis of particular controlling devices. It is interesting to note that such devices sometimes have the structure of neural networks. For instance, the so called linear adaptive filters are the same as two-layer linear networks and can be used in homeostatic control and other applications (Kalman 1960; Widrow & Hoff 1960; Widrow & Stearns 1985; Wiener 1949). However, such an analysis of mechanism is less developed in applications of control theory to animal behavior. Here functional considerations dominate. The need for stable decisions, how to set thresholds for switching between activities and other important questions have been discussed thoroughly, but what mechanism is actually used by animals has been less considered. This has caused some debate over, for instance, whether concepts such as thresholds and decision boundaries are adequate for the study of animal behavior (McFarland 1974b; Toates 1980).

Neural networks allow us to address mechanism in addition to function. We have seen that simple neural networks can take decisions and maintain homeostasis without explicit consideration of thresholds or decision boundaries. Although neural network models of decision making are not well developed, we hope to have shown that they can potentially integrate results from control theory and ethology. Presently it is difficult to see whether a substantial amount of new knowledge will be generated or whether neural networks will simply be a more biologically realistic implementation of already known ideas. It is possible, however, to hint at several areas where neural network models could be particularly useful.

Reactions to stimuli and decision making have often been considered separately. Note, for instance, that the control models in Figure 3.33 do not consider how the animal knows whether water is there or not. Neural networks can integrate mechanisms of reactions to stimuli with decision making. For instance, we may study why some stimuli are more effective than others. Such an integrated model of the processes from stimulus reception to decision making might also help us to understand why the nervous systems sometimes commands nonfunctional behavior.

By providing a concrete implementation, neural networks also allow rigorous tests of intuitive ideas. For instance, one mechanism suggested to prevent dithering (rapid oscillation between activities) is a kind of short-term positive feedback. In other words, when an activity gains control, it is assumed to receive a "boost" in motivation to counter the possible decrease in relevant motivational factors (McFarland 1999). However, we have seen in Section 3.7.3 that decisions taken by mutual inhibition (another traditional idea) are already rather stable without any additional mechanisms.

3.8 MOTOR CONTROL

By *motor control* we mean how the nervous system and the skeletomuscular apparatus contribute to behavior. It is well established today that motor control is exercised by circuits within the central nervous system, called *central pattern generators*, that produce autonomous motor patterns, usually in combination with central and peripheral control based on sensory feedback (Cohen & Boothe 2003; Dean & Cruse 2003; Hinde 1970; Lorenz 1981). The idea of central pattern generators seems to have been suggested first by Graham Brown in 1918, who showed that cats deprived of sensory inputs could still produce coordinated stepping and even suggested a possible neural mechanism (Hill et al. 2003). However, some time passed before this view was accepted because early ideas about motor control were inspired by the view of animals as reactive, reflex-driven machines. Motor patterns were explained as chains of reflexes in which each movement creates appropriate sensory stimuli to elicit the next movement. For instance, walking was explained as a succession of reflexes triggered by changes in leg position, loading, and equilibrium signals during the step. We are not aware of any attempt to verify if such an arrangement could actually work, and empirical research has rejected this idea (Dean & Cruse 2003). Erich von Holst was among the first to establish neurological evidence for central pattern generators (Von Holst 1937, reprinted in von

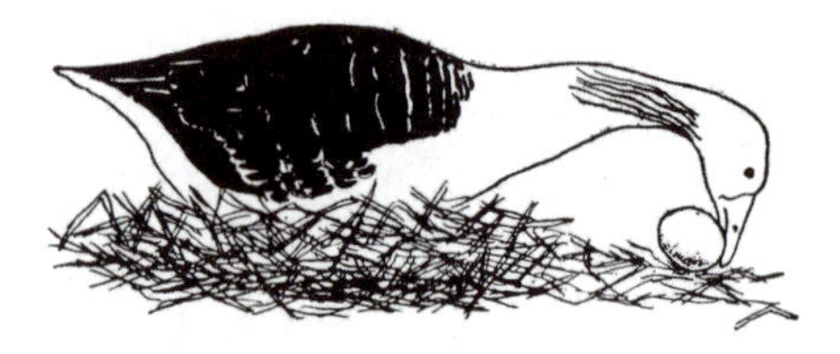

Figure 3.34 Egg retrieval in the greylag goose (*Anser anser*) exemplifies the two concepts used in classical ethology to analyze motor patterns (Lorenz & Tinbergen 1938). The stretching out of the neck and the movement back are stereotyped and rigid (they are completed even if the egg is removed), corresponding to the traditional concept of a "fixed action pattern." The fine adjustment movements that keep the egg from rolling away from the beak correspond instead to a "taxis," an orienting movement guided by sensory information. Reprinted from Tinbergen (1951) by permission of Oxford University Press.

Holst 1973; another early work is Adrian 1931). In a study of earthworm locomotion, he isolated the dorsal ganglia of an earthworm, cutting all afferent nerves and placing the ganglia in a physiological solution. Despite the absence of stimuli that could trigger reflex chains, von Holst recorded rhythmic electrical activity similar to that observed during actual locomotion. He also showed that isolated earthworm segments actually contracted when the wave of activity passed through them.

3.8.1 The analysis of behavior patterns

Classical ethological theory of motor control was based on two concepts: *fixed action patterns* and *taxes* (Hinde 1970; Lorenz 1935, 1937; Tinbergen 1951). A fixed action pattern is, once elicited, generated within the animal and not guided by any sensory information. A taxis, on the other hand, is an orienting motor component controlled by sensory information. In terms of control theory, these concepts separate motor mechanisms that use sensory feedback from those that do not (McFarland 1971; see also Figure 1.6 on page 13). The classical example of a behavior with both a fixed action pattern and a taxis component is egg retrieval in the greylag goose (Figure 3.34). Ethologists also observed that fixed action patterns seem to be inborn, i.e., to develop without any specific experiences (Berridge 1994). In his famous study of courtship behavior in ducks, Lorenz (1941) even showed that such patterns seem as constant and species specific as skeletal bones and that they could be used to detect phylogenetic relationships. A remarkable example of such constancy is the "head throw" display of the goldeneye duck (*Bucephala clangula*), having a mean duration of 1.29 s and a standard deviation among individuals of only 0.08 s (Dane et al. 1959).

Classical ethology defined fixed action patterns solely by the absence of a role of external stimuli in controlling the form of the behavior (Hinde 1970). This did not exclude the possibility of sensory feedback from receptors inside the animals, such as proprioceptors that provide information about muscle tension (Figure 3.35). In other words, although the idea of a centrally generated pattern got most attention, the importance of sensory feedback was also acknowledged. The view that

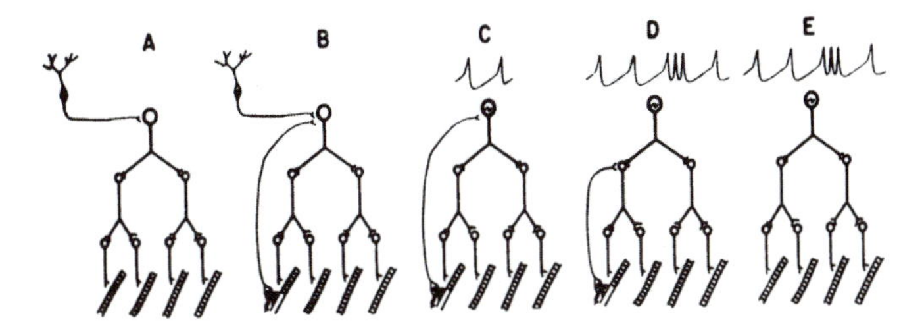

Figure 3.35 An early account of possible nervous organizations in motor control (Bullock 1961). The top node represents a neuron able to trigger the execution of a motor pattern. The other nodes represent neurons that determine the shape of the motor pattern. The diagrams combine in different ways three factors involved in motor control: input from outside the motor mechanism (A and B), proprioceptive feedback (B: feedback determines whether the pattern is executed or not, C: feedback changes the shape of the pattern) and pacemaker neurons (C, D, E). Reprinted by permission of the McGraw-Hill Companies from Hinde (1970). *Animal Behavior*. McGraw-Hill Kogakusha.

emerged was that behavior is made up of combinations of fixed action patterns that are elicited and oriented by sensory information (Hinde 1970; Tinbergen 1951).

Whether the ethological distinction between fixed action patterns and taxes is a useful and correct description of motor phenomena has generated some debate (Hinde 1970). Today the issue is outdated in light of much better knowledge about how motor control is exercised by the nervous system (see below). However, the distinction still makes sense at the behavioral level. The extent to which motor components are rigid or flexible determines the extent to which motor learning is possible, as well as whether the behavior patterns we observe appear precise or clumsy. The "style" of motor control in any particular species is the outcome of evolutionary processes. For instance, flexibility may be related to the environment a species lives in. Animals such as zebras, which live on smooth plains, have a limited number of gaits. In contrast, goats are much more flexible (the four legs can move almost independently of each other), reflecting their mountain habitat.

The emphasis that classical ethology placed on the rigidity of motor patterns can be partly understood by ethologists' interest in signal evolution. Signals often evolve out of other behaviors, in a process called *ritualization* (Eibl-Eibesfeldt 1975, Hinde 1970, § 27.3, see also Chapter 5). This process is associated with significant alterations in motor control, often resulting in highly stereotyped behavior. For instance, components of preening have been the raw material for some courtship signals in ducks (Lorenz 1941). Preening can be directed toward the whole body, but signals evolved from preening have lost most of this flexibility and are always performed in exactly the same way. Another example is the highly stereotyped head-throw display of the goldeneye duck (mentioned above), which is derived from the takeoff leap. An observer usually can distinguish such behavior patterns from those that are guided by sensory information.

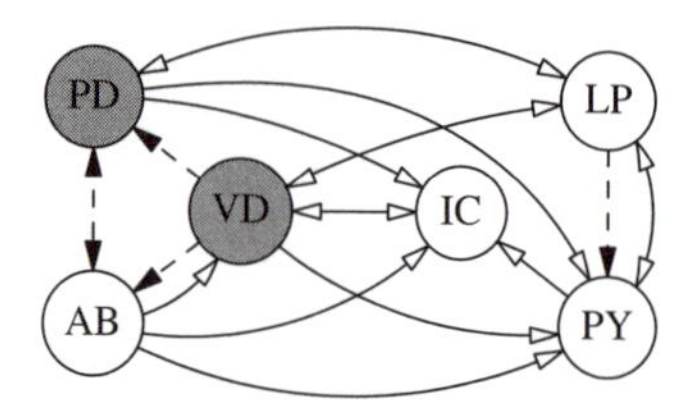

Figure 3.36 Scheme of the neural circuit controlling the pyloric rhythm in the spiny lobster (*Palinurus*; based on Hooper 2003).AB: anterior burster neuron; PD: pyloric dilator neurons (2 cells); VD: ventricular dilator neuron; IC: inferior cardiac neuron; LP: lateral pyloric neuron; PY: pyloric neurons (8 cells). Empty arrows denote inhibitory synapses. Those originating from PD and VD (shaded) cause longer-lasting inhibition. Dashed lines indicate electrical synapses that shift the electrical potential of the postsynaptic cell toward that of the presynaptic cell. Connections that exist in both directions are drawn as a double-headed arrow.

3.8.2 A biological example

To exemplify how biological nervous systems exercise motor control, we turn to the stomatogastric ganglion in the spiny lobster and its control of the digestive tract. There are several rhythms active in the lobster digestive system that are involved in food intake and manipulation and in moving food within the digestive system. A well-studied rhythm is the pyloric rhythm of the spiny lobster (Hartline 1979; Hartline & Gassie 1979; Hooper 2003; Mulloney & Selverston 1974). The pyloric rhythm controls the pylorus, a system of valves and sieve plates that helps to sort out food particles for further digestion. The circuit controlling the pylorus is made up of 13 motoneurons and one interneuron (Figure 3.36). There pyloric circuit has three main phases of activity. The AB neuron serves as a pacemaker. Because of its electrical coupling to the AB neuron, the PD neuron fires together with AB. Together they strongly inhibit the other cells in the network. AB and PD synapse on the pyloric dilator muscles, and when they fire, they open a valve from the gastric mill to the pyloric region. This allows food particles to flow into the pylorus. The LP neuron synapses on the lateral pyloric muscle. It is the first cell to recover from inhibition caused by AB and PD, and it fires causing the valve to close. The IC cell fires in phase with the LP cell and aids in control of the valve. The LP neuron also inhibits the AB and PD neurons, which prevents them from firing again. In the final phase of the cycle the PY neurons recover from inhibition and start firing. These neurons control the pyloric muscles responsible for moving the sieve plates that sort out the food for further digestion. The VD cell often fires in phase with the PY cells, but its firing is not easily correlated with just one movement.

In summary, this example shows how a rhythmic movement is produced through successive activity and inhibition of particular neurons. The valve from the gastric mill is first opened and then closed; then the sieve plates are moved. The pyloric network is controlled by many factors, including the endogenous bursting activity of the AB neuron, inhibitory connections, and the properties of individual cells.

3.8.3 Motor pattern generation in neural networks

We have already seen that neural networks can analyze sensory information and take appropriate decisions. Here we show how they can also model the control of movement. In animals, a behavioral response is the result of coordinated activity in many muscles. Essentially, motor control can be seen as a map from decisions, or central states, onto muscle activities via patterns of activity in motor neurons (Amirikian & Georgopoulos 2003; Flash & Sejnowski 2001). In contrast to sensory processing and decision making, where the mapping is from many inputs to few outputs, motor control is a map from few pattern-generating neurons to many motoneurons. To generate a given behavior, some muscles must contract and others relax. The extent of contractions and relaxations may also vary. Furthermore, motor acts almost always involve sequences of muscle coordinations, which require pattern generation over time. Indeed, much behavior consists of repeating the same elementary motor pattern many times. These behaviors include locomotion, chewing, many displays used in communication, breathing and the beating of the heart. Such cyclicity may be generated by an oscillating central pattern generator (Bullock 1961; Marder & Calabrese 1996), which in nervous systems can be driven by pacemaker neurons (Wang & Rinzel 2003) and/or on reciprocal inhibition among nerve cells (Hill et al. 2003). Many motor patterns are not repeated but still are generated over a period of time. Figure 3.37 illustrates how swallowing involves a centrally generated sequence of tightly control muscle coordinations.

Figure 3.38 shows three simple examples of neural networks able to generate motor patterns. The first network models a simple behavior that consists of one single coordination of muscle activity (no sequences). For instance, stretching a limb involves contracting abductor muscles and relaxing adductors. Bending the limb involves the opposite action. In the network, a command signal (a decision made according to the mechanisms treated above) starts a pattern generator (here consisting of just one node and its connections to motoneurons) that activates two motoneurons. To stretch the limb, the motoneuron connected to the abductor muscle is stimulated, whereas the motoneuron connected to the adductor muscle is inhibited. The second network in Figure 3.38 shows how to extend the mechanism so that one motor command stretches the limb, and another bends it. Lastly, the third network in the figure shows how alternated stretching and folding can be generated by including an oscillator node within the pattern generator. If the oscillating node has a period of two time units, the limb will be stretched in one time unit and bent in the next, and then stretched again and so forth.

From these examples, and with some imagination, one can see how a pattern generator connected to many motoneurons could coordinate the stretching and bending of many muscles, thereby allowing an animal to, for instance, swim or walk. An illustrative example is the network model of human walking developed by Prentice et al. (1998). The model consists of two networks with different functions, as suggested by behavioral data (see references in Prentice et al. 1998). The first network maps a constant input into alternated oscillatory activity of two output nodes. The constant input represents the decision to walk at a given speed, with stronger input meaning higher speed. The frequency of the oscillating output is the frequency of

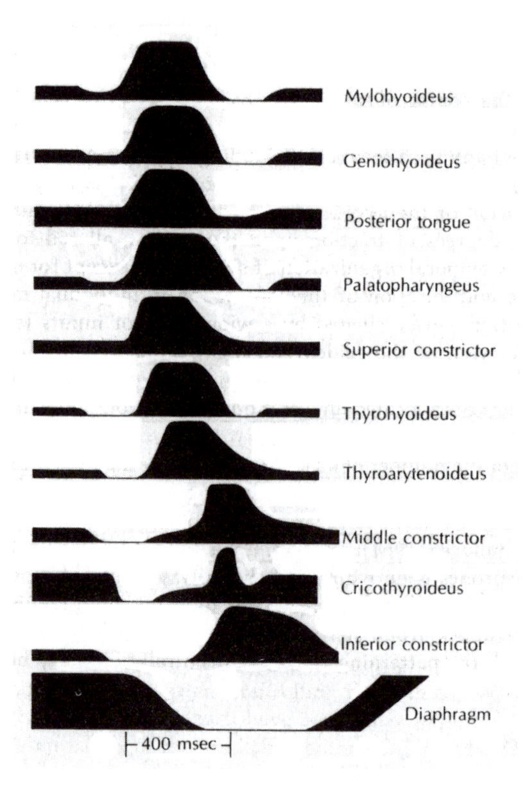

Figure 3.37 The sequence of contractions for some of about twenty muscles that control swallowing in the dog (Doty & Bosma 1956). This sequence can be elicited by many kinds of stimuli (e.g., water or disturbing objects in the mouth) but is performed always in the same way, even when sensory feedback is distorted or abolished (Doty 1951). Reprinted by permission of the McGraw-Hill Companies from Hinde (1970). *Animal Behavior*. McGraw-Hill Kogakusha.

walking that corresponds to the desired speed (measured in, e.g., steps per second). The authors used a simple recurrent network to realize this input-output map. The architecture was similar to the bottom right network in Figure 2.13 on page 56, with four hidden nodes and two nodes to build regulatory states. The second network in the model is a feedforward network with one hidden layer. Its task is to convert the simple oscillatory output of the first network into patterns of muscle activity suitable for walking. The authors used eight output nodes, the activity of each one representing the degree of contraction of one of the main muscles in walking. The network was trained by back-propagation using as desired output the muscle activities recorded from a walking subject. Two sets of recordings were used to train the network, corresponding to the subject walking at speeds of 1.2 and 1.8 m/s. After training, a network with only four hidden nodes could produce realistic walking patterns, although a network with 16 hidden nodes captured better some subtle relationships between muscle activities. In a later test, the model reproduced

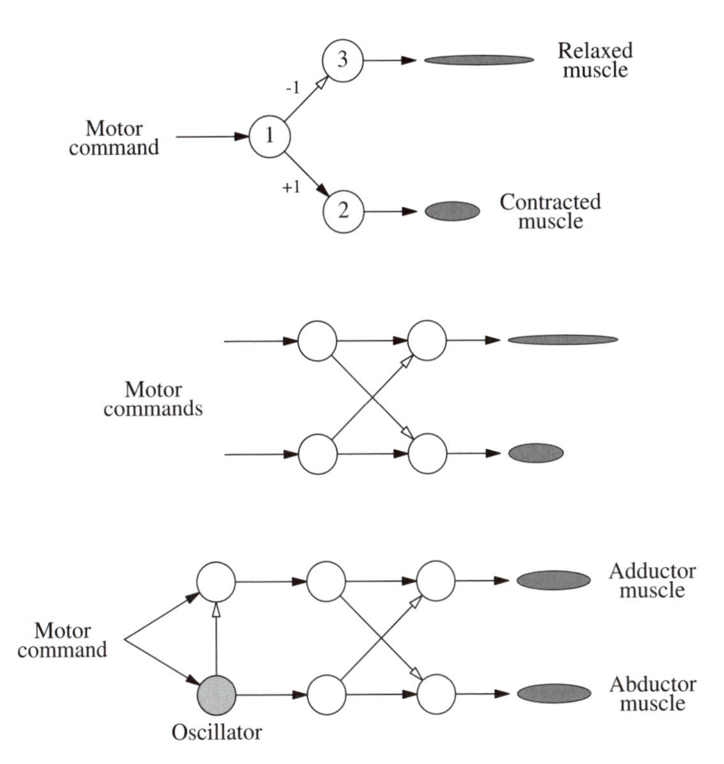

Figure 3.38 Simple networks that show how a pattern generator can map motor commands onto coordinated muscle activities. Open arrows denote inhibition; closed arrows, excitation. In the first network, a motor command results in excitation of one motoneuron and inhibition of another, causing contraction of one muscle and relaxation of another. The second network illustrates how different motor commands (i.e., different patterns of stimulations of the pattern generator) can produce different muscle coordination. If the muscles are the abductor and adductor of a limb, the network can command both stretching and bending of the limb. The third network can generate alternating stretching and bending of a limb. The network is derived from the second one by adding the two leftmost nodes, a regular node and an oscillator node, both excited by a motor command. The oscillator node, provided with constant input, produces an alternation of high and low output. When oscillator output is high, the other node is inhibited, and the limb is stretched (abductor contracted, adductor relaxed). When oscillator output is low, the other node is excited by the motor command, and the limb is folded (abductor relaxed, adductor contracted). In the absence of a motor command, both muscles are relaxed.

accurately the muscle activities recorded during walking at 1.4, 1.6 and 2.0 m/s, reflecting the natural generalization abilities of feedforward networks (Section 3.3).

3.8.4 Sensory feedback

The control of motor patterns is of course not just endogenous. Behavior is typically in one way or the other oriented toward the outside world, and even a single motor pattern often includes sensory feedback that partly controls the movement. Such feedback may come from proprioceptors within the body or from receptors sensitive to external stimulation. An important role of feedback from proprioceptors and mechanical receptors is to compensate for errors or unusual load, thereby maintaining orientation or balance (Kandel et al. 2000). External receptors can also have a similar compensatory role, e.g., in maintaining orientation toward a moving prey. In general, empirical work shows that central pattern generation often interacts with sensory feedback (Cohen & Boothe 2003).

An early study demonstrating both external and internal feedback control is prey capture in mantids (McFarland 1971; Mittelstaedt 1957). This behavior can be divided into an orientation phase and a strike phase. The mantid strikes with its forelegs at an angle (relative to the body) that is controlled by the angle to the prey. To determine the aim of the strike, the animal uses sensory information from the eyes, measuring the angle between head and prey, and information from proprioceptors, measuring the angle between body and head. The strike is not controlled by sensory feedback, corresponding thus to a fixed action pattern. The division into an orientation phase using sensory feedback and a strike or attack phase that is generated centrally is common to many species (McFarland 1971; Tinbergen 1951).

Entering sensory feedback into a neural network model of motor control is, at least in principle, easy. For instance, a network that generates the motor pattern necessary for flying could be integrated with a network that analyzes flow fields to control flying (Section 3.5.1). An interesting example is the pioneering study of Beer and Gallagher (1992), who developed a neural network to control walking in a simulated cockroach-like insect, shown in the left panel of Figure 3.39. The six legs were controlled by identical copies of a recurrent network, shown in the middle panel. The right panel shows how the six networks were connected with each other to form the complete walking control system. Important to the present discussion is that each network received sensory feedback from a proprioceptor that measured the angle between the leg and the body. Although simulated insects could walk without such feedback, the latter allowed better coordination and faster walking. Other network studies of locomotion in simulated animals and legged robots are reviewed by Beer et al. (1998); Ijspeert (2003) and Nolfi and Floreano (2000).

The understanding of motor control in terms of neural networks is perhaps the best developed field of neural network research, and we cannot review all relevant studies here. We just want to state that a number of motor systems have been studied in detail and are well understood. These systems range from very simple, such as the three-neuron network that controls aerial respiration in the pond snail (*Lymnaea stagnalis*; Taylor & Lukowiak 2000), to rather complex, such as the control of swimming in lampreys (Grillner 1996; Grillner et al. 1995). The latter is

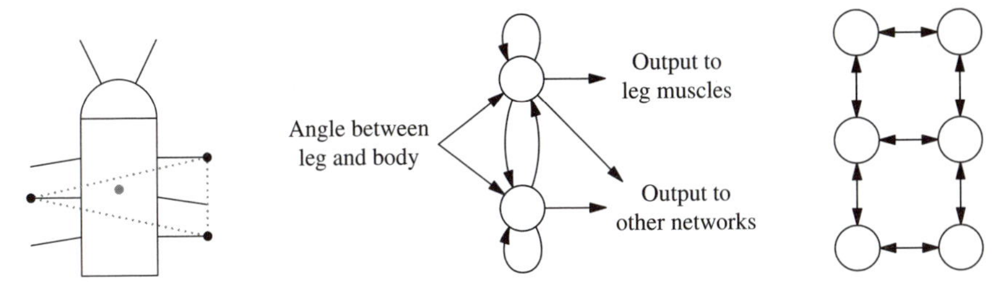

Figure 3.39 Description of the simulation of an insect walking by Beer and Gallagher (1992). Left: Scheme of the simulated insect. The three legs marked with a black dot are touching the ground; the other legs are lifted. The insect is stable if its center of mass (gray circle) falls within the polygon formed by the support points (dotted triangle); otherwise it falls. The insect moves according to the sum of forces exerted by all legs on the ground. Middle: Scheme of the neural network used in to control a leg. The top and bottom circles represent three and two nodes, respectively. The network received input from a proprioceptor measuring the angle between the leg and the body. Network output is sent to leg muscles and to other networks. Right: Scheme of the interconnections between the six networks that control the legs (each one represented by a single circle). Network weights were set by simulated evolution, with fitness given by fast walking. This resulted in efficient walking using the so-called tripod gait, in which the six legs are coordinated into two groups of three in the way shown in the left panel. The two groups push alternatively, one group pushing while the other is lifted. This gait is actually used by fast moving insects.

of particular importance because it seems to contain the basics for locomotion in vertebrates in general. Lampreys swim by generating a traveling wave of neural activity through their spinal cord, which causes a front-to-back undulation of the body that propels the animal forward. The generation of such a wave and how it is modulated to turn and move at different speeds are well understood based on a detailed neurophysiological model of the underlying neural network. Other well-understood systems are walking and escaping in the cockroach (Ritzman 1993), prey capture in toads and frogs (Cobas & Arbib 1992; Ewert 1980) and swimming in the leech (Brodfuehrer et al. 1995).

3.9 CONSEQUENCES OF DAMAGE TO NERVOUS SYSTEMS

Often limited damage to nervous systems leaves performance relatively unaffected. This is especially true when we compare nervous systems with the products of human engineering, say, a television. A further example is that of a computer program, which often can be broken by changing a single bit. Such resistance to damage is reached by several routes. Small biological neural networks can be severely affected by lesions but often can regenerate. The pond snail (*L. stagnalis*), for instance, breathes by a central pattern generator of just three neurons (Taylor & Lukowiak 2000). The circuit stops working properly if any of the connections is cut

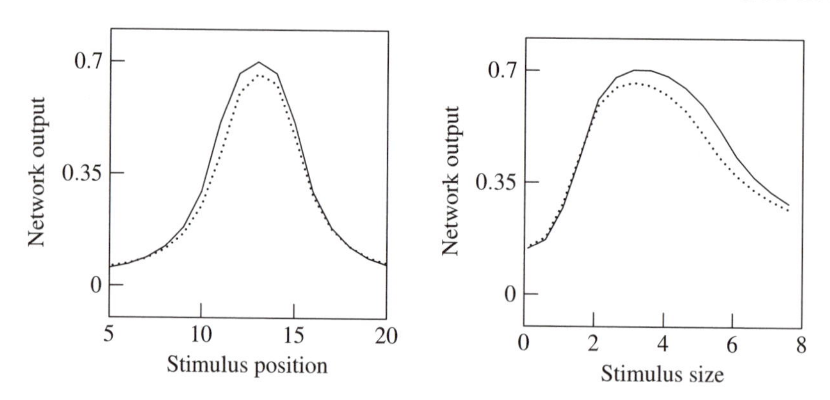

Figure 3.40 Consequence of damage on rearrangement and size generalization in a two-layer neural network (Section 3.3.1). The network is tested before (continuous line) and after (dotted line) severing 4 of 25 weights (number 1, 7, 14 and 18 in Figure 3.10).

but can grow back even after *all* connections have been cut! In general, small nervous systems are sensitive to damage because each individual neuron and synapse makes a substantial contribution, indeed often a unique one. For instance, only one neuron in the pattern generator of the pond snail is capable of self-generated activity (see also Figure 1.8 on page 18 and Figure 3.36 on page 118). Large nervous systems, in contrast, exhibit various degrees of redundancy; i.e., workload is shared among many neurons with similar function. This is crucial to their robustness. Additionally, nervous systems do not passively accept damage, even when they cannot replace dead cells. Extensive reorganization (learning) may occur that can often compensate for the smaller amount of resources available after a lesion.

Neural networks and nervous systems share many similarities in their behavior following damage. This essentially stems from their common network structure. We should emphasize at this point that so far we have dealt with minimal models for simplicity. Clearly, these models, like small nervous systems, show little resistance to damage. Any damage to the networks in Figure 3.21 or Figure 3.25, for instance, would render them not functional. In these and other cases, however, we can use larger networks with the same patterns of connectivity. For instance, each node in Figure 3.21 could be replaced by many nodes. Such a model could function in essentially the same way while being more robust to damage. Resistance to damage improves quickly with network size. For instance, severing one or two connections in a two-layer network with 25 input units has only minor consequences on its ability to generalize (Figure 3.40). The reason is that, as in large nervous systems, the function of each node partially overlaps with the function of other nodes.

The comparison of damage consequences in neural network models and nervous systems has proven very rewarding (e.g., Plaut & Shallice 1994). Network models of particular abilities such as memory or aspects of language in humans have been set up and then lesioned so as to mimic actual lesions. The networks often exhibit deficits comparable with those seen in nervous systems.

CHAPTER SUMMARY

General

- The nervous system controls behavior based on both internal and external causal factors. Both are sensed through specialized receptors, which can be directed toward either the inside or the outside of the animal.
- The ability to command functional behavior is greatly enhanced by the nervous system building regulatory states (e.g., persistent patterns of neuron activity) and memory states (connections between neurons).
- Through its activity, the nervous system influences both the body and the external environment. Such influences often feed back onto the system itself in the form of changed input at some later time.
- Neural network models of the preceding processes can be built by having input nodes that model receptors (internal and external), inner nodes that process input patterns and output nodes whose activity is interpreted as behavior or muscle control.

Stimulus control

- Many stimulus control phenomena emerge naturally in simple networks, provided input patterns are built by taking into account basic properties of receptors (receptor space approach). Networks account for the shape of generalization gradients and the nature of response biases along different stimulus dimensions (rearrangement and intensity dimensions), as well as other phenomena such as external inhibition and the effect of contextual stimuli.
- Considering overlaps between stimulus patterns is a powerful hueristic tool to analyze how both networks and animals generalize to novel stimuli.
- Networks share many similarities with earlier models of stimulus control from ethology and psychology but are more powerful in view of their concrete interpretation as models of the nervous system. This eliminates many ambiguities, e.g., what is a stimulus.

Sensory processing

- Neural network models of sensory processes are well developed, starting from modeling of invertebrate sensory systems up to the complexities of visual processing in mammals.
- These models show how raw information from sense organs can be transformed to be more useful for further processing and decision making. For instance, we have shown how
    - Changes in input can be highlighted by lateral inhibition.
    - Receptive fields can build patterns of activity that allow easier realization of functional behavior.
    - Local signals about motion on a small part of the retina can be converted into global information about how the animal moves in the environment.

- Signals extended in time such as mating calls can be analyzed by recurrent networks.

Decision making

- Networks can realize complex decision making strategies, including
  - Integration of external and internal factors.
  - Different priorities for different activities.
  - Hierarchical decisions.
  - Independence between activities.
  - Stable decisions in the face of changing input.
  - Functional behavior sequences.
- Networks can include concepts from control theory and ethological observations of behavior.
- Networks can be used to investigate nonfunctional (wrong) decisions as well as functional decisions.

Motor control

The ability of neural networks to model motor control stems, once again, from their flexibility in mapping one pattern of activity into another and from the basic network structure shared with nervous systems. Network models of motor control can be developed according to the following principles:

- The input to a motor control network represent a decision to perform a given behavior pattern; the decision is mapped onto activity patterns in motoneurons, which, in turn, control stretching and relaxation of muscles.
- The timing of muscle contractions can be generated by pacemaker neurons, successive inhibition and excitation between nodes and time delays.
- Sensory feedback can be used to increase precision, correct accidental errors and track external events.

FURTHER READING

Ewert JP, 1985. Concepts in vertebrate neuroethology. *Animal Behaviour* 33, 1–29. A classic of neuroethology.

Ghirlanda S, Enquist M, 2003. A century of generalization. *Animal Behaviour* 66, 15–36. A recent review of generalization.

Grillner S, 1996. Neural networks for vertebrate locomotion. *Scientific American* 274, 64–69. An example of neurobiological research on motor control.

Hinde RA, 1970. *Animal Behaviour*. Tokyo: McGraw-Hill Kogakusha, 2 edition. An attempt to synthesize ethology and experimental psychology. Covers reactions to stimuli, decision making and motor control, sometimes sketching network-like models of behavior.

Kandel E, Schwartz J, Jessell T, 2000. *Principles of Neural Science*. London: Prentice-Hall, 4 edition. The most extensive book on neurophysiology.

Mackintosh NJ, 1974. *The Psychology of Animal Learning*. London: Academic Press. A classical of experimental psychology, with a review of generalization in Chapter 9.

McFarland DJ, 1999. *Animal Behaviour: Psychobiology, Ethology and Evolution*. Harlow, England: Longman, 3 edition. A key text of modern ethology, covering all aspects of animal behavior.

Nolfi S, 2002. Power and limits of reactive agents. *Neurocomputing* 42, 119–145. A surprising analysis of what neural networks without internal states can do in realistic settings.

Rumelhart DE, McClelland JL, editors, 1986b. *Parallel Distributed Processing: Explorations in the Microstructure of Cognition*, volume 1. Cambridge, MA: MIT Press. A key contribution to renewing interest in neural network models in the 1980s.

Simmons PJ, Young D, 1999. *Nerve Cells and Animal Behaviour*. Cambridge, England: Cambridge University Press, 2 edition. A modern text in neuroethology, discussing some neural network models.

Chapter Four

Learning and Ontogeny

This chapter deals with the ontogeny, or development, of behavior. Many factors and processes interact in behavioral development. Particularly important are those that create the nervous system and those that govern learning. The latter are here given special attention owing to their importance for behavior. We begin by discussing what learning and ontogeny are and how they relate to each other, to behavior and to the nervous system. We continue with a detailed section on neural network models of basic learning processes, such as classical conditioning. Learning is an extensively developed topic that we cannot cover in full. Our only intention is to show how neural network models can be applied to the study of learning. The chapter ends by considering the ontogeny of behavior in its broad sense.

4.1 WHAT ARE LEARNING AND ONTOGENY?

In Chapter 3 we considered what causes behavior to occur at a given moment, or how behavior mechanisms operate here and now. The behavior map was fixed, and we studied how stimuli and other motivational variables cause behavior. Memory (connection weights) and network wiring did not change. We started by asking how the presence of a peahen causes, or motivates, courtship behavior in a peacock. Whether the peacock courts or not depends on stimulation picked up by sense organs, in this case the eyes, on recognition mechanisms that are capable of determining whether a female is present or not and on internal variables such as hormone levels. The question was how all these factors operate and how they interact to cause behavior. When we consider ontogeny, we ask a different question: we want to know how a particular mechanism (behavior map) becomes established during an individual's life and how experience and behavior are involved in this process (Chapter 1). In the peacock example, we want to know how the nervous mechanisms that allow female recognition become established in a peacock's life.

Ontogeny of behavior hosts many phenomena. Most researched is learning in its narrow sense, in particular, classical and instrumental conditioning (Mackintosh 1974; Pearce 1997). Such phenomena, covered in Section 4.2, occur in response to changes in the environment and result in changes in how animals respond to stimuli. Although we dedicate to learning a specific section, it should be noted that learning is a part of ontogeny. *Ontogeny* in general refers to a diversity of processes starting very early in life and creating the behavioral machinery of the adult. There are a number of general issues in the study of ontogeny. How can the genes code for complex behavior, such as navigation in migratory birds? What is the interplay between nature and nurture? Why does behavior sometimes develop even without experience, whereas in other cases learning is crucial? A female frog, for example, spontaneously recognizes the call of conspecific males, but zebra finches (*Taeniopygia guttata*) need to learn what their conspecifics look like.

A tentative definition of *learning* might be "changes in behavior caused by experience." For instance, an animal may hear a particular sound (say, leaves being stepped over) immediately before seeing a predator. If such an experience occurs a number of times, most animals will become wary of the sound. An experience, however, may change later behavior without this being immediately apparent. Merely seeing different objects for some time, for example, can facilitate later mastering of a discrimination between them ("perceptual learning"; see Section 4.4.2; Attneave 1957; Gibson & Walk 1956). Thus a better definition of learning is "changes in memory caused by experience," whether or not such changes are immediately reflected at the level of behavior. In our terminology, we say that in learning models memory is included among state variables (variables that can change), and modifications in memory due to learning are described by suitable state-transition equations. In neural network models, memory and learning can be given a precise interpretations reflecting this definition. Memory is equal to connection weights, the analog of synapses in nervous systems. The memory state of a network is thus described by its weight matrices, and learning amounts to changes in the weights. In such models, learning thus entails the modification of existing connections, the last of the four phases of the development of nervous systems described in Section 4.7. Nervous systems, however, may be plastic in additional ways (e.g., by birth and death of neurons; see Mareschal et al. 1996 for network models), making circuitry also a part of memory. Moreover, in both nervous systems and neural networks, memories with a limited lifetime can be sustained as patterns of activity in the network. We have called these *regulatory states* to distinguish them from memory structurally encoded in weights (Sections 1.2 and 3.7.5).

4.2 GENERAL ASPECTS OF LEARNING

Empirical research into animal learning is extensive (Dickinson 1980; Mackintosh 1974, 1994; Mowrer & Klein 2001). Figure 4.1 illustrates an example of such studies, whereby an animal learns to turn in one direction at the end of a T maze as a consequence of food being available at that end but not at the other. In this section we present a general approach to modeling empirical findings about learning in terms of changes in connection weights in neural network models. We start with a theoretical analysis of different factors that drive learning in nervous systems and neural network models. Based on such analysis, we discuss the abilities and biological realism of some neural network models of learning.

4.2.1 Local and global factors in learning

Let us begin with nervous systems. The factors that can cause synaptic changes may be usefully divided into "local" and "global" (Bienenstock et al. 1982; Intrator

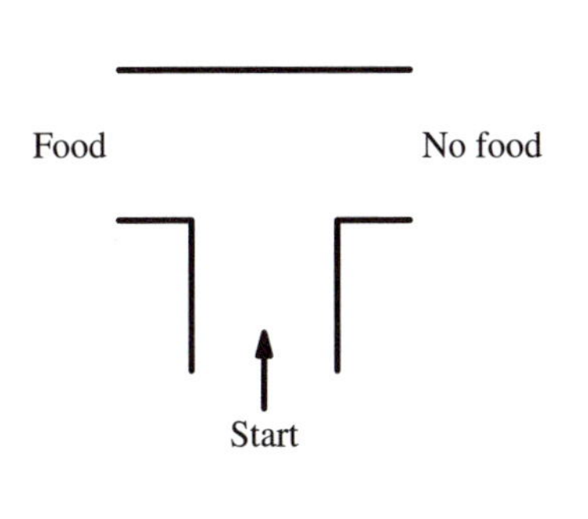

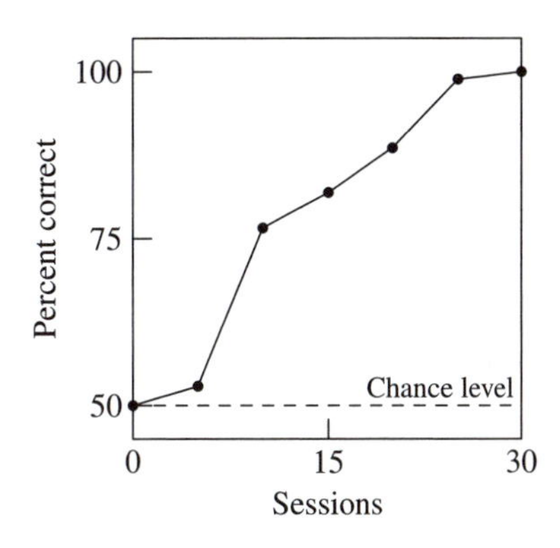

Figure 4.1 An experiment on animal learning. Rats are placed at the start of a T maze that has food at one end but not at the other. Over a number of experimental sessions, the rats learn to turn toward the food. Data from Clayton (1969).

& Cooper 1995; Kandel et al. 2000). By local factors we mean quantities that are naturally available at the synapse between two neurons. These include, for instance, the current strength of the synapse and the activity (or other state variables) of the pre- and postsynaptic neurons. By global factors we mean signals than bring to the synapse information that originates elsewhere in the network. Important aspects of a global factor are what neuron populations are affected and whether the factor affects each weight equally or has a more specific pattern of action (see below). Note that the terms *global* and *local* refer here to the *origin* of learning factors. All factors that influence a synapse ultimately must be "local" in the sense that they must be physically present at the synapse.

The need for global factors arises from the fact that, in many cases, the information needed for successful learning is not naturally present locally. Behavioral data clearly demonstrate that local factors alone cannot be responsible for all types of learning. For instance, in the T maze in Figure 4.1, the crucial information is about which choice of arm leads to food. For learning to occur, the consequence of actions must be evaluated according to whether they lead to food, but this evaluation cannot take place entirely at the synapses that connect stimuli with responses. It is also clear that local and global factors need a suitable time structure. For instance, instrumental conditioning occurs only if rewards closely follow responses (see below).

A major challenge in developing biologically plausible learning mechanisms is to account for global signals in terms of the neural system itself. In neural network theory of learning, this issue is related to the often-made distinction between supervised and unsupervised learning algorithms. The former include global signals that operate outside the network, whereas the latter implement learning as internal to the neural network. Note also that global signals have much in common with the concept of reinforcement, although they are not identical (Section 4.2.3). Neurobiological research has identified several brain parts that seem to generate global (reinforcement) signals and broadcast them to other parts of the brain, e.g., by al-

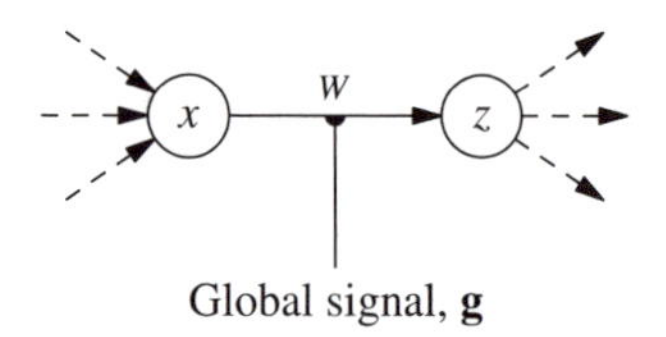

Figure 4.2 In neural network models, the mechanism for updating a connection weight W depends on local factors, usually the activity in the preceding (x) and subsequent (z) nodes, as well as on global signals (**g**, indicated by a line arriving at the connection and terminating in a half circle). Dashed lines indicate incoming and outgoing connections.

tering the concentration of chemicals in the extracellular medium (Martin-Soelch et al. 2001; Pennartz 1996; Schultz et al. 1997; Waelti et al. 2001).

We have seen in Chapter 2 that neural network theory provides an extensive toolbox for achieving memory changes in network models. To what extent do these techniques help us to understand how animals learn? This is a complex issue. Learning has always been the central topic of neural network research (Arbib 2003; Haykin 1999), but biological realism is not always a goal, especially in engineering applications. The analysis of neural network learning algorithms in terms of local and global factors can lead to insights into their biological plausibility. We can formalize the preceding argument by saying that the change ΔW is a function of the current weight value W, the activity or other state variables of the preceding node x and the subsequent node z (the local factors), and global signals $\mathbf{g}$. This is illustrated in Figure 4.2 and formally by

$$\Delta W = l(W, x, z, \mathbf{g}) \tag{4.1}$$

where l is a suitable function. In nervous systems, the most important local factors that influence synaptic change appear to be the states of the pre- and postsynaptic neurons, in particular the state (membrane potential, ionic concentrations, etc.) of the postsynaptic cell at the time the presynaptic cell fires (Dayan & Abbott 2001 and references therein). In neural networks, global factors include all variables that influence weight updates but that do not belong to the two nodes and the connection between them. Global factors can control learning in several ways. The two main ones are what global information enters the local mechanism for weight update and to which weights the information is given (Figure 4.3). A global signal, in principle, can convey many kinds of information. It may originate from nodes in the neighborhood of the connection but also from nodes far away. Furthermore, the global signal may be the output of a single node (including a receptor), or it may result from processing within a whole network that gathers input from many nodes and receptors. The latter yields a very versatile system because it becomes possible to produce almost any map from node and receptor activities to global signals. The specificity of a global signal controls which connections are influenced, adding further flexibility to learning processes. For instance, a global signal may be delivered to nodes that control a particular behavior pattern but not to others.

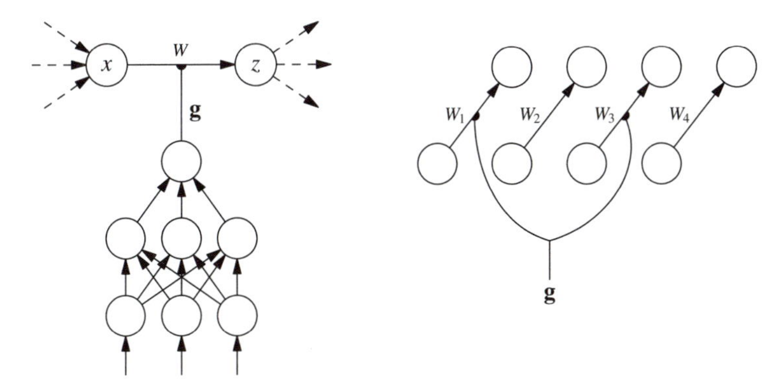

Figure 4.3 Global factors in neural network models of learning. Left: Flexibility of global signals. A global signal may originate from a network with several or many inputs, allowing a diversity of global signals to be generated. Right: Specificity of global signals. A global signal can act on all weights or only, as shown, on particular ones.

The production of a global signal will be very demanding if it requires a large network and needs to send highly specific information to every weight (e.g., a different value to each weight). Networks that communicate global signals to local learning mechanisms are likely to be under genetic control in nervous systems (although it is conceivable that some aspects of these networks may also be changed by learning). We will see in the following sections that the concept of global signals offers considerable potential for understanding genetic predispositions, nature-nurture issues and the evolution of learning mechanisms.

4.2.2 Examples of learning mechanisms

We now turn to some specific examples of learning mechanisms for neural networks. We focus on how these embody the basic principles discussed earlier and how they may account for some fundamental learning phenomena.

4.2.2.1 Hebbian learning

In his pioneering work on the neural basis of learning, Hebb (1949) suggested that a synapse strengthens if the adjoining neurons are often active together. An simple implementation of this suggestion is

$$\Delta W = \eta x z \tag{4.2}$$

where η is a positive parameter that regulates the speed of learning. Equation (4.2) is a paradigmatic example of local learning, based only on the activities x and z of the nodes joined by the weight. The only possible action of global signals is to increase or decrease the rate of learning η. An obvious limit of equation (4.2) is that it allows weights to increase but not to decrease. This leads to unlimited weight growth (Rochester et al. 1956) and also limits what can be learned. For example, a stimulus can become more potent in eliciting a response, but not less. A global signal could be used to make η zero or even negative, allowing weights to decrease. We are not aware of any implementation of this idea, but several modifications of equation (4.2) have been proposed in a similar spirit (see also the next section). One such development is expressed as

$$\Delta W = \eta xz - \gamma z^2 W \tag{4.3}$$

where γ is a positive constant (Oja 1982). The first part of this equation is the same as equation (4.2). The second part limits the increase of W because it subtracts an increasing amount as W gets larger. The multiplication by z^2 has further interesting effects, detailed in Oja (1982) or Dayan and Abbott (2001). In short, equation (4.3) results in a weight vector that matches the "principal component" of the input patterns, i.e., the dimension of the input space along which the input patterns show larger variation. Thus Oja's work is one example of how simple update rules can produce weights that capture statistical features of the input patterns. This has been linked to the ontogeny of receptive fields in nervous systems (Section 4.7). Equation (4.3) is also an example of how additional local factors (here the value of the connection) may enter an update rule. In Section 2.4 we have seen another Hebblike rule of the form

$$\Delta W = \eta(x-a)(z-b) \tag{4.4}$$

The addition of the parameters a and b means that W can decrease as well as increase (Linsker 1986). A simple assumption is that a and b are constants. This can lead to the development of receptive fields of the kind shown in Figure 2.15 on page 59. Further possibilities are gained if a and b are made to depend on past node activities (see, e.g., Dayan & Abbott 2001).

We refer to Shouval and Perrone (1995), Haykin (1999) and Dayan and Abbott (2001) for a more complete discussion of update rules that develop Hebb's original suggestion. We stress again that the hallmark of these rules is that global signals play a limited role. This means that such rules can yield "automatic" learning such as the organization of perception (see Figure 2.15 on page 59; Section 4.4.2) but cannot model learning that relies on feedback from the environment (O'Reilly & Munakata 2000).

4.2.2.2 Reinforcement learning

Instrumental conditioning, as we discussed earlier, requires a global signal that delivers information about the consequences of an animal's actions. So-called reinforcement learning algorithms are based on a similar idea, whereby learning is driven by a global signal about whether a network response has been "good" or "bad" (Section 2.3.4). The global signal is usually the same for many or all weights in the network. A biological example of such a signal may be the nociceptive system, which assigns to stimuli a value along the dimension of pain (Toates 2001). We have already discussed such "value systems" as part of motivational processes in Chapter 3. Figure 4.4 shows a network endowed with a simple value system. The

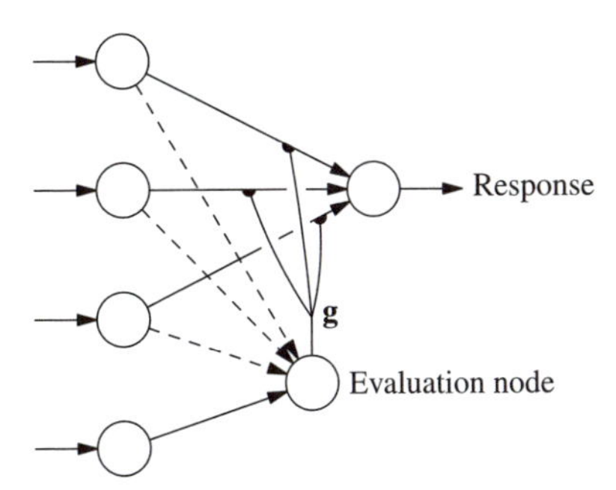

Figure 4.4 A network architecture that uses a global signal to update network weights ("reinforcement learning"; Section 2.3.4). The global signal is evaluated by a node and broadcasted to all weights that link a sense organ with a response. The global signal may come from a different sense organs (e.g., pain receptors, bottom left node) or from the same one (solid lines).

latter is simply a part of the network (only one node in the figure) that receives input from, say, nodes sensitive to temperature or tissue damage. This input is processed and then broadcasted to weights in the network. A weight is then updated based on the global signal and local factors, such as those discussed earlier. We have shown in Section 2.3.4 that such an arrangement is capable of learning input-output maps, e.g., performing a given response to a stimulus if this has lead to positive consequences in the past (as judged by the "value system"). Many simulations of learning in the following are based on the simple reinforcement learning scheme detailed in Section 2.3.4. Reinforcement learning algorithms have been extended to include global signals that are generated by larger networks (see Figure 4.3), but we are not aware of any systematic attempt to apply these models to animal learning.

Although reinforcement learning algorithms for neural networks are usually related to instrumental conditioning, they are also relevant to the study of classical conditioning. Together with instrumental conditioning, classical conditioning is the most studied type of learning (see, e.g., Klein 2002; Mackintosh 1974). A common functional and evolutionary explanation of classical conditioning (Dickinson 1980; Klein 2002; Mackintosh 1975) is that it occurs when one stimulus (conditioned stimulus, or CS) predicts the occurrence of another biologically significant stimulus (unconditioned stimulus, or US). Thus, to account for classical conditioning in a neural network model, we still need a value system that identifies US as biologically significant. One difference is that in classical conditioning the conditioned response (CR, performed to the CS) may differ from the unconditioned response (UR, the response naturally evoked by the US), as discussed in Section 4.2.3.

4.2.2.3 The δ rule and back-propagation

The δ rule and back-propagation (Section 2.3.2) are the most frequently applied methods for training neural networks. These algorithms require exact knowledge of the input-output map to be produced. Based on such knowledge, each weight is modified in a way that reduces the difference between the desired input-output map and the one currently realized by the network Section 2.3.2. These algorithms require the calculation of a different global signal (usually called the *error signal*) for each weight in the network. The calculation of each signal is also complex because it depends on many other weights and node activities in the network. It is such a refined use of global signals that makes back-propagation more efficient than reinforcement learning, especially in network architectures with more layers. At the same time, the complexity of the algorithm makes it less appealing as a model of biological learning. For instance, just to convey the global signals to each weight (let alone to compute it) would require a network as large as the original one. For these reasons, researchers have tried to develop simpler and more plausible algorithms that retain most of the power of back-propagation. One example is the algorithm by O'Reilly (1996). Here the globals signals that serve to update the weights are calculated approximately as a part of network dynamics and broadcast through recurrent connections similar to those seen in real nervous systems (e.g., between different areas of the cortex).

4.2.3 The diversity of learning

The study of animal learning encompasses a number of phenomena and procedures. In addition to instrumental conditioning (see, e.g., Klein 2002; Mackintosh 1974; Pearce 1997; Skinner 1938) and classical conditioning (e.g., Klein 2002; Mackintosh 1974; Pavlov 1927; Pearce 1997), other study areas include perceptual learning (e.g., Gibson 1969; Hall 2001), habituation and sensitization (Horn 1967; Klein 2002), filial and sexual imprinting (Bolhuis 1991; Immelmann 1972; Lorenz 1935), song learning (Slater 2003; Thorpe 1958) and social learning (Hayes & Galef 1996). In addition, these broad categories include subcategories and host a variety findings.

Neural networks may account for most of this diversity, although much work is still needed to explore these potentials. Some neural network applications of learning are based on local factors only (e.g., models of perceptual learning, Section 4.4, and habituation; Horn 1967; Kandel et al. 2000), but most applications also include a suitably designed global signal that interacts with local factors to cause weight changes. An illustrative example is the work on various types of learning in the sea slug, *Aplysia californica*. Figure 4.5 shows a simple network model that summarizes neurobiological knowledge of how three learning phenomena (habituation, sensitization and classical conditioning) arise in this species.

Another example of how network models can be designed to account for different types of learning concerns how connections are updated. A model of instrumental conditioning would include an update mechanism that changes the connections between a given behavior and the stimulus that caused the behavior itself. A model of sexual imprinting, on the other hand, would need to modify the connections between stimuli and sexual behavior when the behavior is not performed, i.e., when the animal is not sexually active. This requires a different design of the conditions that trigger learning. The global factors that control learning, indeed, may be rather complex. Studies of classical and instrumental conditioning, for instance, have

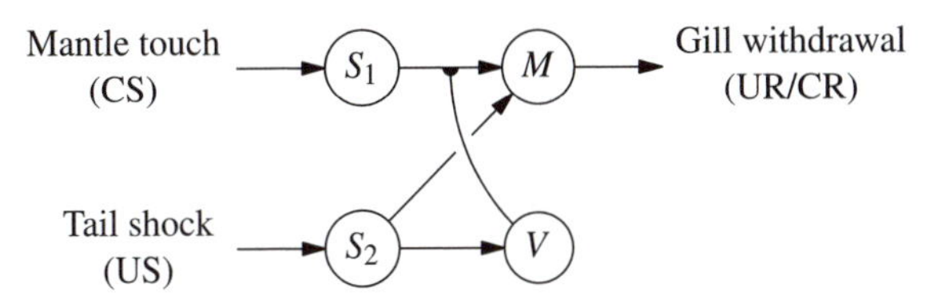

Figure 4.5 Neural network model of the biological neural network that supports habituation, sensitization and classical conditioning in the sea slug, *Aplysia californica* (Dudai 1989; Kandel et al. 2000). Each node or connection in this network stands for many nodes or connections in the animal. In habituation, the mantle of the animal is touched. This initially causes the animal to withdraw its gill, but with repeated stimulation this response progressively fades. Habituation has been shown to derive from a weakening of the $S_1 \rightarrow M$ connection, without the intervention of other neurons. In sensitization, a shock delivered to the tail enhances the ability of a touch to the mantle to elicit gill withdrawal. When the shock is administered, two things happen. First, the gill is withdrawn through the $S_2 \rightarrow M$ connection. Second, activation of the V neuron through the $S_2 \rightarrow V$ connection causes release of chemical near the $S_1 \rightarrow M$ connection, whereby the latter is strengthened. The effect is stronger than the decrease in efficacy caused by habituation. Finally, classical conditioning occurs when the mantle is lightly touched (CS) before administering a tail shock (US). The mantle touch then becomes able to elicit gill withdrawal (UR/CR). Conditioning happens because a mantle touch sets up a transient internal state within the $S_1 \rightarrow M$ connection, lasting about 1 s. If the V neuron is activated within this time, chemicals are released near the $S_1 \rightarrow M$ synapse that cause it to strengthen. Note that sensitization and classical conditioning rely on the same global signal, the "value system" embodied in the V neuron. The difference is that classical conditioning requires an additional (local) state variable that tracks whether the $S_1 \rightarrow M$ connection has been activated in the recent past.

pointed out that learning may depend on the predictive value of stimuli, as well as on the extent to which a given situation may be considered "surprising" (Dickinson 1989; Kamin 1969).

The diversity of learning phenomena has been explored in a number of neural network models. Examples include models of instrumental conditioning (Sutton & Barto 1981; see also Sutton & Barto 1998), classical conditioning (Kehoe 1988; Mauk & Donegan 1997; Pearce & Hall 1980; Schmajuk 1997; Sutton & Barto 1981; Wagner & Brandon 1989, 2001), filial imprinting (Bateson & Horn 1994; O'Reilly & Johnson 1994) and habituation (Horn 1967; Kandel et al. 2000).

4.2.3.1 Constraints on learning and predispositions

Constraints on learning and predispositions refer to factors within the animal that bias learning in certain directions or make some behavior maps more easily established than others (Breland & Breland 1961; Hinde & Stevenson-Hinde 1973; Horn

1985; McFarland 1999; Roper 1983). From a functional and evolutionary point of view, constraints and predispositions may ensure that animals learn the right thing and learn quickly when this is crucial.

Constraints and predispositions are revealed in many studies. For instance, classical conditioning sometimes involves a response to the CS that differs from the response to the US. For example, a rat may freeze to a sound signaling shock but escape the shock itself. It often makes sense to freeze (to avoid detection) or become vigilant (find out were the predator is) in response to a stimulus predicting the presence of a predator rather than to flee. A neural network model of this finding would include a global signal that targets weights that connect the CS with a response not performed to the US.

Other examples include experiments demonstrating that some associations can be learned, whereas others cannot. Garcia and Koelling (1966) exposed rats to water with a novel taste (see also Domjan 1980). Rats from both groups then experienced either illness (induced) or an electric shock. Subsequent testing showed that rats could readily learn associations between taste and illness but not between illness and a compound of a flashing light and a clicking sound. The opposite result was obtained when an electric shock was used instead of induced illness. The results are summarized in Figure 4.6. Similar results were obtained by Shettleworth (1975, 1978), who tried to condition golden hamsters to perform a variety of behaviors using food and nest material as reinforcers (Figure 4.6). Behavior patterns such as "scrabble" and "open rear," which are used in foraging and nest building, could be conditioned with these reinforcers but not "face washing" behavior, which is used to groom the face.

Predispositions have also been seen in filial imprinting (Bolhuis & Hogan 1999; Horn 1985) and in observational learning of snake fear in monkeys (Mineka & Cook 1988). In the latter case, naive laboratory-bred rhesus monkeys were shown a video of another monkey behaving fearfully in the presence of a variety of stimuli. The first monkey is subsequently exposed to these stimuli to look for evidence of acquired fear. These studies have indicated that monkeys who are not initially afraid of snakes will rapidly acquire an intense fear when they have watched another monkey behaving fearfully in response to a toy snake, whereas fear of other stimuli is acquired more slowly.

A different example shows how learning sometimes is highly specialized to deal with particular problems. Generally, associative learning proceeds faster when the delay between the events to be associated is small; e.g., less than one minute is usually required for effective instrumental conditioning (Roper 1983). There is a functional reason for this. In most cases, the likelihood of a causal link between behavior and the reward quickly diminishes as time goes on. Bad food, however, is an exception because it may cause illness first several hours after ingestion. This fact seems to be reflected in taste-aversion learning, in that association between taste and illness is established even though the interval between them is several hours (Figure 4.7).

Network models can encompass constraints on learning and predispositions in similar ways as they can account for different types of learning. One possibility is that different senses have access to different responses. Recall the experiment

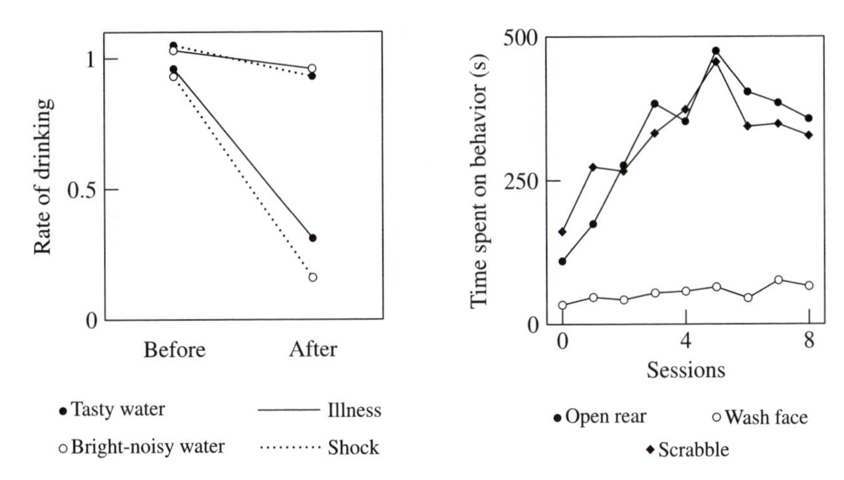

Figure 4.6 Predispositions in learning. Left: Rate of drinking of tasty (flavored) water and bright-noisy water (presented together with a noise and a light) before and after exposure to illness-inducing X radiation or electric shock. Exposure to irradiation caused a reduction in consumption of tasty water only; exposure to shock caused a reduction in consumption of bright-noisy water only (data from Garcia & Koelling 1966). Right: Effect of food reinforcement on scrabbling, rearing and face washing in golden hamsters; only scrabbling and rearing increase when reinforced with food (data from Shettleworth 1975, 1978).

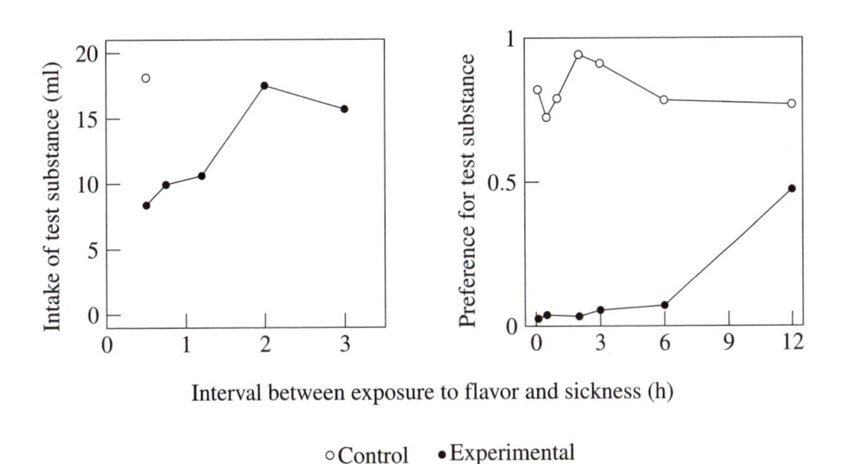

Figure 4.7 Strength of taste-aversion learning as a function of interval between tasting a food and experiencing illness in two separate experiments (left data from Garcia et al. 1966; right data from Smith & Roll 1967). Control subjects were all allowed to taste the food (saccharine solution) but were not made ill. Note that learning involves avoiding to eat the food, so the lower the points in the graph, the stronger is the learning.

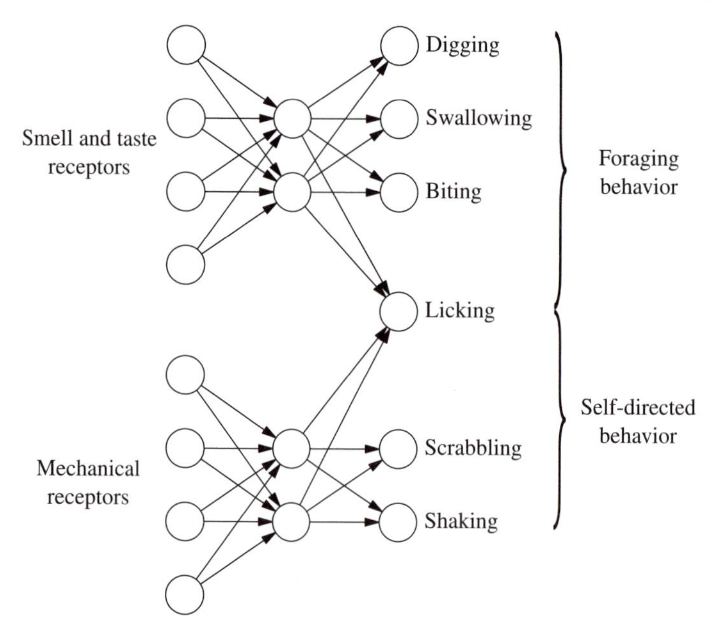

Figure 4.8 A network with predispositions. Learning may occur on all connections, but only some pathways exist from particular sense organs to particular responses. This particular network can reproduce the data in Figure 4.6 (right).

earlier showing that foraging behavior could be conditioned with food but not self-grooming behavior. Figure 4.8 shows a network with such a predisposition. A way of programming such constrains that allows more flexibility is to let global signals target only certain connections (specificity; Figure 4.3).

4.2.3.2 Evolutionary flexibility

That learning results form suitable combinations of network architecture and local and global factors may yield some insight about the evolution of learning. Indeed, the same flexibility that applies to behavior maps also applies to mechanisms of learning, in particular to global signals (Figure 4.3). Thus, given that the network architecture is under genetic control, evolution can mold the global signal to produce a variety of adaptive learning mechanisms. The fact that many observations of learning can be categorized as either classical or instrumental conditioning is explained by functionality. An intriguing perspective is that global signals for learning may arise from networks that are capable of learning themselves. This can be important in order to account for such factors as "expectations" and "surprise." It is an exciting possibility that most types of animal learning can be modeled with a proper combination of local and global factors, as discussed here. However, much more work is needed to fully evaluate this promise, and many biological learning phenomena have so far received little interest from neural network modeling.

4.3 NETWORK MODELS OF GENERAL LEARNING PHENOMENA

We now consider how learning phenomena can be accounted for by neural network models. We focus on a number of fundamental findings in the area of associative learning, i.e., classical and instrumental conditioning (for extensive reviews, see Klein 2002; Mackintosh 1974). We consider the following phenomena, building on a similar list by Roper (1983):

- **Acquisition:**
    - When a reinforcer is suitably applied, the learned response gradually increases in frequency and/or intensity.
    - In classical conditioning, the speed or strength of learning increases with the intensity of the CS.
    - The speed or strength of learning increases with the size of the reinforcer.
- **Extinction:**
    - When the reinforcer is withheld, the learned response declines gradually in frequency and/or intensity.
    - If an acquired response is extinguished and then trained a second time, it is usually acquired faster than the first time.
- **Overshadowing:** When conditioning occurs with a compound stimulus made of stimuli of different intensity, conditioning occurs mainly to the more intense stimulus.
- **Blocking:** Prior conditioning of one element of a compound stimulus impairs subsequent conditioning of the other element.
- **Time and sequence of events:** Associative learning occurs only when the time interval between events is small and when the events occur in a particular order. In classical conditioning, the interval between CS and US must be small (typically a few seconds at most), and the onset of the CS must precede the onset of the US. In instrumental conditioning, the interval between the response and its consequence must be small (up to about one minute), and the latter must follow the former.

4.3.1 Acquisition

In Figure 4.1 we considered a simple experiment in which rats learned to turn in one direction at the end of a T maze as a consequence of food being available at one arm of the maze but not at the other. Figure 4.9 reproduces the empirical acquisition curve as well as the outcome of a network simulation with the network in Figure 4.4. Stimulation arising from the experimental setting are modeled as an input pattern of the kind in Figure 3.7 on page 77. At each time step the network is presented with this stimulus, and it can react (interpreted as turning in the correct direction) or not react (Section 2.2.1.2). Weights are then updated by "reinforcement learning" as follows (technical details are in Section 2.3.4). If the network reaction is followed by a "good" signal from the value system, weights are updated in a way that increases the likelihood of performing the same reaction again in the

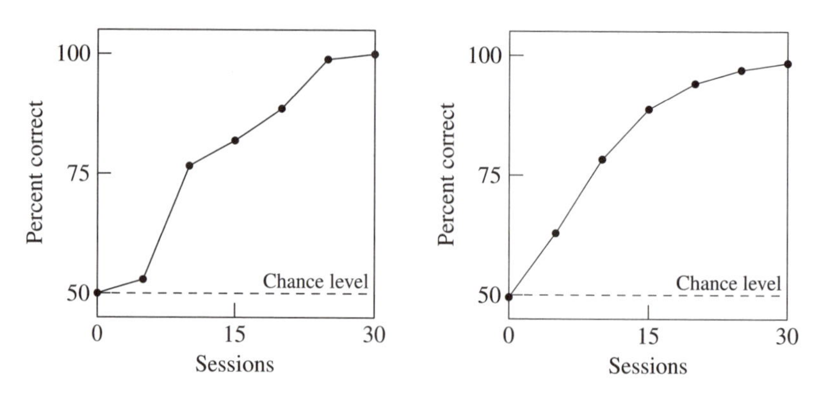

Figure 4.9 Acquisition of a behavior in animals and networks. Left: Reproduction of the data in Figure 4.1 (T-maze learning in rats). Right: Simulation with the network in Figure 4.4 using "reinforcement learning" as described in Section 2.3.4. Network output is interpreted as chance of turning in a particular direction (transfer function of the output node is $f(y) = 0.5 + y$).

same stimulus situation. If a "bad" signal comes, weights are modified so that the likelihood of the action being performed again decreases. Weights linked to more active nodes are updated by larger amounts. The latter leads to faster learning for more intense stimuli, as shown in Figure 4.10, and also leads to a weight vector where larger positive weights correspond to the nodes mostly activated by the stimulus. Such a weight vector has already been considered in Chapter 3 (Figure 3.10 on page 81). The effects of reinforcer size on learning is covered by the assumption that larger reinforcers lead to larger global signals in the network, which increases both the rate of acquisition and the ultimate strength of the response.

4.3.2 Extinction

If a conditioned response stops being reinforced, its strength gradually decreases, a finding referred to as *extinction* of the response. Figure 4.11 shows a simulation of withdrawing food from the T maze in the experiment discussed earlier; preference for turning in the direction previously associated with food declines. This is achieved by lowering the weights linked to nodes activated by the stimulus. The figure also shows that if reinforcement is resumed, response strength increases again. Such "reacquisition" is predicted to occur at the same rate as the original acquisition, but this is at odds with animal studies. The latter show that animals learn faster when reinforcement is introduced again (e.g., Scavio & Thompson 1979). This is called a *savings effect*, and it nicely illustrates the general point that behavior may not tell the whole story about memory changes. In the model used so far, extinction more or less reverses the effects of acquisition; hence there is no savings. A neural network model that exhibits savings is depicted in Figure 4.12. A group of inner nodes has been added between input and output nodes. The weights between the inner and output nodes change as before, based on the occurrence of reinforcement. However, weights connecting the input and inner nodes change without relation to

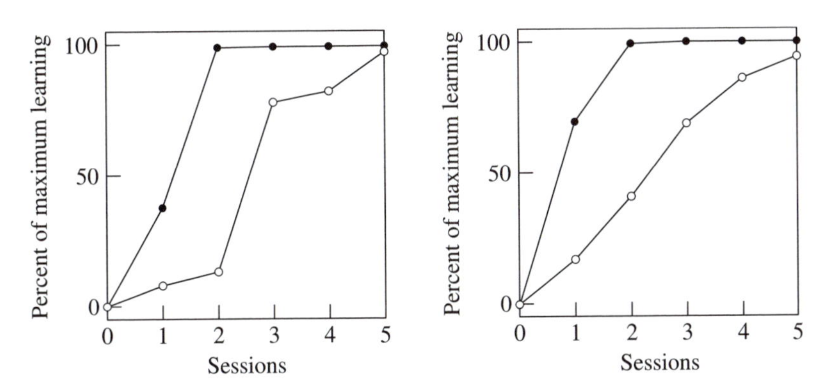

Figure 4.10 Speed of learning as a function of stimulus intensity. Left: Rat data from Kamin and Schaub (1963). Strong CS (closed circles) was an 81-dB noise; weak CS (open circles), a 49-dB noise. Right: Network simulations with the same model as Figure 4.9, with each session consisting of 60 learning trials. The stimuli are like those in Figure 3.8, with maximum activity of 0.7 for the strong stimulus (closed circles) and 0.35 for the weak one (open circles).

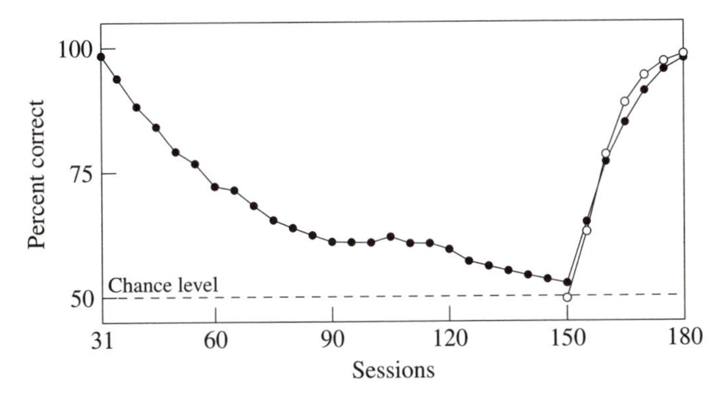

Figure 4.11 Extinction and reacquisition of a conditioned response (closed circles). This simulation is an ideal continuation of the one in Figure 4.9. Sessions from 31 to 150 simulate absence of food in the T maze (neither left nor right turn is reinforced). Reinforcement is resumed at session 151, leading to the reacquisition of the response. Open circles replicate the acquisition curve in Figure 4.9, showing that the second acquisition is not faster than the first. For unreinforced trials, the learning rate is reduced to one-tenth the value for reinforced trials, reflecting empirical observations (discussed in, e.g., Blough 1975).

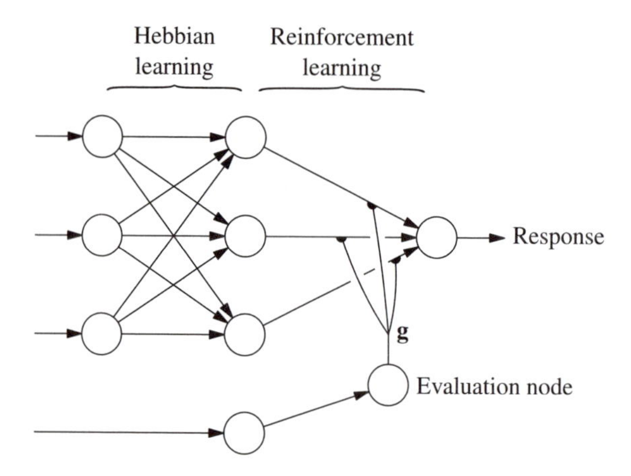

Figure 4.12 A network that shows "savings," obtained by adding one layer to the network in Figure 4.4. The corresponding weight vector changes according to a correlation (Hebbian) rule, namely $\Delta W_i = \eta_1 x_i z_i - \eta_2 W_i$. In the simulations reported in Figure 4.13 we have $\eta_1 = 5 \times 10^{-3}$, $\eta_2 = 5 \times 10^{-4}$. The second matrix changes by reinforcement learning as in Figure 4.4.

the occurrence of reinforcement. Rather, they change by means of a Hebbian rule (Section 2.4) that increases weights joining nodes that are strongly coactive. Spontaneous weight decay is also assumed, so weights connecting nodes that are seldom active, or weakly active, tend to decrease. The net effect of these two factors is to increase the network's sensitivity to oft-seen patterns and is a crude example of perceptual learning (see also Section 4.4.2). These weights will not decrease during extinction, and this facilitates the reacquisition of a previously learned response (Figure 4.13, left).

Another possible savings mechanism is suggested by a different consideration. In nervous systems, each neuron makes either excitatory or inhibitory synapses with other neurons, but not both. Each neuron thus can be classified as excitatory or inhibitory. It follows that two excitatory neurons can inhibit each other only via an inhibitory neuron, as shown in Figure 4.14. The figure also shows a simplified circuit where the inhibitory neuron is dropped while retaining the separation between inhibitory and excitatory synapses. This can be modeled by having two separate weight matrices, say, W_E and W_I, whose weights are constrained to stay, respectively, positive and negative (Dayan & Abbott 2001; Pearce 1987).

To see how this bears on savings, consider acquisition and extinction in such a network. The only change in the learning algorithm is that it cannot cause a weight to change sign. In such a network, acquisition derives necessarily from growth of excitatory weights. Extinction, however, may occur via two routes: decrease of excitatory weights or increase of inhibitory ones. If none of these processes is privileged, as we assume, they will go on at the same time. Thus, after extinction, there will be residual W_E weights as well as nonzero W_I weights. Reacquisition will result from growing W_E weights, starting from their residual value, and simultaneous decrease of W_I. This is faster than starting from null W_E and W_I weights

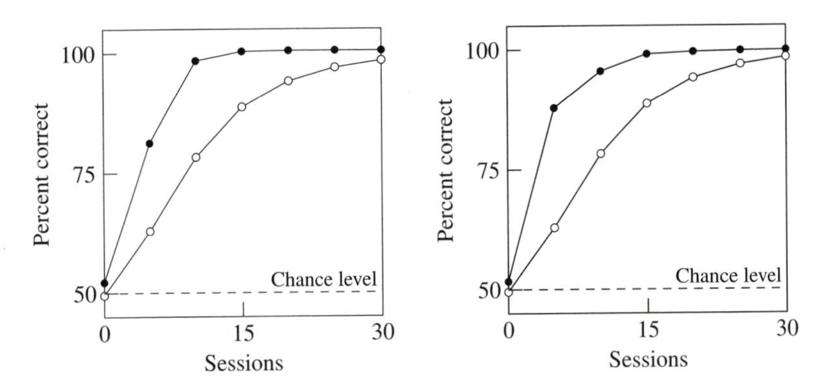

Figure 4.13 Simulation of two network models of savings. Both panels compare rate of learning in the first (open circles) and second (closed circles) acquisition stages in simulations of an acquisition-extinction-acquisition experiment. Left: Simulation with the three-layer network in Figure 4.12. Right: Simulation with a two-layer network similar to the one in Figure 4.4, but where two weights connect each input node with the output node. One weight is constrained to be positive; the other to be negative (Figure 4.14). The learning rule on all weights is the basic "reinforcement learning" algorithm used in the previous simulations.

Figure 4.14 Models of excitation and inhibition. Left: Realistic circuit where excitation is direct and inhibition goes through an inhibitory node (bottom node). Right: Simplified model where a node can make both excitatory and inhibitory connections with other nodes, but the sign of each connection is fixed.

(Figure 4.13, right). It is intriguing that an apparently unrelated anatomical feature such as the separation between excitatory and inhibitory synapses leads to savings, which in many psychological models is handled via ad hoc assumptions (Balkenius & Morén 1998). Still other mechanisms for savings are conceivable, e.g., the existence of two sites of synaptic plasticity that change at different rates (Mauk & Donegan 1997). Clearly, the mechanisms discussed are not mutually exclusive.

4.3.3 Overshadowing and blocking

Table 4.1 presents the typical design of *overshadowing* and *blocking* experiments (Kamin 1968; Pearce 1997). Consider two stimuli *A* and *B* and the compound stimulus *AB* obtained presenting *A* and *B* together. Overshadowing, in its most general sense, refers to the fact that when trained to respond to *AB*, animals generally respond differently to *A* and *B* presented alone, to the point that one of them may be almost ignored (i.e., it has been "overshadowed" by the other). The blocking

Table 4.1 Blocking and overshadowing.

	Pretraining	Training	Typical test result
Overshadowing	–	AB	$R(A) \neq R(B)$
Blocking:			
Group 1	–	AB	$R_2(B) < R_1(B)$
Group 2	A	AB	

Note: A, B: stimuli; $R(A)$: reaction to A. Training also includes non-reinforcement in the absence of the training stimuli.

design is slightly more complex. Responding to B is compared in two groups of animals trained in different ways. For the first group, training consists of reinforced presentations of AB. For the second group, training with AB is preceded by training with A. When B is presented after training, the animals in the second group respond much less than animals in the first group. The extent to which blocking and overshadowing occur varies. An intense B stimulus, for instance, is less affected by blocking than a weaker one (Feldman 1975). Likewise, it is usually the more intense of A and B that dominates responding in an overshadowing experiment (Mackintosh 1976). Animals may also show predispositions for learning about some stimuli rather than others, as seen in Section 4.2.3.

Let us explore these phenomena in a simple two-layer network. It matters little what learning algorithm is used, but in the case of blocking it is crucial to reproduce the succession of events, i.e., to include a period of A-only training before training with AB. We start with very different stimuli (little overlap on the sense organs; Section 3.3) because this is the most common situation in experiments (e.g., a light and a tone). In the network model, overshadowing follows readily from the fact that weights linked to more active nodes are modified more because input node activity x_i is a multiplicative factor in weight update rules, such as reinforcement learning and the δ rule, that are used to model associative learning (Section 2.3). In Section 4.3.1 we saw that this leads to faster acquisition of responses to more intense stimuli. Here we note that when learning is completed (i.e., the response has reached the maximum possible for a given reinforcer), weights linked to more active nodes will have grown most. Thus the nodes that were more active during training end up controlling the response to a larger extent. In other words, if A is more intense than B, overshadowing of B is predicted.

In blocking, the first part of training raises the weights linked to nodes activated by A, producing a weight profile such as the one in Figure 3.10 on page 81. When we add B, responding drops at first (because if B is different from A, it stimulates nodes with negative weight; see Section 3.3.3). This triggers further learning that causes the response to AB to grow. Nodes excited by A, however, start with large positive weights, whereas nodes excited by B with small negative weights. Thus responding to B will remain weak, even if the relevant weights grow somewhat.

Varying the similarity and relative intensity of A and B, we can derive a more thorough prediction about the occurrence of blocking and overshadowing, as shown

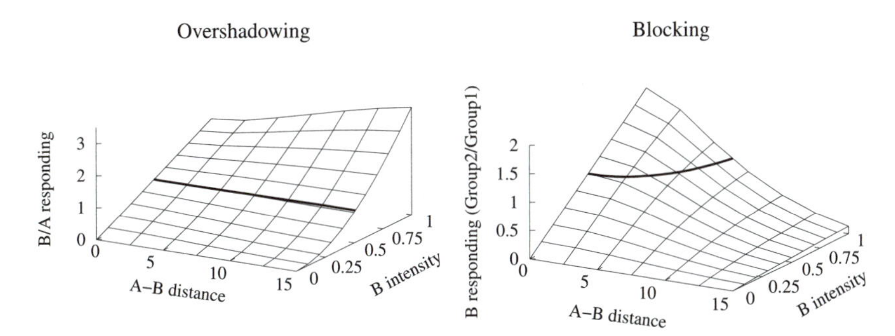

Figure 4.15 Results of training a two-layer network on the overshadowing and blocking designs (Table 4.1), showing the predicted effects of stimulus intensity and similarity on these phenomena. The thick line highlights points for which responding to *A* and *B* (for overshadowing) or responding to *B* in the two experimental groups (for blocking) is predicted to be equal. Stimuli of the kind shown in Figure 3.8 on page 79 are used.

in Figure 4.15. The effect of intensity agrees with empirical data: less overshadowing and blocking occur when *B* is more intense. Unfortunately, we have little data on the effects of similarity (the experiments typically involve stimuli from different sensory modalities). The network model predicts that if *B* is similar to *A*, it will be less affected by blocking. For overshadowing, the model predicts a much smaller effect of stimulus similarity. These predictions await testing.

4.3.4 Time and sequence of events

So far we have neglected several issues concerning the temporal relationship and duration of stimuli. Here we consider the basic findings that associative learning takes place only when external events occur in a particular order and within a short time interval from each other (see page 141; for some exceptions to the latter, see Section 4.2.3.1 and Mackintosh 1983; Roper 1983). It is obvious that a successful model of these findings must embody some form of time representation. Reasoning similarly to Section 3.4, we see that information about temporal relationships and duration of stimuli is, by definition, not available at the sense organs at a single instant of time. Thus some form of memory is necessary to make such information available to the learning mechanism. In earlier chapters we saw that feedforward networks of nodes without internal states cannot implement such memory. We thus need to introduce recurrent connections and/or more complex dynamics for nodes and connections.

In this section we focus on classical conditioning for simplicity, but our arguments generalize to other types of learning. We first describe some major empirical findings and then continue with two examples of how different kinds of regulatory states may underlie such findings. For further discussion of temporal factors in learning, including neural network models, we refer to Balkenius and Morén (1998), Buonomano and Karmakar (2002), Mackintosh (1974), Medina and Mauk

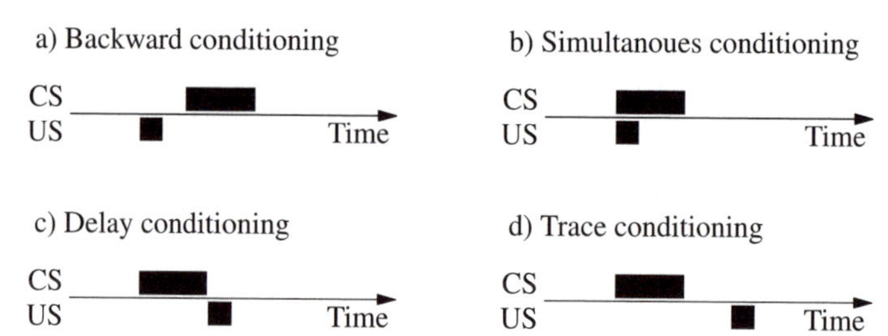

Figure 4.16 Temporal arrangements of the CS and US in classical conditioning, ordered in terms of increasing inter stimulus interval (ISI, the interval between US and CS onset). The black bars indicate the time when a stimulus is on. In backward conditioning, the US precedes the CS (negative ISI). In simultaneous conditioning, CS and US are delivered at the same time (null ISI). In delay conditioning, the US is delivered at the offset of the CS (positive ISI equal to the duration of the CS). In trace conditioning, the US is delivered some time after CS offset (positive ISI longer than the CS). Empirical results on trace conditioning are similar to delay conditioning (see Figure 4.17).

(2000) and Schmajuk (1997). Figure 4.16 shows the arrangements of the CS and US used to study temporal factors in classical conditioning. The main variable is the *interstimulus interval* (ISI), i.e., the interval between CS onset and US onset. The ISI is a crucial determinant of the extent to which learning occurs, as shown in Figure 4.17. A first necessary condition for learning is that the US occurs after the CS by at least a short time (about 50 ms in the particular case portrayed in the figure). Figure 4.17 also shows that the maximum amount of learning is observed for a limited range of ISIs and declines if the ISI is too long. Similar findings have been observed for many species and behaviors (reviewed in Mackintosh 1974).

Our first example concerns extensive research on classical conditioning in the sea slug, *Aplysia californica*. We have sketchily summarized this research as a network model in Figure 4.5 on page 137. In this network, classical conditioning depends on strengthening the $S_1 \to M$ connection based on two factors. The first factor is a global signal delivered to the connection by the "value" node V. This signal is delivered every time the US occurs. The second factor is a state variable of the $S_1 \to M$ connection itself, say, q. The rule for learning is that whenever the global signal is delivered (i.e., whenever US occurs), the $S_1 \to M$ connection is strengthened by an amount proportional to the value of q at that moment. The ISI curve emerges from the particular dynamics of q. Normally, q has a value of 0. When an external stimulus activates S_1, however, q raises rapidly up to a maximum value, and then it starts to decay again toward 0. Thus, if the US is delivered before the CS, then $q = 0$, and no learning follows. The same is true if CS and US are delivered simultaneously or if the US is delivered so shortly after the CS that q has not yet increased significantly. Substantial learning will occur if the ISI is such that q has a large value when the US is delivered. Longer ISIs are less and less effective because q decays toward 0. A more detailed exposition of similar ideas, including a formal specification of the dynamics for q, can be found in Gluck and Thompson

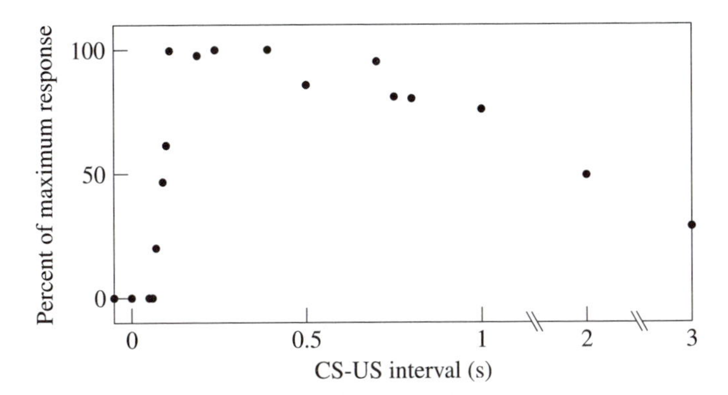

Figure 4.17 Example of an ISI (interstimulus interval) curve showing how the amount of learning in classical conditioning depends on the interval between CS and US (note the breaks at the right of the horizontal axis). With reference to Figure 4.16, the leftmost point refers to backward conditioning (ISI < 0), the next point to simultaneous conditioning (ISI $= 0$) and the remaining points to delay conditioning (ISI > 0; CS duration is here equal to the ISI). Data from Medina and Mauk (2000) relative to rabbits learning to blink an eye (CR) when hearing a tone (CS) that signals an unpleasant stimulus (mild shock to the eye, US).

(1987). In the animal, what we have summarized as a single variable q actually corresponds to the abundance of various chemicals within the S_1 and M neurons (see Chapter 63 in Kandel et al. 2000).

Our second example comes from recent research on the neural basis of eyeblink conditioning in the rabbit (revewied in, e.g., Medina & Mauk 2000). One function of the eyeblink is to protect the eye from noxious stimuli, and this response can be classically conditioned with standard procedures. In most experiments, a sound (CS) is played some time before the delivery of a mild shock to an eye (US), whereby rabbits learn to blink that eye (CR) after hearing the sound. The data in Figure 4.17 refer to such an experiment using delay conditioning (Figure 4.16c). The biological neural network that underlies such learning has been identified in the cerebellum. In highly simplified form, it can be described as composed of two subnetworks and a number of pathways, as shown in Figure 4.18. Information about external stimuli (potential CSs) reaches subnetworks 1 and 2 by a large number of connections originating in other parts of the brain. Each particular CS activates about 3% of these connections for the entire duration of the CS (an additional 1% of connections is activated only briefly at CS onset). The US commands closure of the eye by exciting subnetwork 2. Occurrence of the US also controls plasticity in two sets of synapses, labeled **W** and **V** in the figure. Changes in these synapses are the material basis of learning. The rules governing such changes appear to conform well to ideas about reinforcement learning. First, occurrence of the US has been shown to deliver a global signal to both sets of synapses. Second, there is good evidence that the change in each synapse depends on the activity of the adjoining neurons around the time the global signal is delivered. Thus, despite differences in network architecture and the great evolutionary distance, we see some important

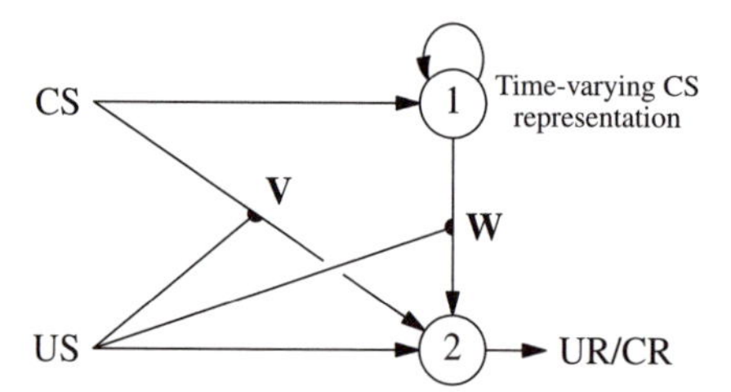

Figure 4.18 Highly semplified scheme of the cerebellar network that underlies eyeblink conditioning in the rabbit. The CS (usually a tone) can elicit the CR (eye closure) via a direct pathway and an indirect pathway mediated by a recurrent network (subnetwork 1, shown as a single node with feedback connection). The latter builds a time-varying representation of the CS that underlies the timing abilities of the network. The US (mild electric shock to the eye) elicits the UR (eye closure) via a direct pathway but also influences two sets of plastic connections via specific pathways (connections ending in half circles). See text for further details. In the animal, the recurrent network consists of two separate cell populations, called *Golgi* and *granule cells*. The output network consists of cerebellar nuclei, Purkinje cells, basket cells and stellate cells. See Medina and Mauk (2000) and Medina et al. (2000) for further details.

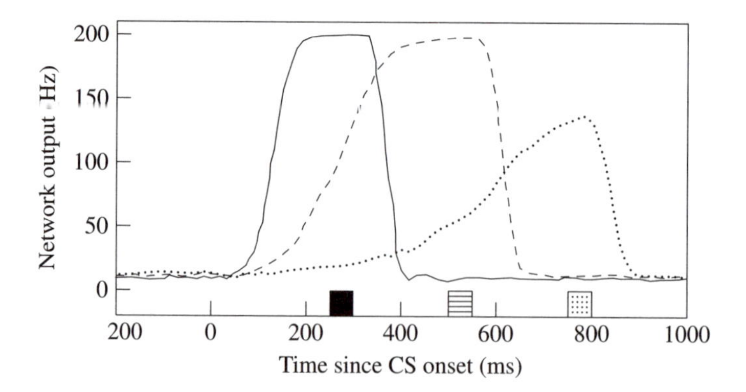

Figure 4.19 Network output after simulated classical conditioning in a model of eyeblink conditioning in the rabbit (Medina & Mauk 2000). The network learns to respond at the appropriate time, i.e., the time when the US was delivered during training (no US was delivered during the test trials shown in the figure). The three lines refer to training with three interstimulus intervals (ISIs): 250 ms (continuous line), 500 ms (dashed line) and 750 ms (dotted line). The box with matching style indicates the time during which the US was on during training. The CS starts at time 0 and lasts until the US is delivered (delay conditioning; see Figure 4.16). Note that the response amplitude is smaller with the longer interval of 750 ms, in agreement with the empirical data in Figure 4.17.

common elements in mechanisms of conditioning in *Aplysia* and the rabbit.

The most important difference is perhaps the presence of extensive recurrent connections within subnetwork 1. These connections allow the network not only to learn to respond to the CS but also to learn to respond at the right time. Indeed, to give any protection from the noxious US, the eyeblink CR must not be performed when the CS is played, but just before the US is delivered. In brief, the mechanism of timing in this network is as follows. The recurrent connections in subnetwork 1 cause the pattern of activity in this network to change with time, even though the CS input is constant. The activity of subnetwork 1 encodes thus, implicitly, the time since CS onset as well as the identity of the CS. Such a time representation may be "read out" by strengthening or weakening specific connections. Suppose, for instance, that we wish the network to respond at time t_0 after CS onset. Owing to the recurrent connections, subnetwork 1 will be in a particular activity state, say, $\mathbf{z}_1(t_0)$, at this time. We can then look up what nodes are strongly active in this state and strengthen the connections between these nodes and subnetwork 2. This increases the likelihood that a response is produced whenever subnetwork 1 is in state $\mathbf{z}_1(t_0)$, i.e., at time t_0 after CS onset. A sharper timing may also be achieved by lowering the weights from nodes in subnetwork 1 that are strongly active at other times. This decreases the likelihood that the CS is produced at an incorrect time. Indeed, there is physiological evidence that both increase and decrease of synapses occur in the rabbit's cerebellum: synapses between nodes that are active when the US occurs are strengthened, and synapses between nodes that are active when the US does not occur are weakened. That this mechanism can produce accurately timed responses is shown in Figure 4.19, reproducing simulation data from Medina et al. (2000). These authors used a much more detailed (but also less easily understood) network model of the rabbit's cerebellum, from which we have derived the simplified architecture in Figure 4.18. We refer to this paper as well as to Buonomano and Karmakar (2002) and Medina and Mauk (2000) for further discussion of timing models of this kind, including the idea that the presence of two sets of plastic connections (rather than one) may produce savings (Section 4.3.2).

4.4 BEHAVIORALLY SILENT LEARNING

We have already noted that not all memory changes are immediately apparent at the behavior level. This was clear, for instance, in our discussion of the memory changes that underlie extinction. Furthermore, not all memory changes can be related to the occurrence of meaningful events in the environment ("reinforcers"). We mentioned, as an example, that merely seeing some stimuli for a period of time can help to discriminate between them later on (Attneave 1957; Gibson & Walk 1956). Below we consider two further examples.

4.4.1 Latent inhibition, novelty and "attention"

The basic observation about latent inhibition is that learning an association between a stimulus and another event is slowed down by preexposing the animal to the stim-

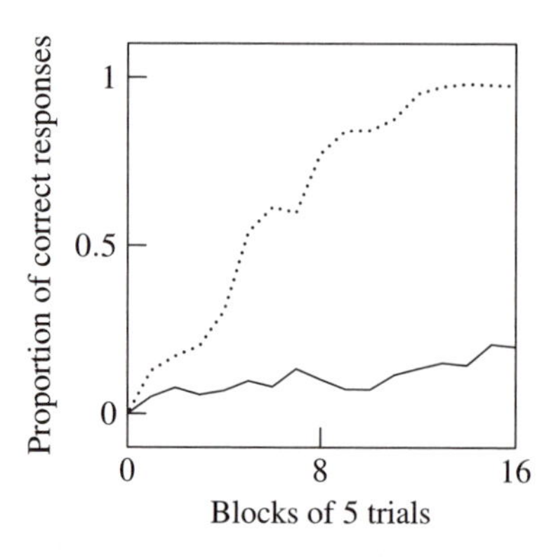

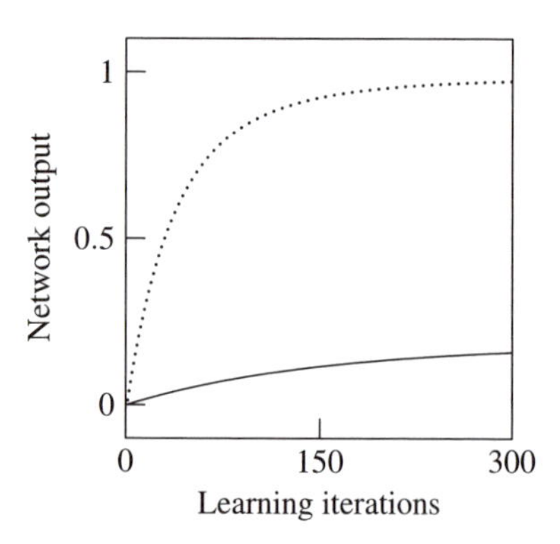

Figure 4.20 Latent inhibition in animals and networks. Left: Data from Lubow et al. (1976) comparing acquisition of a response to a novel stimulus (dotted line) and a stimulus that had been previously shown, unreinforced, to the animal. Right: Simulation of the same experimental paradigm using the neural network model described in the text (equations 4.5 and 4.7).

ulus alone (Figure 4.20; reviewed in Lubow 1989). This simple phenomenon has been investigated thoroughly and has been suggested to play important roles in both perceptual (McLaren & Mackintosh 2000) and associative learning (Lubow 1989). Different proposals have been advanced to explain latent inhibition (Mackintosh 2000; Schmajuk 1997; older ideas reviewed in Lubow 1989). Most assume that a learning rate parameter is lowered during the preexposure phase. In addition, learning rates may be stimulus-specific, which allows different learning speeds for different stimuli. The details, however, vary considerably. Here we discuss the main lines of Lubow's *conditioned attention theory* (Lubow 1989). The concept of attention is used frequently in theories of latent inhibition and other phenomena (Lubow 1989), but it is not an explanation unless one specifies what attention is and how it is directed to one stimulus or another (Johnston & Dark 1986). Lubow has proposed a concrete interpretation of attention simply as an internal response that is not directly observable. Such an *attentional response* is initially performed to all stimuli, but, according to Lubow, it cannot be reinforced by any event. Consequently, attention is predicted to decline to any stimulus, although it may do so at different rates (and sometimes temporarily increase; see Lubow 1989). The link between attention and learning is that the magnitude of the attentional response to the stimulus influences how much an animal learns about it. In our terminology, *attention* is simply a regulatory state elicited by stimuli, and the extent to which a stimulus is capable of eliciting such a state affects learning about the stimulus.

The interpretation of attention as a response is particularly appealing within the neural network paradigm because it makes available all machinery for generating responses from neural network models. To see how Lubow's theory can be implemented, consider a two-layer network with two output nodes. Output node 1 generates the behavioral response, whereas the activity of node 2 is interpreted as the attention elicited by a stimulus. Both activities are calculated as usual: given that

$\mathbf{W}_i$ is the weight vector that conveys stimulation to output node i, the activity of the latter node is $z_i = \sum_j W_{ij} x_j$ (we use linear nodes for simplicity). For node 2 to implement Lubow's attentional response, it should be highly active for novel stimuli and should lower its activity with mere exposure to the stimulus. It is surprisingly simple to achieve this. We assume that all weights W_{2i} between the input nodes and the second output node are initially high, say, equal to 1, and subsequently change according to

$$\Delta W_{2i} = -\eta_2 z_2 x_i \tag{4.5}$$

A system with these properties has been called a "novelty detector" by Kohonen (1984) in the context of signal analysis. For a constant input pattern, the activity of node 2 is shown Figure 4.21. Note that equation (4.5) is identical to a δ rule with 0 as desired response (see Section 2.3.2). This is a simple translation of Lubow's idea that attentional responses are never reinforced and also means that we do not need to postulate a special learning rule for changes in attention. Note also that the rule does not need any global signal. If we add a positive term to equation (4.5), we get a model in which novelty recovers if a stimulus is not experienced for some time. One example is

$$\Delta W_{2i} = -\eta_1 z_2 x_i + \gamma(1 - W_{2i}) \tag{4.6}$$

where $\gamma > 0$. The term $1 - W_{2i}$ approaches zero as W_{2i} approaches 1, which sets an upper limit to the novelty of a stimulus. Now we must specify how novelty or attention affects learning. We consider learning by the δ rule for simplicity. We modify the standard δ rule by including novelty as a global signal able to affect the rate of learning as follows:

$$\Delta W_{1i} = \eta_1 z_2 (\lambda - z_2) x_i \tag{4.7}$$

This is consistent both with Lubow's theory and with the learning rate modulation present in most psychological models of latent inhibition. According to equation (4.7), learning proceeds faster for stimuli of higher novelty, as required to reproduce the phenomenon of latent inhibition (Figure 4.20).

Several considerations can be made at this point. First, we refer the reader to Lubow (1989) for further details of conditioned attention theory. Second, we note that using a network model yields automatically a number of features observed in latent inhibition, e.g., generalization of latent inhibition and greater inhibition for more intense stimuli (because novelty decreases faster; see Lubow 1989). Lastly, we are not aware of specific physiological evidence bearing on our assumption that novelty affects learning as a global modulating factor. However, this hypothesis could be a starting point for research into animal behavior in response to novelty. For instance, the novelty node could inhibit output nodes to model the initial caution that most animals show in novel environments (Russell 1973).

4.4.2 Sensory preconditioning

The phenomenon of sensory preconditioning is observed in the following two-stage experimental design. In the first stage, animals are exposed simultaneously to two

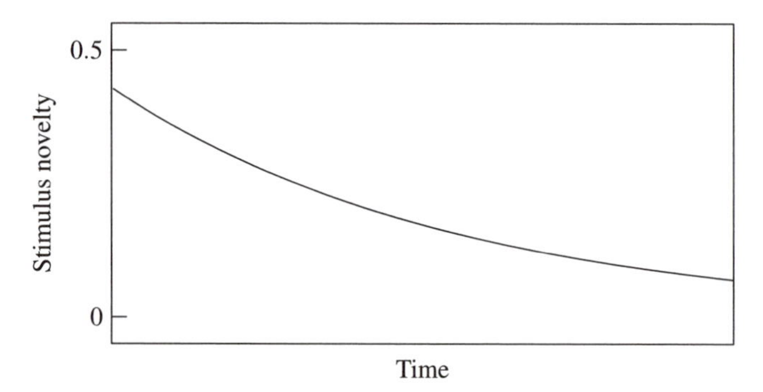

Figure 4.21 Simulation of equation (4.5), showing how the ability of a stimulus to elicit a global signal that regulates learning ("attentional response," see text) decreases with exposure to the stimulus.

stimuli A and B in the absence of reinforcement. In the second stage, a response is trained to one of the two stimuli, say, A, shown alone. The interesting finding is that the other stimulus, B, can also acquire some ability to elicit the response, although only responding to A was reinforced. Sensory preconditioning has been reported in different systematic groups, such as in bees (Müller et al. 2000), snail (Kojima et al. 1998) and rats (Rizley & Rescorla 1972). It differs from "ordinary" generalization (as discussed in Section 3.3) in two crucial ways. First, the stimuli A and B are typically very different, and indeed they often relate to different sensory modalities. Kojima et al. (1998), for instance, used a sucrose solution and a vibratory stimulus (for a design that controls for generalization, see Müller et al. 2000). Second, and most important, preexposure to the compound stimulus AB is necessary for B to elicit any response after training with A. Generalization of the kind discussed in Chapter 3 instead would be observed irrespective of preexposure to AB. This second observation means that exposure to AB must trigger memory changes that become capable of affecting response to B only after a response to A has been acquired.

Figure 4.22 shows a simple network model of sensory preconditioning, similar to a suggestion by Hebb (1966). It differs from a feedforward network by the addition of recurrent connections between nodes activated by A and B (these may be directly the input nodes or, more realistically, nodes in further stages of processing). The recurrent connections change via a local (Hebbian) learning mechanisms that increases the weight between nodes that are often coactive (see Figure 4.12). At the beginning of the simulation, both the recurrent connections and the feedforward ones are weak (signified by the dotted arrows in Figure 4.22a). Exposure to A and B together leads, via the local learning mechanism, to strengthening of the recurrent connections (Figure 4.22b). At this stage B is capable of activating to some extent nodes that initially could be activated only by A. This is not apparent in behavior because none of the two stimuli can elicit an overt response at this stage. If the network is trained to respond to A, however, the recurrent connections become part of a pathway that connects nodes activated by B with the output

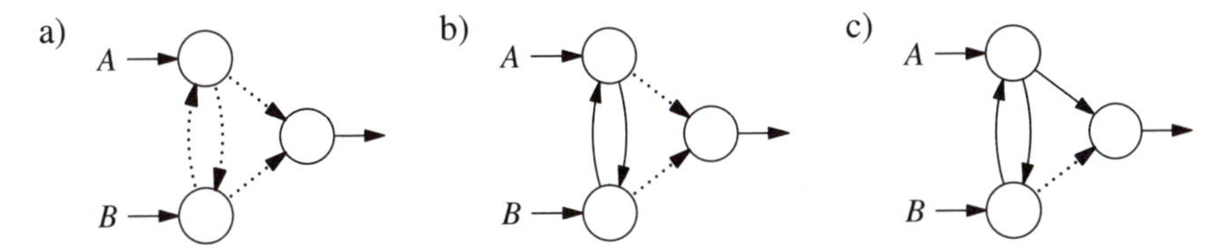

Figure 4.22 A simple network model of sensory preconditioning. It is supposed for simplicity that the stimuli *A* and *B* activate different subset of nodes, each indicated with a single circle. The dotted arrows signify connections that exist but are very weak. The three panels represent the network at three different times: (a) at the start of the simulation; (b) after unreinforced presentations of *A* and *B* together; and (c) after a further stage consisting of reinforced presentations of *A*. See text for further details.

node, as shown in Figure 4.22c. This means that *B* will also be able to elicit the response. Figure 4.23 shows the response elicited by both *A* and *B* throughout the experiment. Responding to *B* remains weaker than responding to *A* because we have assumed that the recurrent connections cannot grow beyond a certain value; see the legend to Figure 4.12. In terms of animal learning theory, sensory preconditioning involves the establishment of stimulus-stimulus associations, as opposed to the stimulus-response associations that are most commonly assumed to underlie classical and instrumental conditioning (Mackintosh 1974; Pearce 1997). This issue will be considered briefly in the next section.

4.5 COMPARISON WITH ANIMAL LEARNING THEORY

We have already seen in Chapter 3 some similarities and differences between neural networks and other models of behavior. With respect to learning, neural networks have many similarities with psychological theory of animal learning, but there are also important differences. Animal learning theory is one of the most developed subjects in animal behavior studies. It has developed almost exclusively based on observations and experimentation at the behavior level. However, learning involves also processes and states within the animal (learning mechanisms, memory). Consequently, learning theory holds a number assumptions about internal factors that have been inferred from observations of behavior. We now review some of these assumptions and relate them to corresponding ones in neural network models.

4.5.1 Stimulus representation

How stimuli are represented is important because it can influence the predictions of models of learning and reactions to stimuli. In studies of animal learning, stimuli are identified and represented based on their physical nature, such as a circle or a tone. They are often referred to with capital letters such as *A* or *B*. Compounds of stimuli are referred to as a series of letter, e.g., *AB*. This holds in most learning models, including the classical Rescorla and Wagner (1972) equation. We recalled

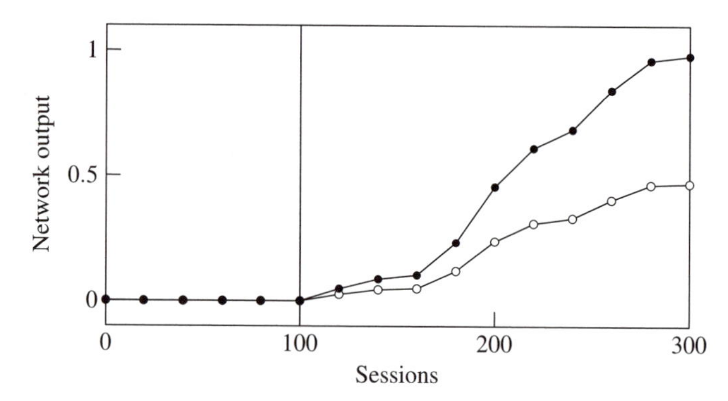

Figure 4.23 Simulation of a preconditioning experiment using the network in Figure 4.22. The simulation involves two stimuli, *A* and *B*. Closed and open circles show network output that *would be* observed if presenting *A* or *B* alone at any time, respectively. In a first phase (training sessions 0–100) *A* and *B* are shown together, unreinforced. Responding to *A* or *B* is not modified. In a second phase (sessions 101–300) responses to *A* are reinforced, whereas *B* is never presented. Closed circles show that network output to *A* grows as expected, and open circles show that *B*, if shown, would also elicit a response. See text for details.

in Section 3.2.5 that such models assume that a particular stimulus is represented as activating one "stimulus element." Such activities are typically restricted to be 0 or 1, corresponding to the stimulus being absent or present, respectively. This way of representing stimuli is adequate for some purposes but comes with shortcomings. One problem is that there is no clear rule for deciding what is a stimulus and what is a combination or compound of stimuli and to define precisely similarity between stimuli. Furthermore, there is no natural way of dealing with continuous changes in stimulation, e.g., in intensity. Another issue is "context." The stimulation reaching the animal during an experiment is not limited to the stimuli that the experimenter is interested in because the animal will also perceive other aspects of the testing environment. Context is sometimes taken into account as an additional stimulus but, as for other stimuli, unambiguous procedures are lacking (e.g., to calculate the similarity between different contexts).

In neural network models, we can represent stimuli more realistically using knowledge of receptors and sense organs (or further stage of processing) to build network input. As we saw in Chapter 3, this approach offers a more precise representation of stimulation that allows us to model both continuous and discrete changes in stimulation (i.e., varying a stimulus versus adding or removing one) and to clearly define similarity and difference between stimuli. A potential drawback of this approach is that it is more complex. The transformation from the physical stimulus to receptor activities, for instance, is not always trivial. However, we have seen in Chapter 3 that taking into account some general and simple features of perception is often enough to account for many findings. The same appears to hold for learning phenomena, as shown earlier.

4.5.2 Learning

Learning refers here to the processes that change the behavior map. Research into animal learning has produced refined and successful theories of learning mechanisms based on both external and internal factors. Many theories share similarities with neural network models. For instance, the models in Atkinson and Estes (1963), Blough (1975), Bush and Mosteller (1951), McLaren and Mackintosh (2000) and Rescorla and Wagner (1972) can all be cast within the neural network framework. As seen in Chapter 3, many popular models predict responding based on a weighted sum of "stimulus elements" that are assumed to be activated by environmental stimuli. In these models as in neural networks, learning proceeds by changes in the weights (often called *associative strengths* in psychology). The influential model by Rescorla and Wagner (1972) prescribes a rule for weight change identical to the δ rule (equation 2.38), although the activity of "stimulus elements" was restricted to the values 0 and 1 (see the preceding section):

$$\Delta W_i = \eta \left(d - \sum_i W_i x_i\right) x_i \tag{4.8}$$

Rescorla and Wagner were the first to apply equation (4.8) to behavioral modeling, showing that it can reproduce a number of learning phenomena such as acquisition, extinction, blocking and overshadowing. The first application to stimulus representation with continuous elements was given by Blough (1975); see Section 3.3.3. The Rescorla-Wagner equation was originally aimed at classical conditioning, but Blough's model addressed instrumental conditioning. Similarly, Pearce's (1987, 1994) models of classical conditioning can be applied as well to instrumental conditioning. It may then come as little surprise that the same network model can support both classical and instrumental conditioning without the need of different assumptions. Consider, for instance, the reinforcement learning model we have been using in most of this chapter. Signals provided by a "value system" drive learning in the same way independent of whether they arise from an action performed by the animal (instrumental conditioning) or from an event unrelated to the animal's behavior (classical conditioning). It is only necessary that reinforcement, behavior and a particular stimulus situation occur in temporal contiguity.

4.5.2.1 The influence of stimuli and internal factors on learning

Many learning phenomena relate to experiences with one or two external stimuli. Habituation and sensitization involve only one stimulus, whereas classical and instrumental conditioning require a first stimulus that elicits a behavior and a second stimulus usually called the *reinforcer* (also US in classical conditioning). Empirical data show that the nature of these stimuli influences learning (Section 4.3.1). For example, learning usually proceeds faster for more intense stimuli (Mackintosh 1974). To account for this, it is often assumed that the learning rate η in equation (4.8) has a higher value for more intense stimuli. By interpreting "stimulus elements" as model receptors (or model neurons, in general), this can be derived from neurophysiology rather than being assumed. In fact, intense stimuli will activate receptors more strongly (higher values of x_i in equation 4.8), yielding larger weight changes (since x_i is a factor in determining ΔW_i). A similar remark was made in Section 4.3.1 relative to the strength of the reinforcing stimulus and in Section 4.3.3 about the causes of overshadowing.

Animals (unsurprisingly) take more time to learn discriminations between more physically similar stimuli. Making predictions about the speed of learning a discrimination obviously requires modeling the similarity between stimuli. This, as we have seen earlier, is traditionally a problem for many theories of animal learning. Additionally, Pearce (1987) showed that a discrimination between the compound stimuli *AC* and *BC* is solved faster in the (Rescorla & Wagner 1972) model than a discrimination between *A* and *B* despite the former pair of stimuli having more "elements" in common than the latter. Neural networks and models such as Blough's (1975) that can be interpreted as networks do not suffer from these shortcomings, provided input patterns are realistically built based on knowledge of animal perception (see Figure 5.5 and Section 3.3.2).

Earlier we also saw how to include within neural network models internal factors that influence learning. Our simple model of how novelty of a stimulus influences learning showed how an abstract concept such as "attention" could be implemented in a simple neural network. Because the amount of attention paid to a stimulus was a function of previous experience, the model was also a simple example of a "value system" that can learn.

4.5.2.2 What changes within the animal?

Learning results in changes of the animal's memory that may cause a change in responding to stimuli. A fundamental question is what is stored in such memories. In learning theory it is common to refer to associations or associative strengths (Mackintosh 1974; Rescorla & Wagner 1972). Many associations are considered to link a stimulus and a response (Guthrie 1935; Hull 1943; Mackintosh 1974), with stronger association yielding stronger responding. However, associations between stimuli have also been suggested. Classical conditioning, for instance, has been modeled as the building of an association between the CS and US rather than between the CS and the response (Mackintosh 1974; Pearce 1997; Tolman 1949). There are indeed empirical findings that support both types of associations (see, e.g., Pearce 1997). We will not review this debate here; our purpose is rather to compare some of the existing ideas with what neural networks models suggest.

In neural network models, all learning consists of changes in connection weights. Some of the connections can be interpreted as stimulus-response connections. For instance, in the model of classical conditioning in *Aplysia*, only stimulus-response connections are modified (Figure 4.5). However, more complex network models of classical conditioning may contain also stimulus-stimulus connections that are modified by learning (e.g., McLaren & Mackintosh 2000). It is clear that we still do not understand fully what connections are required for neural networks to explain the empirical findings. Additionally, in networks we find changes to connections that are difficult to interpret as either stimulus-response or stimulus-stimulus associations. For instance, during perceptual learning (see Sections 2.4 and 4.4.2) connections may be modified to provide more useful representations for decision making (see Sections 3.4.4 and 3.7). There are also some difficulties in relating the concept of an association in the sense of learning theory with connections between nodes in a network. Networks use distributed processing, whereby a single connection may contribute to responding to many stimuli. On the other hand, learning theory typically considers one-to-one associations between stimuli or responses.

In learning theory the concept of associations is also used to distinguish between different learning phenomena. For instance, habituation is considered "nonassociative learning" (because the stimulus is not associated with any other event in the environment) and is contrasted with "associative learning," such as classical and instrumental conditioning (so labeled because they involve an "association" between two events). However, habituation can be described as the extinction of an initial reaction within the same model as other learning phenomena. We have also seen that in models of classical conditioning deriving from research on *Aplysia*, the neural network that supports sensitization (a form of "nonassociative" learning) coincides to a large extent with the network responsible for classical conditioning. Thus neural network models of learning do not compel us to make any strong difference between associative and nonassociative learning.

4.5.3 Responding after learning

At the behavior level, the results of learning are studied by exploring changes in responding to stimuli. In this book we have drawn a clear conceptual distinction between the behavior map (how responses are generated from stimuli) and learning (how the behavior map is changed by experiences). It is important to note, however, that features of the behavior map are also important when we study learning. For instance, we saw earlier that assuming different representations of stimuli can lead to different predictions about learning phenomena. This suggests that we cannot study learning separately from representation, as most of learning theory does.

4.6 TRAINING ANIMALS VERSUS TRAINING NETWORKS

It is interesting to compare methods developed to train animals with methods used to train networks. To our knowledge, these two subjects have developed independently of each other. Animal training is based mainly on knowledge gained during the behaviorist era, particularly by Skinner and his followers (Skinner 1938; Staddon 2001). An example of the practical efficacy of this knowledge is the animal training firm Animal Behavior Enterprises, led by Keller Breland, Marian Breland Bailey and Bob Bailey. Applying the principles of instrumental conditioning, its staff has trained about 15,000 animals of 140 different species in 45 years for purposes ranging from military applications to entertainment. In a seminar in Stockholm in 2002, Bob Bailey summarized the art of training animals with the following line: "Training is simple but not easy." He meant that successful training results from the application of simple principles to the complex interplay of the trainer's aims in combination with the animal's own wants and needs. His overall advice is to organize training in many steps and to simplify each step so that its goal can

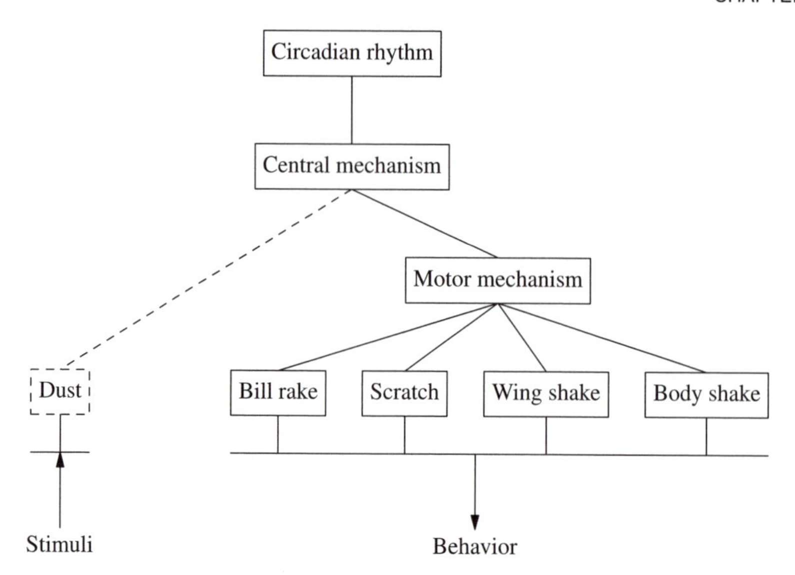

Figure 4.24 The dustbathing system in the fowl. Boxes represent putative subsystems: a sensory mechanism responsible for recognizing dust; a central dustbathing mechanism responsible for integrating sensory and internal influences; several motor mechanisms responsible for specific actions as well as a higher level motor mechanism responsible for the patterning of actions during dustbathing bouts. Solid lines indicate mechanisms and connections among them that develop without specific experiences (prefunctionally); dashed lines indicate mechanisms and connections that develop as the result of specific functional experience. Adapted from Vestergaard et al. (1990) and Hogan (2001).

be made unambiguous to the animal, which can be achieved by precise administration of reinforcement. It would be interesting to apply these principles to training neural networks. Perhaps engineers and others interested in practical applications of neural networks will be able to cut training times or even reach by small steps complex skills that are difficult to train with more standard techniques.

4.7 ONTOGENY

Ontogeny refers here to the development of an individual's behavior, or its behavior system, during the individual's whole life span. This process starts very early in embryonic life and includes a diversity of processes, many not well understood. From a behavioral perspective, ontogeny is particularly challenging because at any time important changes may occur within the animal that are not expressed in behavior until a later time.

Dustbathing in the fowl is an illustrative example of behavioral ontogeny (Figure 4.24; Hogan 2001; Larsen et al. 2000; Vestergaard et al. 1990). We have already discussed the motivational system regulating dustbathing in Chapter 3. Dustbathing

Table 4.2 Ontogeny of dustbathing in the fowl.

Age	Behavior
Day 1	Pecking
Day 3	Learning of dust stimuli
First week	Some other movement occur
Day 8	Signs of central regulation
Day 12	All movements occur
Day 14	Central regulation fully developed
Second week	Coordination of movements
First month	Full integration reached
Up to fifth month	Changes in amount of dustbathing

consists of a sequence of coordinated movements of the wings, feet, head and body that serve to spread dust through the feathers. The function is to remove lipids and to maintain good feather condition. The sequence of behaviors in a dustbathing bout begins with the bird pecking and raking the substrate with its bill and scratching with its feet. These behaviors continue as the bird sits down. Then the bird tosses dust into its feathers with vertical movements of its wings and also rubs its head in the dust. It then rolls on its side and rubs the dust thoroughly through its feathers. These sequences of movements may be repeated several times. Dustbathing ends with the bird standing and shaking its body.

Dustbathing does not appear fully formed in the young animal. Instead, it develops during the first months of life. Pecking appears already on the first day after hatching, and in the first week some of the specific movements of dustbathing appear. Around day 3, the chick learns the external stimulus that will elicit dustbathing. Genetic predispositions determine that under normal conditions such a stimulus is indeed dust. After a week, one can observe some central regulation of the amount and timing of dustbathing. In the second week, all movements occur, and they become coordinated, and central regulation is well developed. However, it is not until the end of the first month that the perceptual mechanism responsible for recognizing dust is integrated with the other parts. This sequence of events is summarized in Table 4.2. The movements, their coordination and central regulation develop without specific experiences, whereas the development of perceptual mechanisms and their connections with central mechanisms require specific experiences in the context of dustbathing, e.g. experiences with the "dust" stimuli.

Most of the general points about the ontogeny of behavior can be made about dustbathing. First, behavior develops through a series of events, eventually reaching its adult form. Different components of a behavior system develop partly independently of each other and only gradually become fixed in the normal adult form. A consequence of this is that behavior seen during this process may make little functional sense and does not fulfill any goals. For instance, before the perceptual mechanism is integrated into the dustbathing system, dustbathing occurs on any surface regardless of the presence of dust. The functional behavior emerges once all its perceptual, central and motor components are fully integrated.

Second, for development to occur in a "normal" fashion, particular experiences are often needed (Hogan 2001). Often only some aspects of the behavior machinery need such experiences, whereas other parts develop spontaneously (Figure 4.24). However, considerable variation among species and behaviors exists. In many "simpler" species most behavior mechanisms seem to just be ready when needed. For instance, a butterfly emerging from metamorphosis can fly as soon as its wings are fully extended, without any need for practice. When learning occurs, genetic predispositions are vital to ensure that appropriate things are learned (Hogan 2001; Horn 1985).

Third, many changes that occur inside the animal during ontogeny are not revealed in behavior until later in life. Observability of internal variables is a general problem of behavior studies, but it is particularly acute in the case of ontogeny. As a result, it is not always easy to discover what drives ontogeny, and observations of behavior are often difficult to interpret. Below we will use a general framework that has emerged from extensive ethological research (Bolhuis & Hogan 1999; Hogan 1994a, 2001). It consider behavior as emerging from "building blocks" consisting of various kinds of perceptual, central and motor components, all of which can exist independently. Development can then be understood in terms of changes in the components themselves and in their relationships (Hogan 2001). This picture is, of course, a simplification (e.g., it may be difficult to draw a line between "perceptual" and "central" mechanisms), but it is helpful to understand ontogeny, in particular at the level of behavior. Before we consider neural network models of ontogeny, we will summarize what is known about the ontogeny of the nervous system. Some of these issues, in particular the relationships between genes and nervous system structure and functionality, are also discussed in Chapter 5.

4.7.1 Ontogeny of the nervous system

The ontogeny of behavior has its material basis in the ontogeny of the nervous system. All the organization and specificity of behavior is due to the nervous system being highly structured rather than randomly assembled neurons. Our question is, then, what this "structure" consists of and how it is established during ontogeny. In essence, the structure, as discussed in Chapter 3, consists of a variety of cell types, including receptor and effector cells, as well as patterns of connections between cells. The latter are often intriguingly complex. There are connections within clusters of neighboring cells but also long-distance connections between different parts of the brain and connections with sense organs and muscles. Connections can be of different kinds, and neurons may also be sensitive to or produce hormones or neuromodulators with longer-lasting influence on other neurons or organs.

Our knowledge about the ontogeny of nervous systems has improved considerably in recent years. We can roughly identify four steps in which the nervous system develops (Blass 2001; Kandel et al. 2000; Whatson & Stirling 1998; Wolpert et al. 1998). First, neurons of various types are generated and migrate to particular

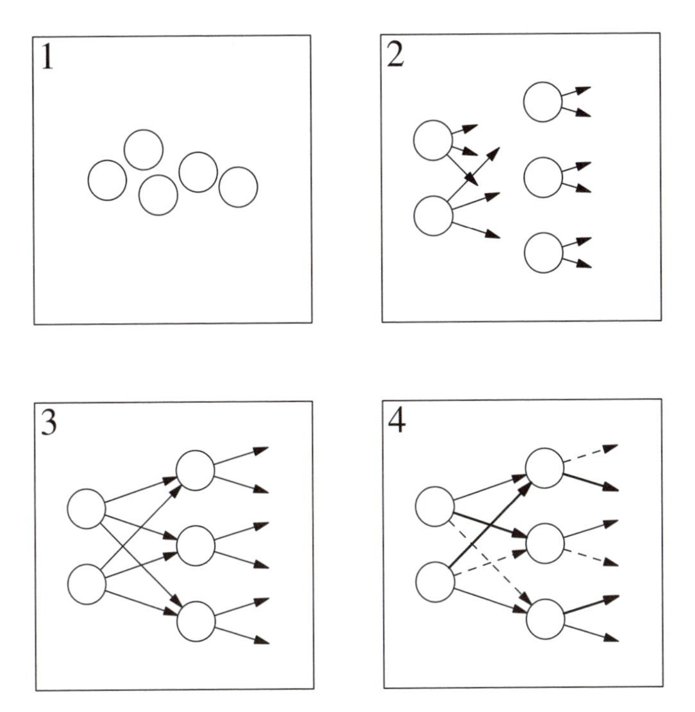

Figure 4.25 Stages of nervous system development, illustrated in a simple neural network. Stage 1: Neurons are born and start migrating to their final locations. Stage 2: Axons start to grow toward target cells. Stage 3: Connections are established, often in large numbers and an unspecific pattern. Stage 4: Connectivity is refined by weakening or removing some connections (dashed lines) and strengthening others (thick lines). External stimuli are most important in the last two phases. See Kandel et al. (2000) and Wolpert et al. (1998) for details.

locations, resulting in spatial organization. Then follows outgrowth of axons and dendrites. A variety of processes guides the axons to their targets. These growth processes lead to synaptic connections being established between cells. Finally, there are processes that refine these connections. These phases are illustrated in Figure 4.25. It is also important to the ontogeny of animal behavior that different parts of the brain mature at different times and different speeds and that the development of the nervous system is not limited to the time before the individual is born. One example is the development of the neural machinery behind song learning and the production of other sounds in song birds (Zeigler & Marler 2004). Here, birth of new neurons, migration of neurons and establishment and refinement of connections occur well after birth and even in adults. The process follows a definite temporal organization that reflects in behavior as critical periods for sensory learning and motor learning.

The development of the nervous system is controlled both by genes and external

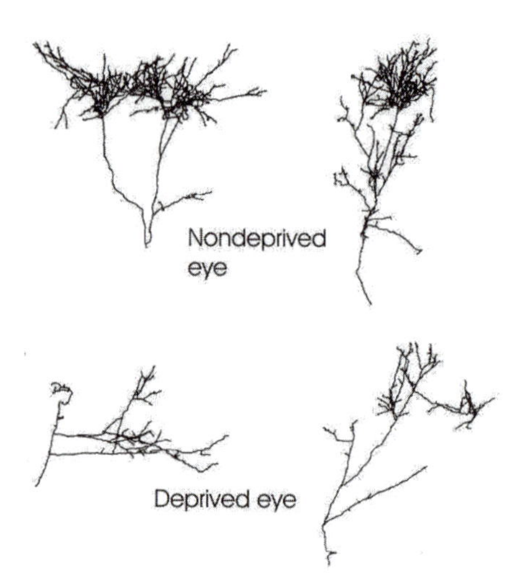

Figure 4.26 Effect of visual experience on the development of neural complexity. The example shows the terminal arbors of geniculocortical axons in the visual system of the cat. Deprivation of normal visual input during development reduces dramatically the complexity of these axons. Reprinted with permission from Antonini A, Stryker MP, 1993. Rapid remodeling of axonal arbors in the visual cortex. *Science* 260, 1819–1821. © 1993 AAAS.

events (see also Section 5.3). In addition, there are various processes of "self-organization." In organisms with very simple nervous systems, genetic control dominates, as if the genes contained a "blueprint" of the networks to be realized. One well studied example is the nematode worm, *Caenorhabditis elegans*, whose nervous system contains precisely 302 neurons (falling into 118 classes), about 5000 chemical synapses and 600 electrical ones (Riddle 1997; Thomas & Lockery 1999). This nervous system develops according to a strict plan of cell divisions and cell death. However, not all the network properties of *C. elegans* are specified by the genes because the worm is capable of learning (Hedgecock & Russell 1975).

In species with larger nervous system, made up of millions or billions of nerve cells, there is no exact "blueprint." Instead, much of the development depends on nerve cells being genetically instructed to make selective contact with other nerve cells. This leads to a kind of "self-organization," in the sense that nerve cells are born with properties that ultimately cause network organization to emerge. For instance, one type of nerve cell may grow axons that find their way to another group of cells and connect to them. Exactly which cells will make contact with each other is not genetically specified, nor which connections will survive among those initially made. Refinement of connections depends sometimes on spontaneously generated neural activity, independent of external stimulation. For instance, in humans, the segregation of retinal inputs to the thalamus occurs before birth and is driven by synchronized bursts of activity in retinal axons (Galli & Maffei 1988; Kandel et al. 2000; see Linsker 1986 for a related neural network model). However, many developmental processes are open to influences from sensory experiences and behavior.

Not only the efficacy of synapses but also network organization can depend on experiences. A classical kind of experiment that demonstrates how the environment can influence brain ontogeny consists of raising animals, e.g., rats, in two different environments (see, for instance, Rosenzweig & Bennett 1996). An "impoverished" environment may consist of plain cages in a dimly illuminated, quiet room, with nothing special for the animals to do. A "rich" environment would contain running wheels, ladders, slides and toys. The latter were changed daily. Rats raised in the rich environment had a thicker cortex, better blood supply to neurons, more glial cells, more cholinergic neurons, more dendrites in the cortex and more dendritic branching. Figure 4.26 shows the result of a related experiment in kittens comparing nerve fibers from an open eye and an eye deprived from sensory input. The latter show a considerable decrease in complexity. Sensory input also results in refinement of synaptic connections; i.e., changes in the efficacy of synapses. This includes learning in its more narrow sense, which was covered earlier.

4.7.2 Neural network studies of ontogeny

Learning, in the sense of refinement of connections and thus the last stage of ontogeny, has always been the core area of neural network research. In comparison, only a few studies exist about the ontogeny of behavior in a broader sense (for reviews, see Elman et al. 1996; Nolfi 2003). Most of these studies, moreover, are limited to tracking changes in network behavior that arise from changing the value of connections within a fixed network architecture. Thus they do not consider the birth and growth of neurons, how connections are formed and what controls such events. Problems studied with this approach include human language acquisition (Elman et al. 1996; Rumelhart & McClelland 1986a), perceptual learning (Kohonen 2001; McLaren & Mackintosh 2000) and imprinting (Bateson & Horn 1994; O'Reilly & Johnson 1994).

Rumelhart and McClelland (1986a) present one the earliest neural network studies of a complex ontogenetic process. Their aim was to model how children learning to speak English acquire the ability to form the past tense of verbs. A feedforward network received verb stems in input and was trained by back-propagation to produce the corresponding past tenses as outputs. Figure 4.27 shows the acquisition curves for regular and irregular verbs separately. Mastering of regular verbs displays a steady growth in performance. The ability to handle irregular verbs, on the other hand, shows a sudden drop after a number of training sessions. At this time the network seems to have "grasped" the relatively simple rule for constructing the past tense of regular verbs and uses the same rule, incorrectly, to handle irregular verbs. The performance on irregular verbs increases again as the network learns each verb on a case-by-case basis. This study has been discussed widely, and also criticized, as a model of grammar acquisition (Elman et al. 1996; Pinker & Ullman 2002). In the context of this book, the example is interesting because it suggests that complex learning sequences can arise from learning based on weight updates within a fixed network architecture (see below).

Another neural network model that builds only on weight changes has been suggested for filial imprinting, the process whereby a young bird such as a chick or

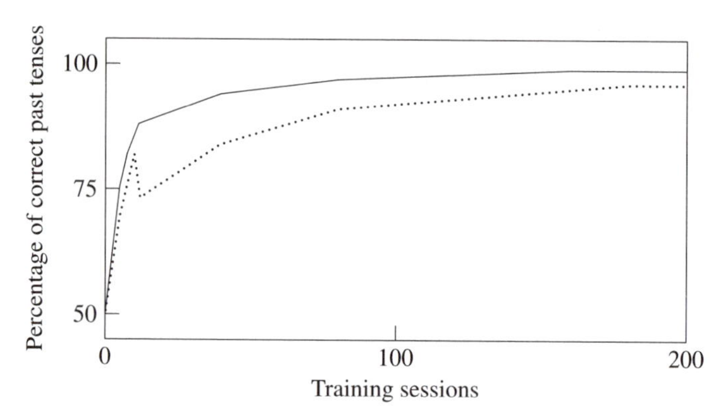

Figure 4.27 Performance on regular and irregular verbs of a network that learns to form the past tense of English verbs (data from Rumelhart & McClelland 1986a).

duck learns to approach and follow its mother while avoiding other objects. Learning the characteristics of the mother occurs during a sensitive period that starts almost directly after hatching and terminates after a certain amount of experience. Bateson and Horn (1994) have developed a neural network model of this process that includes genetically preset weight values and two kinds of learning mechanisms for updating connection weights. The network, shown in Figure 4.28, has an analysis layer (compare with Figure 1.11 on page 26 and Figure 3.13 on page 87), a recognition layer and an executive layer. Although the terminology is different, the model is a feedforward network. The two executive nodes command approaching a stimulus (normally the mother) or withdrawing from it. Initially, all weights in the network are small, except for the weights that directly connect the analysis layer and the executive layer. The consequence of these connections is that the newly hatched chick will approach any stimulus. A Hebbian learning rule strengthens any connections that contribute to the approach of a stimulus encountered. In parallel, owing to a second learning rule that lowers weights, the potency of other stimuli (those not encountered) to elicit an approach response declines and eventually drops to zero (Figure 4.28). This effectively ends the sensitive period; any new stimuli encountered subsequently will elicit withdrawal.

O'Reilly and Johnson (1994) have developed a model of imprinting with similar abilities but a different architecture based on the connectivity of brain regions that underlie imprinting in chickens. In particular, the authors speculate that the organization of the IMHV region, including recurrent connections and spatially organized receptive fields, may allow the imprinting object to be recognized robustly even when its retinal image varies (see Figure 3.3 and Section 3.4.3). They also argue that major features of imprinting derive from how such robust recognition is learned. These hypotheses are studied with a recurrent network organized similarly to the IMHV region and whose connections are modified by Hebbian learning. Repeated experiences with an object lead to the development of strong feedback loops within the network such that the network ultimately responds only to the experienced object (or similar ones—the network is capable of generalization). As

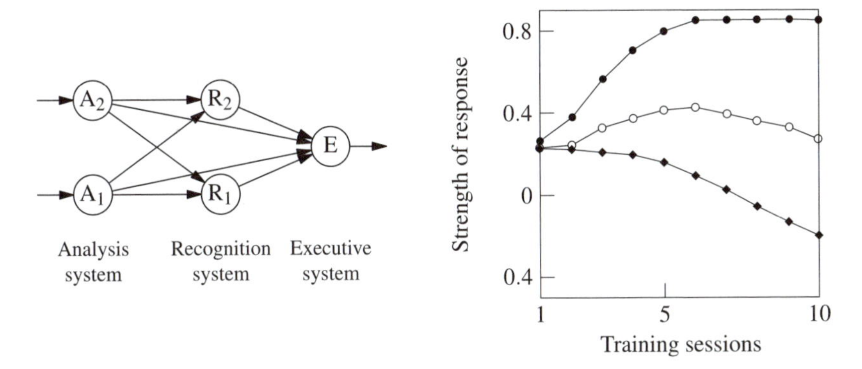

Figure 4.28 A neural network model of imprinting. Left: Model architecture (redrawn based on Bateson & Horn 1994). Right: Results of simulations, showing the response to the imprinted stimulus (closed circles), a somewhat similar stimulus (open circles) and a very different stimulus (closed diamonds).

anticipated, the experienced object is recognized when projected at different positions on the retina. Crucial to the interpretation of the network as a model of imprinting is that the learning process is one-way. This follows from the fact that Hebbian learning depends on the activity of nodes: after an object has become able to elicit the response, different objects cannot trigger learning because they are unable to excite the network. Another interesting feature is that the model learns to respond to moving objects more easily than to still ones, a well-known phenomenon in studies of imprinting (Sluckin 1972). In the model, movement is used to develop a single representation (pattern of node activity) for the different retinal images projected by an object based on the fact that retinal images at two successive instants in time likely arise from the same object (Földiák 1991). Whether movement plays the same role in animal learning, and in particular in imprinting, is not known. In summary, the model accounts for several aspects of imprinting (more phenomena are reviewed in O'Reilly & Johnson 1994) and integrates the study of imprinting with the development of object recognition. Perhaps its major drawback is that, by appealing to general ideas about perceptual learning, it does not fully clarify the differences between imprinting and other learning processes (i.e., why not all learning terminates after a small amount of experience).

We have thus reviewed three studies that model aspects of behavioral ontogeny by weight changes in a fixed network architecture. At least some observations of behavioral ontogeny may depend on such reorganizations of memory. In particular, the strategy of acquiring general information first and specific information later, exemplified by the Rumelhart and McClelland (1986a) study on the English past tense, seems a general feature of perceptual learning. To an important extent, perceptual learning takes place early in life and seems to tune the nervous system to general characteristics of the environment the animal lives in. When later in life the animal is faced with various recognition and discrimination problems, the nervous system is already equipped with important information, e.g., the ability to construct and use particular stimulus dimensions (Section 3.4). It is very likely that predispositions exist that facilitate perceptual learning in particular directions, but it is interesting that even simple learning rules in a network without predispositions can achieve considerable sensory organization relevant to functional behavior.

Perceptual learning, as other ontogenetic phenomena, usually occurs in the absence of clear behavioral expressions, as well as in the absence of feedback from the environment. The latter is reflected, in network models of these phenomena, in the frequent use of learning algorithms that operate without global signals (see Section 4.2.2). In Section 2.4 we have seen some simple examples in which spatially organized receptive fields develop by Hebbian learning (Figure 2.15 on page 59). Those simulations started with a model retina in which all nodes sent input to an output node with equal weight. During the simulations, however, receptive fields of different kinds could develop, depending on the value of a parameter in the Hebbian rule. One example is center-surround receptive fields (Figure 2.15, center), also observed in visual systems. In such a receptive field, the output node is excited by input nodes located in a well-delimited portion of the model retina, approximately circular in shape. Other input nodes, arranged in a disc around the circular region, provide inhibitory input. Nodes outside this disc affect the output node negligibly. The latter is noteworthy because it can be interpreted as pruning of connections during the ontogeny of receptive fields. In another simulation, we saw that a network with many output nodes linked by inhibitory connections and subject to the same ontogenetic process could develop receptive fields that span the whole retina (Figure 2.16).

There are today many studies on organizing network weights without explicit feedback from the environment (see, e.g., Kohonen 2001 and articles in Arbib 2003). These models have been often related to empirical findings about the ontogeny of the nervous system, but applications to the ontogeny of behavior are not well developed.

4.7.3 Toward a general framework

The preceding results are intriguing, but the ontogeny of behavior cannot depend only on refinement of connections within a fixed network architecture. Nervous systems can develop also in terms of, e.g., number and connectivity of nerve cells. For instance, in male zebra finches (*T. guttata*), connections from the HVC brain nucleus grow to the RA nucleus mainly between 25 and 60 days of age (Nordeen & Nordeen 1988), during which period the young male memorizes the song of other males (Zeigler & Marler 2004). Today there is no general framework based on neural networks for modeling the ontogeny of behavior in its broadest sense (but see Chapter 5; Nolfi 2003). Here we sketch some possible research lines.

The goal of a general neural network model of ontogeny is a relatively simple and intuitive account of animal development as it appears at the behavioral level of analysis. Particularly important is to account for how ontogenetic factors, such as sensitive periods and predispositions, emerge and regulate ontogeny. Within a neural network framework, changes in behavior should be explained in terms of how network parts and their interactions develop. The latter can, in general, happen as

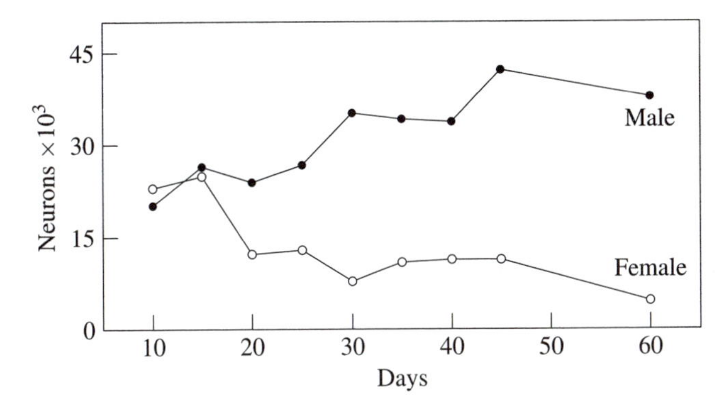

Figure 4.29 The presence or absence of steroid hormones can regulate the ontogeny of the nervous system (Forger 2001). This diagram illustrates how the volume of the song nucleus HVC in Zebra finches increases in males and decreases in females from an age of around 20 days. This has been linked to the development of singing in male song birds (DeVoogd 1991).

a result of automatic maturation processes, behavior and sensory experiences. An example of a mainly internal process is the influence of steroid hormones on sexual differentiation. Figure 4.29 shows how steroid hormones control the proliferation of a song nucleus (HVC) in male and female zebra finches. This is also an example of a process that is difficult to extract from observations of behavior alone. Behaviorally silent ontogenetic processes present an interesting hierarchy that may be important in developing models. That is, the development of motor programs is more likely to find behavioral expression because it is connected directly to muscles. Important development of sensory mechanisms, on the other hand, may occur without any immediate consequence for behavior because connections that link such changes with behavior have not yet been established.

For a complete understanding of ontogeny, existing neural network models must be developed in a number of ways. We need additional state variables that describe, at least, network architecture and properties of nodes. We also need rules for changing such states (formally, state-transition equations) to model changes in real nervous systems, such as the generation of new cells and the establishment of specific connections within or between groups of nerve cells. The construction of a general model, of course, should be guided by knowledge about how nervous systems develop and must reflect the material processes that establish neural networks in animals.

Particular experiences may be important and trigger new developmental phases, resulting in learning in terms of synaptic changes and other changes such as number of nerve cells and connectivity (see preceding examples). To account for both internal process and interaction with external factors, neural network models of ontogeny should take into account the dynamical interaction between organism and environment. One example is network models in which the system itself recruits new nerve cells as part of learning, called *generative connectionist architectures*

(Mareschal et al. 1996; Quartz & Sejnowski 1997). Such models are consistent with observations that number of nerve cells and amount of connectivity increase with experience (see above). A general model of ontogeny must also explain the behavior we observe before the final developmental stage is reached. It is common that behavior patterns make little functional sense when they first appear. For instance, male chickens try to copulate long before they are sexually mature (Kruit 1964). Perhaps such behavior is important because it triggers further development, but perhaps it is simply a by-product of ongoing development.

4.8 CONCLUSIONS

Neural networks are potentially powerful models of learning in animals, but much work is still needed to have a satisfactory theory of learning based on neural networks. Many learning phenomena so far have not been modeled. In several respects, learning in neural network is similar to learning in other successful models, such as the classical learning equation of Rescorla and Wagner (1972). Neural network models, however, offer a few advantages. A first one is that neural networks can smoothly exploit knowledge from both behavioral observations and neurobiology. Most learning theories, including contemporary ones, are based on observations of behavior only.

A second advantage is that internal factors that govern learning can be given a concrete interpretation. Memory is defined in terms of connection weights, and learning is defined in terms of changes to connection weights. The mechanisms for weight change include local factors, depending, for example, on the activity of pre- and postsynaptic nodes, and often also global signals that selectively communicate information from other parts of the network to particular connections. Based on local and global factors, mechanisms of weight change can be tailored to account for various learning process.

A third advantage is that a neural network model of learning inherits all the features of behavior systems that we covered in Chapter 3, including sensory processing, decision making and motor control. The learning mechanism is integrated with the behavior system simply as a different and clearly identifiable aspect of the operation of the same network. For instance, neural network models of learning may easily include a realistic handling of stimuli. One example is the effect of stimulus intensity on learning, which traditionally has been dealt with with ad hoc assumptions (see Section 4.3.1). Another example of this integration is that learning mechanisms can exploit the ability of networks as input-output maps to produce and communicate global signals that guide learning (see, e.g., the model of "attention" sketched earlier). Such networks are not part of the behavior map, in the sense that they do not produce observable behavior.

There are, of course, challenges and problems with neural network models of learning. A major challenge is the diversity of findings and the compexity of learning. A particularly acute problem is retroactive interference, from which networks often suffer to a degree that is unrealistic—old information can be almost totally overwritten when new information is learned (McCloskey & Cohen 1989). Despite considerable progress in recent years, we still have to understand how memory can be better protected in neural networks (French 1999; Mackintosh 1995; McClelland et al. 1995). Neural network models for ontogeny in the broad sense are not developed satisfactorily and require new ideas. We need to model how new nodes are generated and receive their properties, how connectivity is established and refined and how such process can be adapted to generate functional behavior.

Neural network models of learning and ontogeny may be important for understanding the nature-nurture issue as well as the evolution of learning. A problem with understanding the relationship between genes and behavior is that they operate at different levels. However, assumptions about how the genes determine network architecture, local mechanisms and global factors of learning can provide a very concrete model of how ontogeny arises from the interaction of genes and experiences. Neural network models also make the evolution of learning mechanisms more understandable because the latter can be expressed in terms of selection on architecture, local mechanisms and global factors. This gives genes control over the learning process and can give insight into the generality or specialization of learning mechanisms. These issues will be explored in the next chapter.

CHAPTER SUMMARY

- Neural network models offer a concrete interpretation of (long term) memory and learning in terms of connection weights and how they change.
- Neural network learning procedures such as reinforcement learning, the δ rule and back-propagation reproduce major observations of animal learning.
- The local rules for updating connection weights (memory) in neural networks are usually a function of the activity in the connection, the current value of the weight and a global signal. This agrees with available data about the neurophysiology of learning.
- Global signals generated elsewhere in the network allow any (available) information to influence local weight updates. Global signals can be generated by full-fledged networks, which provide evolutionary flexibility about what events can cause learning.
- However, learning is still not fully understood in terms of neural networks:
  - Most learning procedures contain biologically unrealistic features (e.g., back-propagation).
  - What mechanisms underlie behaviorally silent learning is still partly unresolved. We have presented simple models of latent inhibition, sensory preconditioning and perceptual learning that could be developed further. The models embody known features of the nervous system (recurrency, Hebbian synaptic plasticity) and reproduce a few major findings.
- No general framework exists for modeling the ontogeny of behavior based on neural network models, although a number of interesting applications have been developed.

FURTHER READING

Ontogeny of animal behavior:

Blass E, editor, 2001. *Developmental Psychobiology*, volume 13 of *Handbook of Behavioral Neurobiology*. New York: Kluwer.

Bolhuis JJ, Hogan JA, editors, 1999. *The Development of Animal Behaviour: A Reader*. Malden, MA: Blackwell.

Hogan JA, 2001. Development of behavior systems. In Blass (2001), 229–279.

Hogan JA, Bolhuis JJ, editors, 1994. *Causal Mechanisms of Behavioural Development*. Cambridge, England: Cambridge University Press.

Animal learning and memory:

Kandel E, Schwartz J, Jessell T, 2000. *Principles of Neural Science*. London: Prentice-Hall, 4 edition.

Klein SB, 2002. *Learning: Principles and Applications*. New York: McGraw-Hill, 4 edition.

Mackintosh NJ, 1974. *The Psychology of Animal Learning*. London: Academic Press.

Mackintosh NJ, 1983. General Principles of Learning. In Halliday & Slater (1983), 149–177.

Roper TJ, 1983. Learning as a biological phenomenon. In Halliday & Slater (1983), 178–212.

Squire LR, editor, 1992. *Encyclopedia of Learning and Memory*. New York: Macmillan.

Learning in neural networks:

Arbib MA, 2003. *The Handbook of Brain Theory and Neural Networks*. Cambridge, MA: MIT Press, 2 edition.

Haykin S, 1999. *Neural Networks: A Comprehensive Foundation*. New York: Macmillan, 2 edition.

Neural network models of ontogeny:

Elman JL, Bates EE, Johnson MH, Karmiloff-Smith A, Parisi D, Plunkett K, 1996. *Rethinking Innateness: A Connectionist Perspective on Development*. Cambridge, MA: MIT Press.

Chapter Five

Evolution

In this chapter we tackle the evolution of behavior and nervous systems. All multicellular animals, with a few exceptions, have a nervous system. Phylogenetic studies suggest a single origin, which consequently must date back more than 500 million years. The nervous system offers an all-purpose machinery for receiving information and producing responses, allowing the organism to be much more flexible toward its environment. With a nervous system, organisms can move, perform all sorts of activities, communicate, generate internal responses such as heartbeat, gut movements and hormone secretions, and so on. Other solutions for doing some of these things exist, e.g., in bacteria, but never on the same scale. Clearly, the nervous system is a success story in evolutionary terms.

We have shown in earlier chapters that neural network models naturally share many properties with nervous systems, essentially by virtue of a shared network structure. This suggests that neural networks can also help us to understand the evolution of behavior, as we attempt to show below. This chapter contains two parts that use neural networks in different ways. We consider first the general question of how behavior systems evolve. We show that neural networks fulfill many requirements for an evolving behavior system. We also discuss the material basis for such evolution and biases and constraints arising from the concrete implementation of behavior systems in terms of neural networks. In the second part of this chapter we show how neural network can be used to study specific evolutionary questions such as the evolution of communication and social behavior.

Some of the issues discussed herein are poorly understood, and we explore ideas rather than provide established facts. Recall, for instance, that the relationship between genes and the nervous system is still not well understood (Chapter 4), nor do we know to what extent neural networks are helpful for understanding this and other issues of behavior evolution.

5.1 THE EVOLUTION OF BEHAVIOR SYSTEMS

How does a behavior systems evolve, and what features does evolution favor? This question ranges from the basic operations and abilities of animals to the adaptation of specific behavior patterns in a given species. Behavior evolution arises from the evolution of the nervous system, its receptors and sense organs and effectors such as the skeletomuscular apparatus. It is the flexibility and limitations of these systems that, together with selection pressures, determine the evolution of behavior.

Before discussing what insights neural network models can provide, we ask what general characteristics we expect to find in behavior systems produced by evolution. We start from the concept of a "look-up table" introduced in Chapter 1. The look-up table is an idealized behavior mechanism that stores the response to each possible stimulus as the value of a table cell (see Figure 1.3 on page 9). This is equivalent to the approach of game theory and optimization theory, which traditionally seek an independent, best response to each possible situation (Binmore 1987; Maynard Smith 1982; McNeill 1982; Section 1.3.2). We now ask: can such an idealized mechanism evolve? The answer is clearly no.

A first reason is that an impossibly large memory would be required to store the table. Consider, for instance, a sense organ with just 1000 receptors (an insect eye can have up to 200.000, a human eye has about 100 million), with each receptor capable of being in only one of two states, "on" or "off." Then there are 2^{1000} possible patterns of stimulation, each needing its own entry in the table. In constrast, estimates of the storage capacity of the human brain vary between 2^{33} and 2^{52} bits (Crevier 1993), and the storage capacity of the genome is even more limited. The problem becomes even more acute if, as it is often the case, sequences of stimuli rather than single stimuli must be stored to take functional decisions.

A second difficulty concerns how entries in a look-up table could be optimized. Evolution proceeds by trying out new solutions and retaining those innovations that are improvements. However, there are limits to the number of parameters that evolution can optimize because the combinations to be tried out are too many even with a modest number of parameters. The same argument applies to responses established through learning. A third problem is that reactions to novel stimuli are completely undetermined because corresponding table entries have not been under any selection. This is important because genuinely novel stimuli may appear during evolution and also because of the extreme variability of sensory data that arise from existing stimuli (Figure 3.3).

From these arguments we conclude that only behavior mechanisms with a much coarser memory structure (fewer parameters) than a look-up table can evolve. For instance, we could have a small table where only the responses to some stimuli would be stored. Responses to other stimuli could then arise from stored responses according to some rule. In such a system, each memory parameter would affect responding to more than one stimulus, and responding to each stimulus would be governed by more than one parameter. Reactions to novel stimuli would be determined by the current content of the memory and its structure, i.e., the particular rules that produce responses based on stimuli and the content of memory.

Besides making a system possible, limiting the number of memory parameters has other advantages. Selection favors those mechanisms that show some "intelligence" vis-à-vis reality. For instance, stimuli that are similar to one another often share some causal relationship with events in the outside world. An "intelligent" mechanism can detect and use such regularities, generalizing knowledge about familiar situations to novel ones. A look-up table is incapable of generalization because the response to each stimulus is independent of responses to other stimuli, however similar they might be. On the other hand, a behavior mechanism whose output is a continuous function of the input automatically displays generalization. That is, if S' is a slight modification of S, then output to S' will not be too different from output to S. Lastly, limiting the number of parameters can also simplify the ontogeny of the system because fewer parameters have to be set.

To summarize, we can view the problem of behavioral evolution as analogous to a fitting problem. When we fit a function to a set of points, we adjust function parameters to achieve the best fit according to some criterion. How well we can fit the points depends on what function we use. A polynomial of first degree ($a_0 + a_1x$) can describe perfectly only straight lines (fitting a_0 and a_1), whereas a second-degree polynomial ($a_0 + a_1x + a_2x^2$) can describe both straight lines (with $a_2 = 0$) and quadratic shapes. In the case of evolution, the "points" to be fit are requirements on an animal's behavior, and we can see selection as a fitting process that attempts to set genetically determined variables of the nervous system so that such requirements are satisfied as closely as possible.

5.2 REQUIREMENTS FOR EVOLVING BEHAVIOR MECHANISMS

Clearly, what behavior can evolve depends partly on the raw material evolution works with, i.e., on properties of nervous systems. Below we expand the remarks we made earlier by listing some properties of behavior systems that we think have been and still are crucial for the evolution of behavior. By *crucial*, we mean that these properties have allowed nervous systems to evolve from the simplest forms to the human brain, with a flexibility that makes it possible to fit any species' needs. We also discuss to what extent these properties are found in neural network models and nervous systems.

5.2.1 Flexibility and efficiency

Perhaps the most important requirement of a general behavior mechanism is that it should be able to implement almost any behavior map, including maps that require regulatory states (Chapter 3). Only a system with such a flexibility would allow behavior to be tailored continuously by natural selection to fit prevailing circumstances. We have already seen in previous chapters that neural network models are very flexible. Memory parameters (weights) can be fitted to practically any needs, and network architecture can grow, shrink and alter patterns of connections. It has been proven mathematically that neural networks have excellent fitting properties; almost any map can be described with these models (Funahashi 1989; Haykin 1999; Hornik et al. 1989; McCulloch & Pitts 1943). The potential to build appropriate regulatory states allows the network to detect temporal patterns in sensory data and to organize behavior in time (Sections 3.5, 3.7.5 and 3.8).

Flexibility also derives from the potential to add, modify or remove receptors, muscles and other effectors. We have already noted that network models, as well as nervous systems, can deal equally well with any kind of sensory information, provided that receptors exist that can transform stimulation into neural activity (Section 3.6; Wang et al. 1999). If a new receptor type becomes available during evolution, for instance, a neural network capable of learning can immediately exploit any new information that the receptor may provide, thus exploiting the po-

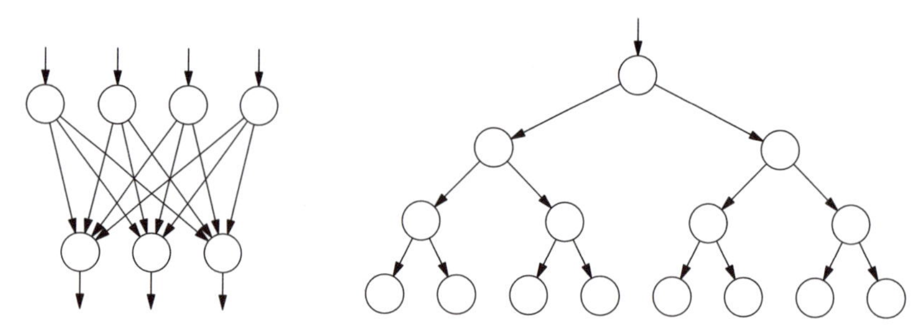

Figure 5.1 Speed and economy of a feedforward neural network and a decision tree (used in, e.g., many computer algorithms). A feedforward neural network makes a decision in a limited number of steps (equal the number of weight matrices), whereas searching in a tree may require many steps. Neural networks also tend to be more economical because the same nodes and connections participate in the processing of many stimuli.

tential of improved sensory abilities. This would be much more difficult if large changes were required in the machinery downstream from the sense organs.

In addition to flexibility, evolution also requires efficiency, meaning that the behavior mechanism should be fast while being economical with resources such as receptors, nerve cells and connections. This is crucial to both simple and complex nervous systems and seems to follow naturally from a network architecture. Note that it is not obvious that a biological system should be fast. Physiological process are slow or very slow in comparison with, say, electronic processes. Despite this, behavior systems are usually fast and can outperform computers in many tasks. The reason for this is that neural networks operate in parallel, i.e., each node or neuron operates simultaneously with many others. Standard digital computers, by comparison, typically operate in a sequence, performing only a small number of operations at any given time. Consider, for instance, a three-layer feedforward network. The response to an input is determined in just two steps: the activity of input nodes causes the activity of inner nodes, which in turn causes output activities. Each step comprises many operations, depending on the number of nodes and existing connections. Increasing the number of nodes or connections does not increase response time, but rather it increases the number of operations that are performed in each step. The result is still calculated in two steps. Compare this with, for instance, a decision tree, which operates sequentially and requires more time as the complexity of the problem (number of tree nodes to be examined) increases (Figure 5.1). In summary, parallel processing allows animals to react quickly to a range of problems and to recognize most objects at a glance. This is obviously an important benefit in situations that require rapid reactions. Similarly, economy with respect to resources is achieved by limiting the number of parts (connections and cells), using the same parts for many stimulus-response relationships (Figure 5.1). Despite this flexibility, neural networks and, it seems, behavior systems also have some limitations, addressed in Section 5.3.3.

5.2.2 Genetic control

For genetic evolution to occur, behavior mechanisms must be under genetic control to some extent. Note that this does not exclude that learning and other experiences influence behavior. The genes may, for instance, specify a learning mechanism rather than exactly determine behavior. Behavior systems have the potential of being guided by both genetic information and experiences, and it appears that this control can shift rather quickly during evolution. For instance, some song birds develop their songs without any specific experiences, whereas others need a particular temporal arrangement of learning events (Catchpole & Slater 1995; Slater 2003). That neural networks can learn is well established, but only a handful of studies have explored the issue of genetic control of network features. Some of this work is discussed in Chapter 4 in the context of the ontogeny of behavior. Below we discuss some further studies investigating genetic control in the context of behavior evolution. This area, however, is not yet well developed.

5.2.3 Already simple systems are powerful

Even animals with small nervous systems, such as nematode worms, are capable of functional behavior. Neural network models clearly demonstrate, as seen in previous chapters, that already very simple networks have fundamental abilities such as generalization and decision making. If this were not the case, it would be difficult to explain how simple nervous systems (and thus also complex ones) could have evolved. It seems impossible that a complex system such as the fruit fly brain or crow brain could emerge without a more or less continuous evolution starting from much simpler system.

5.2.4 Possibility of smooth incremental changes

In the course of evolution, some simple nervous systems have evolved toward greater complexity. As just recalled, it is easier for evolution to proceed if there is no or only small phenotypic gaps owing to developmental constraints (Futuyma 1998). Neural network models suggest that nervous systems can change smoothly, for instance, by progressively adding more nodes or slowly changing connection patterns. For example, a genetic mutation causing the production of slightly more neurons could lead to slightly greater memory capacity. The evolution of predispositions such as those examined in Section 4.2.3 (see especially Figure 4.8 on page 140) could occur by gradual changes in the likelihood that particular groups of nodes are connected with each other. Moreover, the possibility for incremental changes makes it easier to evolve efficient behavior from crude behavior. The learning abilities of neural networks may also contribute to this by bringing the behavior of certain architectures closer to the optimal solution, thus amplifying the effects of genetic variation and exposing it to stronger selection—the Baldwin effect (Maynard Smith 1987).

It is also much easier to imagine a biological system that has some fault tolerance to evolve. A system without such tolerance would be rendered useless after any disturbance, genetic or environmental. As we have seen in Section 3.9, neural network models have fault tolerance similar to nervous systems both with respect to damage per se and in their ability to compensate for damage by learning. Contrast this with, for example, a computer program that can be rendered completely ineffective after only minor faults such as a 1-bit change or an unexpected input. In summary, it appears that networks can often evolve by smooth changes, although constraints on this ability cannot be ruled out.

5.2.5 Handling of novel situations

We have already stressed the importance of responding properly to novel stimuli. Without such ability, an animal could not survive. This problem includes both stimuli that are similar to experienced ones and those that are very different. For instance, stimuli that are similar to one another often share some causal relationship with events in the outside world. Most neural network models generalize spontaneously, trying in novel situations responses that were appropriate in similar familiar situations. As analyzed in Section 3.3, these abilities are present even in very simple networks and stem from the distributed character of memory and input processing. In addition, animals have evolved general methods to cope with novelty, including exploratory behavior and avoidance behavior (Russell 1973). However, exactly how novel situations are detected and handled by nervous systems is not satisfactorily understood. We are not aware of any neural network model addressing this issue from a functional or evolutionary perspective (but see Section 4.4.1).

5.3 THE MATERIAL BASIS OF BEHAVIORAL EVOLUTION

5.3.1 Evolution of nervous systems

To study the evolution of behavior, it seems important to know what modification of behavior mechanisms mutations or new combinations of genes can produce. The question is complex because genes do not code for behavior directly. Rather, they control the ontogeny of the nervous system (including receptors and effectors) that, in turn, controls behavior. Comparative anatomical, physiological, biochemical and molecular studies of the nervous systems have revealed a number of changes whereby nervous systems evolve (Butler & Hodos 1996; Ghysen 2003; Nieuwenhuys et al. 1998; Nishikawa 1997). In brief, these are as follows. On the basis of a highly conserved molecular machinery of embryonic development, the nervous system evolves through changes in both the number and structure of its parts, e.g., by duplications and subsequent differentiation. The pattern of connectivity between parts also evolves, e.g., when axons from one part "invade" another part or by loss of connections. The parts themselves evolve in a variety of ways. Most obvious are changes in size and number of neurons. Finer details of neural circuitry also evolve. Finally, nerve cells evolve through changes in their physiology, for instance, by changes in the properties of ion channels or synapses. This may result in the creation of new cell types, including new receptors. It has been suggested that many of these changes can be influenced by relatively simple mutational events that control the development of the nervous system (Chapter 4; Butler & Hodos 1996; Katz & Harris-Warrick 1999; Nishikawa 1997, 2002).

The nervous system is evolutionary conservative compared with other traits, including behavior, and major reorganizations are rare (Katz & Harris-Warrick 1999; Nishikawa 1997). Anatomical differences are often small, even within large taxonomic groups, and often this holds also for circuit organization (Katz & Harris-Warrick 1999). For instance, the basic neuronal organization for locomotion has been conserved in distantly related chordates (Kiehn et al. 1997). Thus evolution of behavior seems not to be based primarily on anatomical changes. Recent advances suggest instead that the material basis for evolution of species-specific behavior often consists of small changes to neuronal circuitry, potentially resulting in dramatic changes to behavior (Katz & Harris-Warrick 1999; Nishikawa 1997, 2002). These changes may involve details of how cells are connected but also properties of neurons. Of considerable interest here is the discovery of "multifunctional" networks (Katz & Harris-Warrick 1999), circuits that are able to produce different behavior patterns by small changes to some parameters, such as the number of ion channels. On the other end, anatomically different neural circuits seem to have evolved several times to solve similar problems, perhaps because there are only a few ways of solving a given problem within a network structure (Nishikawa 2002).

The latter point suggests that referring to structural changes is not sufficient: we must also understand how these relate to behavioral function (on which selection operates). Such a link has been evidenced in a number of studies. For example, food-storing birds have larger hippocampal volumes than nonstoring species (Basil et al. 1996; Krebs et al. 1989), and in song birds, there is a correlation between size of song repertoire and volume of the HVC song control center (DeVoogd et al. 1993). The evolved complexity of motor pathways for singing in song birds is another example (Catchpole & Slater 1995; Nottebohm et al. 1976; Slater 2003). But it is evolutionary changes on a finer scale that should interest us most, e.g., changes in information processing, decision making or learning. Recent studies, for instance, of electric fish and prey capture in frogs (Nishikawa 1997) have demonstrated that adaptation of novel behavior, absent in ancestors, can be caused by small changes to neural circuitry, e.g., patterns of connections.

Similar points to those we just made about the nervous system can also be made about receptors (sense organs) and effectors such as the skeletomuscular apparatus. For instance, pigments in photoreceptors can shift their sensitivity to other wavelengths as a consequence of single genetic mutations and thus give rise to improved color vision (Goldsmith 1990; Katz & Harris-Warrick 1999; Wang et al. 1999).

5.3.2 Evolution of neural networks

To study the evolution of behavior using neural network models, we need to make assumptions on how genes code for properties of a neural network. Network properties that may be put under genetic control are, for instance (Nolfi 2003; Nolfi & Floreano 2000; Rolls & Stringer 2000):

1. Transfer functions and other properties of nodes (neurons).

2. Properties of the weights (synapses), e.g., their initial value and rules for weight change.
3. Network architecture, including size and number of layers, as well as patterns of connectivity between layers or node groups; an alternative and more biologically realistic assumption would be that genes code for developmental rules that build network architecture.
4. Properties of receptors, their numbers and organization into sense organs.
5. Properties of the skeletomuscular apparatus and other effectors.

Notwithstanding all these possibilities, most studies of neural network evolution have been concerned so far only with model genotypes that determine individual weights within a fixed architecture (Enquist & Arak 1993, 1998; Huber & Lenz 1996; Kamo et al. 1998; Nolfi & Floreano 2000; Phelps & Ryan 1998). From the point view of the evolution of behavior mechanisms, this is perhaps the less interesting case because it limits evolution to developing behavior maps within a given structure rather than developing receptors, architecture, learning abilities and so forth. That the genes can code for individual network weights is often even unrealistic (see also Section 5.1). Organisms seem to have fewer than 100,000 genes, which must build the whole organism and not just the brain (Maynard Smith & Szathmáry 1995). The number of synapses in the human brain, with 10^{10}–10^{11} neurons and perhaps an average of 10,000 synapses each, is of the order of 10^{14}–10^{15} (Rolls & Stringer 2000). Clearly, the human genome cannot specify individual synaptic weights (Nolfi 2003; Rolls & Stringer 2000). This applies to many other organisms. In small neural networks, however, genes may determine individual synapses to a large extent.

The literature offers a number of interesting studies of behavioral evolution based on neural networks. In Section 3.8 we described a study by Beer and Gallagher (1992) showing that a walking pattern actually seen in insects evolved spontaneously in a population of simulated insects whose locomotion was controlled by a neural network (see Figure 3.39 on page 123). Ijspeert and coworkers provide further valuable examples (Ijspeert 2001, 2003; Ijspeert et al. 1998; Ijspeert & Kodjabachian 1999). One example concerns changes in locomotion associated with the transition from water to land habitats. Starting from a neural network model of swimming in the lamprey (page 122), Ijspeert (2001) added front and hind legs, each pair controlled by a central pattern generator similar to the one present in the swimming model. This arrangement is found in salamanders, which can both swim (in the same style as lampreys) and walk. Moreover, salamanders are believed to be the living vertebrates closest to the ancestral animals that came on land. The network model was subject to simulated evolution with fitness given by efficient swimming and walking, after which it could swim and walk at different speeds (depending on the amount of input to the central pattern generators) and turn (if opposite sides of the body were given different input). The evolved network would swim, with limbs held close to the body as salamanders do, if the limb pattern generators were off and would walk with a salamander gait if all three pattern generators were switched on.

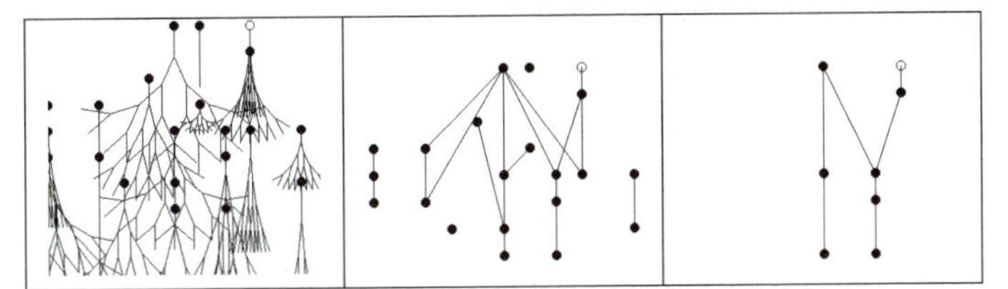

Figure 5.2 Development, in three stages, of a feedforward network resulting from simulated evolution (Nolfi et al. 1994). Left: Each node grows an "axon" following genetic instructions, with amount of growth also influenced by node input. Center: The nodes whose axons touch are connected together. Right: The final network is obtained by eliminating isolated and nonfunctional nodes. The bottom nodes represent receptors, the top ones effectors. Reprinted from Nolfi and Floreano (2002). *Evolutionary Robotics*. The MIT Press.

A few studies have considered genetic coding of network parameters other than the weights. For instance, Rolls and Stringer (2000) simulate evolution of network parameters in three kinds of networks that can learn on their own by modifying the weights: *pattern-association networks* (e.g., classical conditioning), *autoassociative networks* (e.g., short-term memory) and *competitive networks* (e.g., perceptual learning). A number of genes were assumed to control specific network properties, such as rules for the formation of connections, operation of neurons and learning. All three networks were subject to evolution with mutations and recombination. In each generation, the fitness of networks was calculated based on their ability to learn particular tasks. The learning ability of all three types of networks improved significantly during evolution.

Nolfi et al. (1994), studying the evolution of navigation abilities in robots controlled by neural networks, encoded network architecture as follows (see also Nolfi 2003; Nolfi & Floreano 2000). The genotype explicitly coded the number of nodes and their position in a two-dimensional space, but connections resulted from growth of axons based on genetic instructions (Figure 5.2). The latter specified how long an axon would grow before branching and the direction of growth after branching. Stimulation received by a node could also affect growth. After a while, growth was stopped, and nodes whose axons touched were connected (there was no distinction between axons and dendrites in this model). Finally, the network was obtained by eliminating isolated nodes or node groups.

In conclusion, understanding how evolution changes behavior is difficult because behavior is a complex product of gene action, developmental processes and nervous system operation. Our short discussion of brain evolution is not intended to provide an overview of this field but to show that neural network models can describe many of the structural and functional changes that are believed to be important in the evolution of nervous systems. Using neural network models, therefore, can lead to insight along several lines. First, we can gain intuitive insight by considering the evolution of behavior in simple model systems. Unfortunately, only a few studies have explored these issues, although the results suggest that neural network models are helpful in understanding the evolution of behavior and the nervous system.

Second, neural network models can be used in simulations of evolution, as shown later in this chapter.

5.3.3 The role of behavior mechanisms in evolution

Earlier we saw that the basic design of behavior mechanisms provides a lot of potential for the evolution of behavior, but it is also likely that such mechanisms have biases and constraints that can influence evolution (Binmore 1987; Dukas 1998a; Maynard Smith 1978; Maynard Smith et al. 1985; Stephens & Krebs 1986). There are at least two ways in which this may happen. First, mechanisms able to implement efficient behavior may not come cheap: if a mechanism is too costly to build and maintain, it will not be favored in evolution (Kamil 1998). Second, and the subject of this section, limitations and biases in information processing may make some behavior impossible or difficult to implement, or they may bias evolution in certain directions rather than others (Dawkins 1982; Dukas 1998b; Maynard Smith et al. 1985; Stephens & Krebs 1986). There are several reasons to expect such constraints and biases. A behavior system is subject to many demands. Selection is likely to favor a system that handles many situations efficiently, perhaps at the expense of perfection in particular cases. In addition, some shortcomings may derive from basic features of design that in practice cannot change (e.g., that the nervous system consists of interconnected cells). This argument may also apply to particular levels of organization. For instance, a reptile brain may lack some innovations that prevent reptiles from developing a behavior that is possible in, say, birds. Unfortunately, the importance of any of these factors in evolution of behavior is not fully understood. We now consider some features of neural networks that may contribute to understand how biases and constraints of nervous systems may influence evolution.

The most general biases and constraints are likely to emerge from the basic principles of network operation. Perhaps the most important principle is about how external stimuli, other motivational factors and memory interact to produce responses. In our models, the response of a node is an increasing function of the weighted sum of inputs:

$$z = f\left(\sum_i W_i x_i\right) \tag{5.1}$$

We referred to this weighted sum earlier as the *overlap* between the weight vector $\mathbf{W}$ and the input vector $\mathbf{x}$ (Section 3.3). This calculation is a cornerstone of neural network models. The simplest network (receptors connected to one output node) involves only one overlap calculation, whereas in more complex networks (more outputs or layers, recurrence) responses are generated from many such calculations. Important features of the overlap are that it maps an arbitrary number of inputs onto one output and that it generalizes spontaneously (similar inputs yield similar outputs; Section 3.3). Moreover, by adjusting the weights, even a single overlap calculation can implement many input-output maps. Indeed, we saw in previous chapters that simple networks based on this computational principle can reproduce a host of behavioral phenomena. The latter often make apparent what

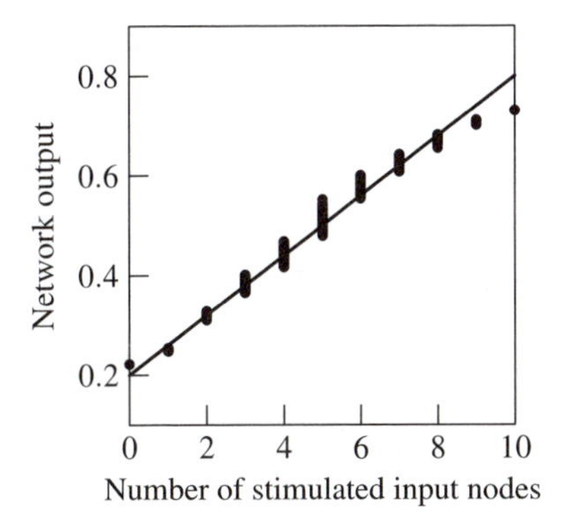

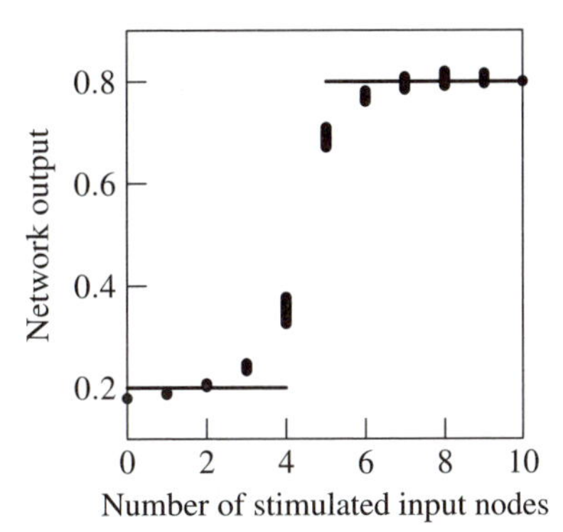

Figure 5.3 A three-layer network can implement a smooth response profile (left) more easily than an abrupt change in response (right). The network had 10 input nodes, 5 hidden nodes and 1 output node. Input nodes could be stimulated (value of 1) or not (value of 0), and the network had to respond as a function of the number of active nodes ("cardinality"), as indicated by the solid line in each panel. Fitting consisted of 5000 iterations of simulated evolution (Sections 2.3.3 and 2.5). All patterns of zeros and ones were used, but the fitting algorithm compensated for the different frequency of different cardinalities (e.g., there is only one pattern with 10 zeros or 10 ones but 252 patterns with 5 zeros and 5 ones). Parameters of the node transfer function (equation 2.7 on page 36), including its slope, were also subject to evolution, to avoid any limitations in fitting ability that might have derived from an unsuitable choice of such parameters.

biases and constraints derive from overlaps (Section 3.3). These can be understood by considering how overlaps vary when changing the input. In general, we can increase (decrease) the overlap of **x** with **W** by increasing any input x_i corresponding to a positive (negative) weight. Moreover, the effect of changing x_i depends on the magnitude (absolute value) of the weight. Thus, although the overlap is an idealization, it captures the key fact that a presynaptic neuron's ability to influence the postsynaptic one depends on the strength and sign (excitatory or inhibitory) of the synapse. We continue with a number of particular examples of biases and constraints that can be traced back to properties of overlaps.

5.3.3.1 Sharp thresholds versus smooth changes

Abrupt changes in responding (thresholds) are harder to implement than smooth changes. This is true in all networks in which the output is a continuous function of the input; e.g., multilayer feedforward networks. Figure 5.3 shows that 5000 generations of simulated evolution can accurately fit the network to a straight line, whereas the fit to a step function is considerably worse. This constraint may be important because optimization theory often predicts step functions as solutions or even more difficult shapes such as stepwise functions. Note that a threshold could be built in the output node (Section 2.2.1.2), but such a network could not fit smooth functions at all and thus would have much narrower behavioral abilities.

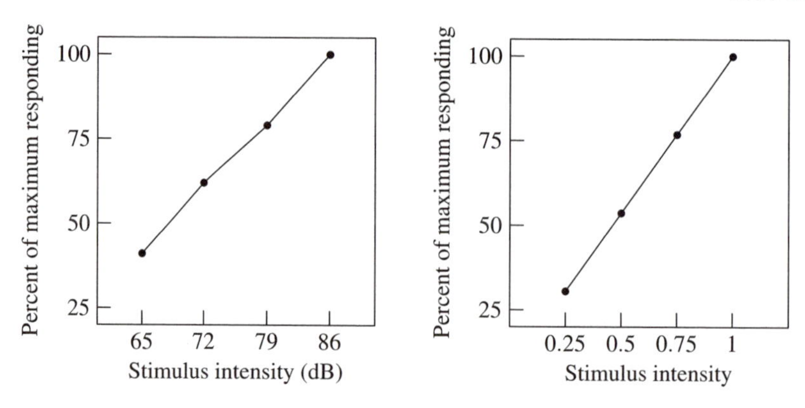

Figure 5.4 Reactions to equally reinforced stimuli of different intensities in animals and networks. Left: Rabbits were trained to react to tones of different intensities. Despite the fact that all stimuli were equally reinforced, the more intense ones elicited more responding (data from Scavio & Gormezano 1974). Right: Simulation of the same experiment with a two-layer feedforward neural network. The stimuli employed are like the ones in Figure 3.8 (right). Stimulus intensity is measured as maximum node activity. The network was trained with reinforcement learning (Sections 2.3.4 and 4.2.2).

5.3.3.2 Effects of stimulus intensity

The ability to form input-output maps along intensity dimensions is limited in neural networks. We saw in Section 3.3 that neural network models reproduce realistically the reactions of animals to stimuli varying in intensity. For instance, both animals and neural network models usually react more strongly to more intense stimuli. Even after extensive training in which stimuli of different intensities are equally rewarded, animals continue to respond more to the more intense stimuli (Figure 5.4). Thus it is difficult for animals to treat alike stimuli of different intensity. We also saw in Section 3.3 that a two-layer network, which entails a single overlap calculation, can only form a monotonic response gradient over intensity dimensions (i.e., either increasing or decreasing with intensity). A three-layer network can, in principle, overcome this limitation because it can fit with arbitrary precision any smooth function, given a large enough hidden layer. In practice, however, it seems very difficult to eliminate the bias (Figure 3.12). Moreover, there are practical limits to the number of nodes that can be used, and using a very large network is also likely to degrade generalization along other dimensions (see Chapter 6 of Haykin 1999).

5.3.3.3 Discrimination

The discrimination abilities of animals (or any mechanism) are, of course, limited, and discriminating between similar stimuli is more difficult than discriminating between different ones (Figure 5.5; Hanson 1959; Raben 1949). The evolution of some remarkable phenomena, such as camouflage and mimicry, seems to depend

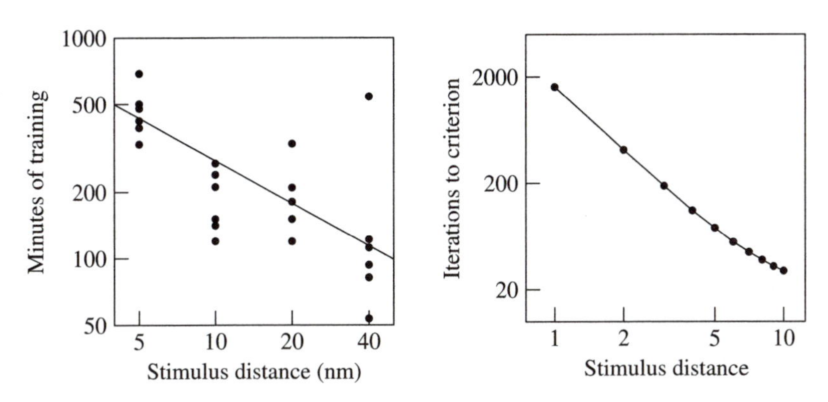

Figure 5.5 Effect of stimulus similarity on discrimination learning in animals and networks. Left: Data from Hanson (1959), who trained four groups of six pigeons to discriminate between a monochromatic light of 550 nm in wavelength and, respectively, lights of 555, 560, 570 and 590 nm. The line is the best fit of the power law $y = ax^b$, which appears straight on a log-log scale (fitted parameters are $a = 1213$, $b = -0.64$). Right: Results of training a two-layer network network on the discrimination in Figure 3.11 on page 84, with different locations of the $\mathbf{s}^-$. The number of δ rule iterations required to reach a total squared error of less than 0.01 decreases approximately as a power law as the distance between the training stimuli increase (log-log scale is used).

critically on limitations to discrimination (Dittrich et al. 1993). Neural network models behave very realistically when faced with discrimination problems. Smaller differences between stimuli to be discriminated result in longer training time or inability to discriminate (Figure 5.5). In addition, adding resources such as receptors or inner nodes often makes discriminations easier. An advantage of neural networks is that such constraints emerge naturally from the network mechanisms rather than being simply assumed or implemented as a discrimination cost (as commonly done in game theoretical models of evolution).

5.3.3.4 Biases among solutions

If there are several alternative ways to gain a selective advantage by modifying behavior, particular features of the behavior mechanism may bias evolution to adopt one alternative rather than another. In neural network models, this corresponds to the ease whereby different solutions are found in simulated evolution (Section 2.3.3). One example is offered by Calabretta et al. (2004). These authors used a three-layer feedforward network to discriminate among a few stimuli that could be projected at different positions on a model retina. The network had three output nodes. Two were used to control a two-segment arm in response to the stimulus. The other one was assumed to command a movement of the eye to the left or to the right, causing a shift of the stimulus image on the retina. The network was trained by simulated evolution to respond with a given movement of the arm to each stimulus. The evolved network solved the task by first moving the eye so that the stimulus was centered on the retina and then reacting in a way appropriate to the stimulus. However, the network could have recognized the stimuli in all positions, as was proved in a simulation in which the eye could not be moved. Nevertheless, this second solution never evolved if the eye could be moved because it is more difficult to reach starting from an unorganized network (i.e., a finer organization of the weights is required).

Another important class of problems where properties of behavior systems may bias evolution in particular directions is the evolution of signal form. The same information is potentially conveyed by a countless number of different signals, and how animals process sensory information seems a major determinant of form. Neural networks research on this problem is summarized in Section 5.4.5.

5.3.3.5 Conclusions

Earlier we provided a few examples of constraints and biases in neural networks. Other examples can be found in, for instance, Minsky and Papert (1969). Biases and constraints may also occur in decision making, receptors and effectors, as well as in the map from genes to networks. However, the existence of constraints should not change our basic conclusions that networks are very flexible and can implement most input-output maps and learning programs. Most constraints and biases are temporary, although some constraints (e.g., those related to stimulus intensity) may be very difficult to overcome.

If real nervous systems have biases and constraints similar to neural network models, we expect a number of consequences for the evolution of behavior. A possible example of puzzling behavior that may be caused by constraints of the type just discussed is reactions to young. Evolutionary theory predicts that parents should only care for their own young or, in some circumstances, the young of kin (Trivers 1972). However, adoption of unrelated young is not uncommon among animals (Redondo et al. 1995; Riedman 1982). Perhaps this is a difficult problem for evolution to solve because it requires opposite responses to very similar stimuli. A more aggressive and less caring behavior toward young in general may be easily associated with decreased care of own young. The alternative is a more caring behavior toward young in general. A telling example concerns gulls that frequently prey on chicks of conspecifics. Sometimes, when they bring to the nest a chick for food, they fail to kill it and end up adopting the chick instead (Graves & Whiten 1980; Parsons 1971). We cannot prove that this behavior has evolved because of constraints in discrimination, but the possibility seems worth considering.

5.4 EXPLORING EVOLUTION WITH NEURAL NETWORK MODELS

We now change focus. So far we have discussed the evolution of behavior and nervous systems in general terms. In the rest of this chapter we will discuss how neural network models may be used to study the evolution of specific behavior programs, usually called *strategies*. Many arguments, as well as the "game of presence" example in Section 5.4.4, are based on Enquist et al. (2002).

There are two common theoretical approaches for exploring the evolution of behavior (Grafen 1991). The first defines strategies in terms of behavioral variables (phenotypic variables) and assumes that evolution acts directly on such variables. For instance, when studying the evolution of fighting behavior, we could define strategies such as "fight if the value of the contested resource exceeds a threshold value; otherwise, give up" (Maynard Smith 1982). Such strategies are usually assumed to transmit directly from parent to offspring and sometimes to change slightly by mutation. The second way of modeling behavioral evolution is to assume a genetic machinery that codes for behavioral strategies. For instance, alternative strategies may be assumed to correspond to different alleles at a given genetic locus, or combinations of alleles at several loci (e.g., Futuyma 1998). Both approaches have provided substantial insight into the evolution of behavior, but applications of these methods generally do not take into account the ontogeny and operation of behavior systems. Neural network models make it possible to investigate how the latter may influence evolution and also to study how ontogenetic programs and behavior systems change during evolution.

Today only a few evolutionary studies have used neural network models as behavior mechanisms. Most of these studies exploit networks as more biologically realistic mechanisms for recognition discrimination (Chapter 3). The problems considered include evolutionary conflicts and manipulation (Arak & Enquist 1995; Krakauer & Johnstone 1995; Wachtmeister & Enquist 2000), the evolution of mimicry (Holmgren & Enquist 1999), the evolution of signal form and responses to signals (Bullock & Cliff 1997; Enquist & Arak 1994, 1998; Enquist & Leimar 1993; Johnstone 1994; Kamo et al. 2002, 1998), reconstruction of signal evolution (Phelps 2001; Phelps & Ryan 1998), host-plant selection (Holmgren & Getz 2000), coy female behavior (Wachtmeister & Enquist 2000), evolution of camouflage (Merilaita 2003), foraging behavior (Holmgren & Getz 2000) and learning abilities (Rolls & Stringer 2000). The evolution of behavior based on neural networks is also studied in robotics, particularity within so-called evolutionary robotics. For more information, readers are referred to Nolfi and Floreano (2000).

Below we first give a brief introduction to the modeling of adaptive behavior. Then we discuss how neural networks can be used to explore the evolution of behavior. The focus is on issues that are difficult to investigate with more standard methods. This is followed by a description of simulation techniques. The chapter ends with three examples describing how neural networks have been used to investigate particular evolutionary problem. The first example considers manipulation in interactions between senders and receivers, the second one is the evolution of signal form, and the third is a case study of evolution of male calling and female responses in a frog species.

5.4.1 Standard approaches

Optimization theory and game theory are today the most common frameworks for investigating the evolution of adaptive behavior (Houston & McNamara 1999; Krebs & Davies 1977, 1987, 1991; Maynard Smith 1982; Maynard Smith & Price 1973; Parker & Maynard Smith 1990). They consider strategies described in terms of behavioral variables or, more generally, phenotypic variables. The purpose of these methods is to find and describe behavior that maximizes individual reproduction or some other measure of fitness. The evolutionary justification for such a focus on fitness maximization is the assumption that natural selection will favor changes in behavioral strategies until an end point is reached in which individuals cannot improve their reproduction by using an alternative achievable strategy (evolutionary equilibrium). Finding fitness-maximizing behavior can be done formally by means of an extensive mathematical toolbox (Grafen 1999; Houston & McNamara 1999; Maynard Smith 1982; Stephens & Krebs 1986), but more intuitive verbal arguments about which strategy is best or produces most offspring are also common.

When an individual's fitness does not depend on the behavior or strategies of others one speaks of *frequency-independent selection*, whereas when an individual's fitness depend on what other individuals do the term *frequency-dependent selection* is used (Futuyma 1998). Frequency-independent selection is well illustrated by the evolution of flying, e.g., in the albatross. The dominating selection pressures for the shape of the body and wings, as well as the control of flying, are the aerodynamics conditions at sea. It seems safe to assume that these selection pressures have remained the same during the whole evolution of albatrosses' adaptation to life at sea. Evolution under frequency-independent selection is usually studied with optimization theory (Houston & McNamara 1999), under the assumption that evolution has been going on long enough to discover an optimal or near-optimal solution.

Evolution under frequency-dependent selection is more complex because selection pressures change as a consequence of evolution. Fighting behavior is an example of behavior that evolves mainly under frequency-dependent selection. Consider, for instance, weaponry such as antlers and horns. Their success depends crucially on what weapons and defense tactics are used by other individuals. The invention of new weapons or defenses often triggers the evolution of counteradaptations that reduce their efficiency. Possible evolutionary end points under frequency-dependent selection are often studied through the application of *evolutionary game theory* (Eshel 1996; Houston & McNamara 1999; Maynard Smith 1982). The aim of such analysis is to find strategies (or sometimes a set of strategies, one for each type of player, such as, e.g., males and females) that will be maintained over time once they have become established in the population (a stable evolutionary equilibrium). Specifically, evolutionary game theory seeks *evolutionarily stable strategies* (ESSs) as solutions to evolutionary problems. According to the definition, a strategy s^* is an ESS if, when used by all or almost all individuals in the population, it cannot be displaced by any alternative strategy $s \neq s^*$ used by a few individuals (Houston & McNamara 1999; Maynard Smith 1982). This requires first that s^* reproduces at least as well as any s. This is usually referred to as a *Nash equilibrium*. To ensure stability, the ESS definition also requires that either s^* has higher fitness than any alternative s, or if an alternative s exists that initially has the same fitness as s^*, its fitness should become lower than the fitness of s^* if the number of individuals using s increases.

If evolutionary processes are indeed at a stable equilibrium, predictions about behavior under frequency-dependent selection can be obtained simply by asking what is the end point of the evolutionary process (Houston & McNamara 1999; Maynard Smith 1982). However, for evolution to occur at all, strategies must exist out of equilibrium at least some of the time (Maynard Smith 1978). If these periods are brief, we may be justified in ignoring them and analyze behavior purely with game theory. However, there is also the possibility that some evolutionary processes never come to rest, with strategies in an almost continuous state of flux (Dawkins & Krebs 1979; Enquist et al. 2002). For instance, the coevolution between a prey species and a predator species may take the form of an *arms race*, i.e., a never-ending series of adaptations and counteradaptations (Abrams 2000; Dawkins 1982; Dawkins & Krebs 1979; Dieckmann et al. 1995; Slatkin & Maynard Smith 1979).

Recent developments in evolutionary modeling have increased and modified our understanding of evolutionary stability (Dieckmann et al. 1995; Eshel 1983; Houston & McNamara 1999; Leimar, in press). We now know that ESS analysis may not find all equilibria and that some potential ESSs are not reached during evolution by natural selection. In addition, it has also been realized that many evolutionary problems lack stable equilibria (Houston & McNamara 1999). This understanding has emerged from a more dynamical approach to evolutionary modeling, which also is necessary if we want to understand what phenotypes emerge when evolution proceeds out of equilibrium. Studying the dynamics of evolution requires tracking the small steps of evolution that occur in each generation. The last decade has seen considerable and exciting development of such dynamical theories, often called *adaptive dynamics* (Dieckmann & Law 1996; Geritz et al. 1998).

5.4.2 How neural networks may contribute

Neural networks may help us to study the evolution of behavior beyond the standard approaches just summarized. A characteristic of both the traditional ESS approach and more dynamical approaches is that they are based either on behavioral variables (decision rules) or on genetic variables assumed to determine behavior directly (the relation between these approaches is increasingly understood; see, e.g., Hammerstein 1996; Leimar in press). Variables related to nervous system operation and ontogeny are rarely considered. The argument for using neural network models in evolutionary modeling is, of course, their biological realism, including features such as flexibility and efficiency but also biases and constraints. Another argument for using neural networks is their generality. The basic design of behavior mechanisms is essentially the same in most animals, and species-specific behavior evolves within a framework that can be covered by neural network models. Here follows a discussion of how neural networks may be used to explore the evolution of behavior.

Let us first consider frequency-independent selection. Under such assumption, the evolutionary dynamics are relatively simple, and we expect improvements to accumulate over evolutionary time. Finding the optimal solution under the assumption that the behavior map is a network means that optimization is performed with respect to network variables such as connection weights and variables specifying the network architecture rather than with respect to behavior-level variables or decisions rules. There are several ways to do this. One option is to use optimization theory to find neural networks that maximize fitness. We are not aware of any studies using this approach to study evolution. Alternatively, one could try to find solutions by an evolutionary simulation (see below) or some other dynamic approach. The two options may not always yield the same result. The former can locate optimal network strategies (given we can solve the optimization problem, of course), whereas the latter is sensitive to, for instance, the initial choice of network architecture, basins of attraction and local fitness maxima (Futuyma 1998). Similar options are available for frequency-dependent selection. We could use standard evolutionary game theory and search for ESS networks, but no one has done this as far as we know. The other option is simulations of evolution, which in addition to evolutionarily stable behavior may also describe behavior that emerges in scenarios with continuous change and without any stability (see below). So what insight could be gained from such analyses? We have seen in this and earlier chapters that basic features of neural network operation entail a number of biases and constraints that potentially can influence evolution. Similar biases are likely to exist in animals, and this is a cornerstone of studies of the evolution of behavior based on neural networks. Let us consider a few examples.

Neural networks show clearly that some input-output maps are more difficult (or impossible) to realize. Even when a map potentially can be implemented, biases in the behavior mechanism can still influence the direction of evolution. One example is offered by Holmgren and Getz (2000), who studied the evolution of host plant preferences in insects. The animals' recognition system was modeled with a three-layer neural network, and an evolutionary simulation was set up in which the networks had to select, by means of sensory cues, those host plants that offered better nutrition to larvae. Although many host plants had the same quality, the networks often evolved to recognize only one kind of plant. This study suggests that specialist strategies are more common because they are more easily formed by nervous systems and therefore may evolve more easily than generalist strategies.

Neural network models generalize spontaneously, responding with similar behavior to similar stimuli (Section 3.3). Generalization may also be important in many evolutionary problems; e.g., problems involving high-dimensional stimuli such as visual stimuli that are received via many receptors. Biases in behavior systems may have additional consequences for the evolution of social interactions. In this case, natural selection favors strategies that are sensitive to the behavior (or physiology or biochemistry) of others, allowing the individual to respond based on more information. However, if there are evolutionary conflicts of interests between individuals, selection would also favor social strategies that can exploit others, e.g., by taking advantage of biases in behavior mechanisms. This has consequences for the evolution of both manipulative strategies and signal form (see examples below).

Neural networks may also provide insight into how learning may affect the evolution of behavior. To analyze such problems, one may use neural networks as more realistic models of learning and memory (Chapter 4). An example is Kamo's study of how learning influences the evolution of signal form (Kamo et al. 2002). His results suggests that signals will be most exaggerated and costly when receivers learn to recognize them and when each receiver meets the signal only a few times.

If learning occasions are many or recognition of the signal is genetically inherited, less costly and less exaggerated signals evolve.

Another use of neural networks could be to study directly the evolution of the behavior system, letting features of the network evolve more freely as discussed in Section 5.3.2. This could provide insight about the evolution of both behavior and the functional structure of the nervous system. For instance, one could try to understand how mechanisms (i.e., networks) for sensory preprocessing, learning, regulatory states and decision making evolve. Simulation results could be compared with the growing literature on the neurobiology of behavior.

5.4.3 Simulation techniques

How can we study behavioral evolution with neural network models? Sections 2.3.3 and 2.5 provide a brief technical introduction and some examples. Most simulations are built around the following core steps:

1. Create an initial population of networks.
2. Calculate the fitness of all networks.
3. Based on fitness, decide which networks "reproduce" and how much.
4. Mutate some networks.
5. If a stop criterion is fulfilled, end; otherwise, go to step 2.

(We refer only to "networks," but other traits such as morphology can be included.) In addition to this general procedure, a number of details have to be decided before a real simulation can be carried out. One is how mutants should be produced; i.e., what variables can mutate, and how exactly do they mutate? Other relevant details are the size of populations and the mode of reproduction. The simplest choice is a population size of two and asexual reproduction. The general procedure then takes the following form:

1. Select an initial network (network 1).
2. Calculate the fitness of network 1.
3. Make a copy of network 1, called network 2.
4. Mutate network 2.
5. Calculate the fitness of network 2.
6. Compare the fitness of networks 1 and 2.
7. Retain the fitter network and label it network 1.
8. If a stop criterion is fulfilled, end; otherwise, go to step 2.

This simple algorithm can be developed along several lines to suit specific problems. One obvious extension is to use a larger network population. Using a population introduces a natural range of variability that is particularly important when simulating evolutionary games (frequency-dependent selection). In a population, mutations with higher fitness enter slowly, and adaptation to the social environment may require to confront a range of different phenotypes. It is also possible to consider sexual reproduction by defining how offspring networks are created from parent networks. The latter is not trivial because, as discussed in Section 5.3, we do not fully understand how genes code for nervous systems. Lastly, we may include individual and social learning in evolutionary simulations. This requires us to calculate fitness based on how the network behaves during a simulated life in which it can learn (e.g., Kamo et al. 2002).

5.4.4 Example 1: Coevolution between signal senders and receivers

This example describes how sender-receiver coevolution can be investigated with neural network models and also illustrates some of the general points discussed earlier regarding the interplay between behavior mechanisms and evolution.

An important feature of social evolution is evolutionary conflicts of interest. Such a conflict exists when the expected reproduction or fitness of one individual could increase at the expense of other individuals (Enquist et al. 2002; Parker 1979). For instance, a potential prey would benefit from the predator not attacking, a contestant from the opponent giving up or a busy female from males stopping courting her. These evolutionary conflicts may appear at first as just of theoretical interest because it is difficult to imagine how an individual can influence what others decide to do. However, many examples of successful manipulation of others' actions are known in nature. Larvae of the blue butterfly are a spectacular example, being able to persuade ants to carry them to the ant nest, where they feed on ant larvae and pupae (Hölldobler & Wilson 1990).

How can manipulation evolve and persist? To get some insight into this problem, we will consider a simple case of evolutionary conflict called *the game of presence* (Enquist et al. 2002). We first present this game and then analyze it with both standard game theory and simulated evolution of neural networks. There are two players, an actor and a reactor. The actor is either present ($\nu = 1$) or absent ($\nu = 0$) but has no choice of actions. The reactor, based on whether or not the actor is present, decides on an effort x ($x > 0$). This structure is depicted in Figure 5.6. In the presence of the actor, the reactor's payoff is first increasing and then decreasing with x. The payoff to the actor of the reactor's effort is ever-increasing with x. In the absence of the actor, providing an effort $x > 0$ returns a negative payoff to the reactor. A simple formalization of this game is as follows:

$$F_A(x) = x$$

$$F_R(x) = \begin{cases} x - x^2 & \nu = 1 \\ -x^2 & \nu = 0 \end{cases} \tag{5.2}$$

where F_A is the payoff (fitness) to the actor, and F_R is the payoff to the reactor. Note the conflict between the two players concerning the amount of effort x to be made by the reactor. Similar conflicts are widespread. For example, in sexual conflicts (Arnqvist & Rowe 2002; Gavrilets et al. 2001), a female (reactor) must trade off current investment in reproduction with future investment, whereas a male (actor) wants the female to invest maximally in the current reproduction (Chapman & Partridge 1996; Rice 1996). We will now present three theoretical analyses of this game given different assumptions about what strategies are available to actors and reactors.

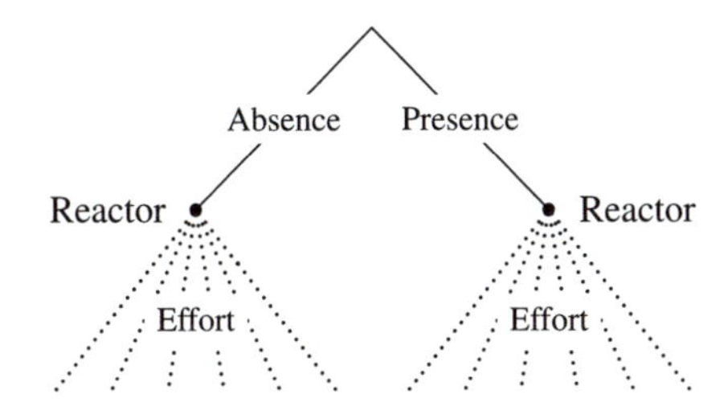

Figure 5.6 A tree description of the game of presence (see text). The dotted lines indicate that the effort can take many possible values. The black dot indicates tree nodes where a player can decide among different courses of action.

5.4.4.1 The standard ESS solution

Let us first solve the game as it stands. We have so far (deliberately) assumed that the reactor is the only player that has a choice of actions. The reactor strategy consists of two levels of effort, one to use when the actor is present and one when the actor is absent. We solve the game simply by finding the effort that maximizes the reactor's payoff in these two cases. If the actor is present, the optimal effort is $x = 0.5$. Actor and reactor then receive in return 0.5 and 0.25, respectively. If the actor is absent, the optimal effort for the reactor is $x = 0$. This solution is an equilibrium (no better response strategy exists), and it is also evolutionary stable because if the reactor's strategy drifts away from the optimum, selection will return it to the optimal levels. We see that manipulation does not exist at the ESS. Indeed, manipulation can never be a part of an ESS because at equilibrium the reactor's response must be the best one. Were it not, the reactor would have an incentive to change its strategy, and the system would not be at equilibrium.

5.4.4.2 Many actor appearances and a look-up-table reactor

The game just analyzed may seem so trivial as to not warrant a formal analysis. However, it nicely illustrates some of the problems of considering only equilibrium solutions (ESSs) and of simplifying too much. In reality, information about the presence or absence of the actor is not provided automatically to the reactor but must be inferred from sensory input. The presence or absence of the actor is detected by a mechanism that reacts to the stimulation or physical energy (e.g., light or sound) that reaches the reactor. The actor must be recognized even when viewed from different distances and angles, in different light conditions and against different backgrounds. We considered such problems at length in Chapter 3.

An additional complication for the reactor is that the actor may take on a variety of appearances (Figure 5.7). Because such appearances give rise to different patterns of stimulation, they may elicit quite different reactions from the reactor. In game theory, it is common to restrict, consciously or unconsciously, the number of strategies considered, e.g., by limiting the number of appearances to the minimum needed to convey the relevant information or by considering variation along a single dimension only. Furthermore, strategies available to players are prescribed

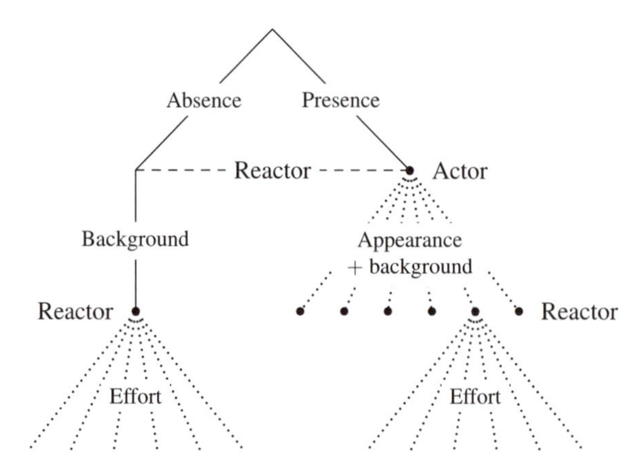

Figure 5.7 A more realistic formulation of the game of presence, where the actor can choose among different appearances (dotted lines stemming from the actor node; see text). The dashed line indicates that the reactor is uncertain about what branch of the game is taken at that point (i.e., whether the actor is present or absent). See Figure 5.6 for additional explanations.

in advance, whereas evolution is an unfolding process in which new strategies become possible as a result of evolution itself. There are several good reasons for making these simplifying assumptions. One is to make mathematical analysis possible. Another motive is to limit the number of solutions to the game or to eliminate those considered implausible. However, as we shall see, these assumptions cannot always be justified from an evolutionary point of view.

To see the implications of removing these assumption, let us now reanalyze the game allowing the actor to have many appearances and the reactor's strategy to be a look-up table storing a response to each actor appearance (Sections 1.3.2 and 5.1). An actor can change appearance by random mutation. Responses are also subject to change by mutation so that over time reactors may increase or decrease the effort made in response to different appearances. Simulations of this game (see Enquist et al. 2002 for details) are highly dynamic, with rapid changes in actor appearances accompanied by reactor counteradaptations (Figure 5.8). In most simulations, appearances evolved that enabled actors to manipulate reactors into producing an effort $x > 0.5$. A stable equilibrium was found only when the number of appearances available to actors was restricted to one, corresponding to our first analysis of the game. As the number of possible appearances was allowed to increase, reactors were less able to resist manipulation and thus produced a greater effort. Note also that actors typically do not use the most effective appearance and thus also behave suboptimally.

Why do not reactors evolve effective countermeasures against manipulation? This is not possible because reactor responses toward appearances not currently used by actors are neutral with respect to selection and subject to mutations and subsequent drift. Thus, at any given time, there is a set of appearances not currently in use that would stimulate reactors to produce efforts larger than the opti-

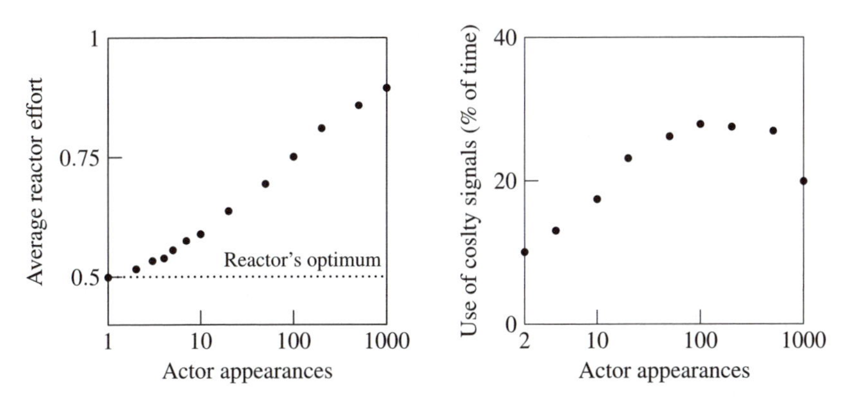

Figure 5.8 Simulations of the realistic formulation of the game of presence, using actor and reactor populations of 1000 individuals each. Left: The more strategies (appearances) are available to actors, the greater effort they are able to elicit from reactors. Each simulation lasted 15,000 generations (we show the average of 10 simulations, with effort calculated on the last 10,000 generations only). Simulations started at a Nash equilibrium; all actors used a particular appearance, and all reactors respond with the optimal effort $x = 0.5$ to any actor appearance. Such equilibrium is stable only with one possible appearance, corresponding to the simplistic version of the game in Figure 5.6. With two or more appearances there is no stability. The effort fluctuates, and signals replace each other (the rate of change of the most common signal varied between 0.004 to 0.014 per generation). Right: The use of costly signals from a new set of 10 simulations in which half the appearances were costly to use. In all other respects the simulations were the same as above. Costly signals were used to a considerable extent. The degree of manipulation was somewhat lower compared with the previous simulations.

mum. When such appearances arise by chance mutation in the population of actors, they are favored by selection and spread rapidly; this has been referred to as *sensory exploitation* (Basolo 1990; Ryan 1990). Selection acting on the population of reactors then tends to desensitize their mechanisms to these new appearances, restoring the effort back toward $x = 0.5$. However, at any time appearances may arise in the actor population that can bypass the reactors' defense mechanisms. In conclusion, our second analysis has a very different outcome from the simplistic model. Stable equilibria do not exist, and the game resembles an evolutionary race in which at least some players are manipulated and thus behave irrationally.

5.4.4.3 Many actor appearances and neural network reactors

In our final analysis, we replace the look-up table of the reactor with a three-layer feedforward neural network. This has some interesting consequences. The strategy space of reactors is not an array of independent efforts but a (smaller) number of network weights. This is a more realistic model of actual reactors (animals); e.g., it generalizes spontaneously to novel experiences from familiar ones (Section 3.3). The network is able to sense (e.g., see) actor appearances that vary continuously in many dimensions. This also means that we can make the actor strategy more diverse. Namely, an actor appearance consists of a pattern of activations of the network input nodes (we still keep the simplification that each actor has only one possible appearance). Figure 5.9 illustrates the outcome of simulated evolution of this game. Parts (a) and (b) both show how the average rector effort changes over time. Shading shows when the reactor is manipulated (effort $x > 0.5$). First, actor and reactor were allowed to coevolve for 150,000 iterations of the simple evolutionary algorithm on page 191 (both diagrams show this part). From this point, the simulation was split in two. In one (a) only the actors were allowed to evolve. In the other (b) only the reactor evolved. After another 150,000 iterations coevolution was again introduced. As shown some degree of manipulation is maintained during coevolution. When the reactor's evolution was stopped the actor became more efficient at manipulation. When instead evolution of the actor was stopped the reactor evolved an optimal effort.

Compared with the previous analysis, instability is still present, with reactors spending more effort than optimal. A major difference between the network and the look-up table is that the network is more "intelligent" towards novel stimuli. At least two factors contribute to this: the reluctance of the network to produce the same response to very different stimuli and the existence of a common output mechanism from which it is difficult to elicit responses that deviate substantially from the reactor's optimum. That is, after some evolutionary time network weights will be set so that it is not possible to elicit an effort much greater than 0.5.

5.4.4.4 Complexity

Based on the comparison between the look-up table and the smarter network, we may ask whether the behavior mechanism can be made so "intelligent" as to effectively abolish manipulation and restore the predictions of game theory. Will a more refined mechanics be better at avoiding manipulation? Perhaps, but the evolutionary consequences of intelligence may be mixed. For instance, to solve a problem with greater accuracy a more complex mechanism is usually needed. For example, eyes with more receptors allow for more accurate discrimination. How does such increased complexity affect reactions to novel stimuli? Are humans less susceptible to manipulation than, say, a snail? One may think that if a more complex mechanism evolves, it will inevitably be better than the mechanism from which it evolved. For example, one may add a bias-correcting device to the original mechanism. But such a new mechanism also will have its own biases. In general, while added complexity may function well in solving currently existing problems, it will not necessarily work better when faced with new problems. In fact, the opposite may be true because as complexity increases, the number of different stimuli the organism can perceive is also likely to increase, as well as the number of physiological and biochemical processes that are open to interference (Figure 5.10). An increase in complexity may also allow qualitatively different stimulation to be effective. For example, the ability of a mechanism to recognize patterns in time may favour signals that are variable in time over constant signals (Wachtmeister & Enquist 2000).

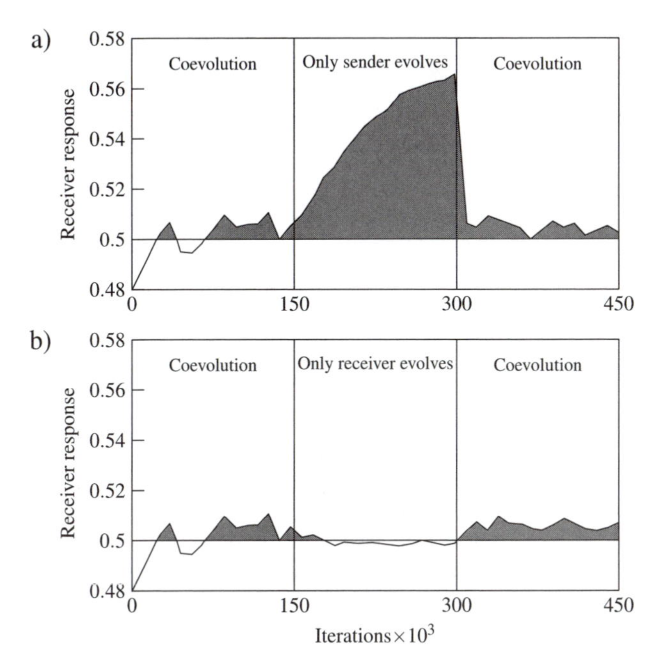

Figure 5.9 Simulation of the game of presence using a three-layer neural network as the reactor. Parts (a) and (b) both show how the average reactor effort changes over time. The reactor's optimal effort is 0.5. Shading shows when the reactor is manipulated (effort > 0.5). First, the actor and reactor were allowed to coevolve for 150,000 iterations of the simple algorithm on page 191 (both diagrams show this part). From this point the simulation was split in two. In one (a) only the actor were allowed to evolve. In the other (b) only the reactor evolved. After another 150,000 iterations, coevolution was again introduced. As shown, some degree of manipulation is maintained during coevolution. When the reactor's evolution was stopped, the actor became more efficient at manipulation. When instead evolution of the actor was stopped, the reactor evolved an optimal effort.

5.4.4.5 Accumulation of costly appearances

In another study we investigated the relationship between evolutionary conflict and cost (Arak & Enquist 1995). The signal consisted of a grid with 5×5 cells. Each cell could take on values between 0 (black) and 1 (white), with 0.5 being optimal for the signal bearer's survival. Thus the cost of the signal in terms of decreased survival was a function of the deviation of cell values from 0.5. Two types of conflicts were considered, one between actors, or signal senders, competing for the favorable response of signal receivers or reactors. For instance, a male wants to be chosen over other males, and a potential prey discarded. The other conflict is between actors and reactors about the reactors' response, as seen above. Reactors were modeled as three-layer feedforward networks. For both kinds of conflict, simulations showed that actors were willing to pay larger signal costs as conflict increased (Figure 5.11).

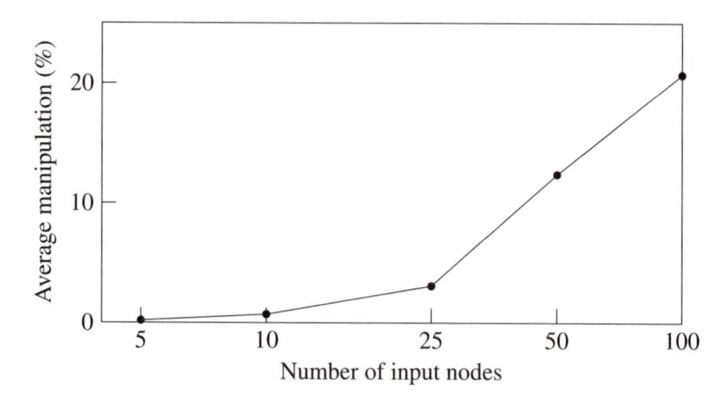

Figure 5.10 Effects of mechanism complexity on manipulation. The diagram demonstrates how varying the size of the reactor's artificial retina influences the opportunity for manipulation in an evolutionary process. On average, the reactor is more susceptible to manipulation when the number of receptors increases. Percent manipulation is defined as $100(r-o)/o$, where r and o are actual and optimal response, respectively. If $r < o$, manipulation is defined as zero.

5.4.5 Example 2: Evolution of signal form

Studies of signal evolution using neural network models have dealt with both visual (Enquist & Arak 1998) and acoustic (Phelps & Ryan 1998, 2000) signals. Some of these results are described below. Neural networks allow us to study aspects of the evolution of signal form that are difficult to investigate with game theory or other approaches such as statistical decision making theory. Applications of standard game theory are often useful for predicting whether signaling will evolve or not and the cost of signals. However, this theory has been much less successful in predicting the form signals take. Why, for instance, are repeated signal elements and long tails so common? Application of neural network models opens up new possibilities for research. With neural network models we can study how sense organs and neural processing influence the evolution of signal form. In addition, we can take into account the many difficult problems of recognition that potentially can influence signal design (Section 3.3 and 3.4). Neural networks can also handle complex signals (many dimensions of variation), whereas game theory is most easily applied to one-dimensional or discrete signals.

5.4.5.1 Signals in nature

In nature we find an enormous variety of signal forms. Among closely related species, signals are usually dramatically more variable in form than nonsignaling traits. For example, plumage coloration in birds often varies to a much greater extent within genera than do other morphological features such as beak, wing or body shape—a most useful fact to bird watchers. In some taxa there is also considerable intraspecific variation in signal form. All this diversity agrees with the preceding remarks, suggesting that instability is common in signal evolution. An interesting

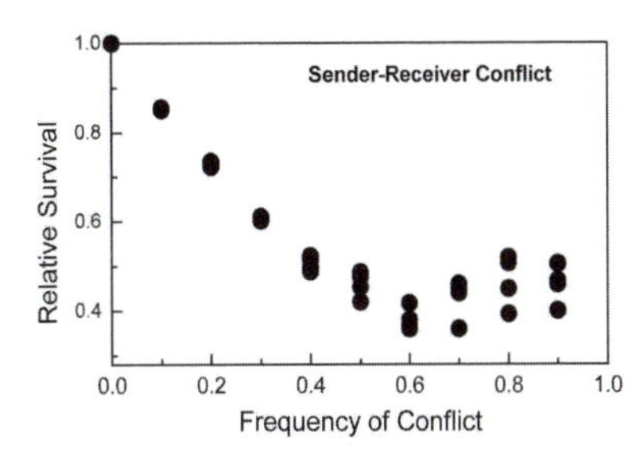

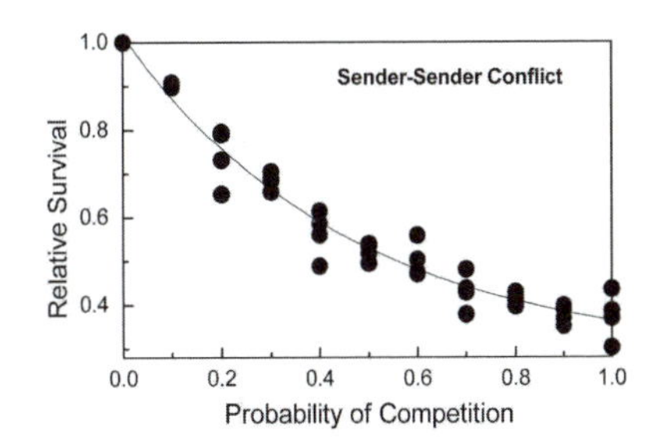

Figure 5.11 The relation between cost of signals (measured as relative survival) and evolutionary conflict. In the left panel conflict is measured as the frequency of sender-receiver conflict; in the right panel conflict is measured as the probability of sender-sender conflict. For each level of conflict, results are shown for four different runs of the simulation; some points overlap. From Arak & Enquist (1995).

observation is that signal form is relatively independent of context and systematic group. The amazingly diverse body coloration of coral reef fish exemplifies the problem of relating function and form. These signals are known to have a number of functions depending on species, sex and age (Andersson 1994; Eibl-Eibesfeldt 1975). Yet, except for certain cases of deceit, their particular appearance seems to provide few clues to their function.

Natural signals, however, also show a number common themes (Figure 5.12). Here we will just summarize the most important ones. For a more detailed discussion about forms of real signals, see Brown (1975), Cott (1940), Eibl-Eibesfeldt (1975), Enquist and Arak (1998) and Smith (1977). A signal is typically made up of a number of distinct elements. It is common to find that some of these elements are *exaggerated* and/or *contrasting* with other signal elements. In addition, elements may be *repeated* several or many times and usually are arranged to form various *symmetries*. Signals also relate to other signals and the background. It is possible for signals to be conspicuous in isolation, but signals are conspicuous also because they are different from the background. Another general feature of signals is their *distinctiveness*. The most extreme situation, noticed by Darwin (1872), is that signals designed to provoke opposite reactions are often opposite in form ("principle of antithesis"; Figure 5.12d).

5.4.5.2 Visual signals

In a number of studies we have investigated the evolution of visual signal form in computer simulations using a neural network as the receiver's recognition and decision mechanism. The basic simulation algoritm is the same as described earlier. In most simulations, a sender's fitness benefited from a particular response from the receiver, whereas the receiver's fitness depended on its ability to discriminate between the appearance of the sender and other stimuli (including background). Sender-sender conflict was introduced by allowing several senders to compete for the favorable responses. Sender-receiver conflict was introduced by not making it profitable for the receiver to respond all the time to the sender's signal. A number of studies based on similar simulations have shown that most common themes of visual signals can be generated, such as exaggerations of size and coloration, contrasts, symmetries and repetitions (reviewed in Enquist & Arak 1998). Two main factors seem to be responsible for all these results: the recognition/discrimination problem at hand and biases in the network (Section 3.3). Figure 5.13 shows one such study that resulted in a blocky or symmetrical arrangement of contrasting elements. Symmetry is an interesting and widespread feature of signals (Figure 5.12). Preferences for symmetries can emerge in neural networks from experiences of asymmetrical patterns if these need to be recognized from several or many directions (Enquist & Arak 1994; Johnstone 1994). If one trains a network on two asymmetrical patterns that are mirror images of each other, certain symmetrical patterns, e.g., a combination of the two, may elicit a stronger response than the training patterns (Enquist & Arak 1994; Enquist & Johnstone 1997; Ghirlanda & Enquist 1999; Johnstone 1994). Such preferences have also been demonstrated in real animals (Jansson et al. 2002). It is important to note that spectacular signals, according to this explanation, are not necessarily for accurate transfer of information. For instance, a network (and real animals?) can be trained easily to recognize asymmetrical patterns from all orientations.

5.4.6 Example 3: Calling and female responses in the Túngara frog

Our final example is a case study of the evolution of male calling and female responses in the Túngara frog, *Physalaemus pustulosus*, conducted by Steven Phelps and Michael Ryan (Phelps 2001; Phelps & Ryan 1998; Ryan 2001). Readers are referred to these papers and references therein for details. The first step was to design model receptors and a network structure that could respond to a call. The "artificial ear" of the network was an input layer of 15 units, each sensitive to a range of sound frequencies. Since calling has an important time dimension, a feedforward network would not do. Instead, a recurrent network with state or context nodes was chosen (Elman 1990; see also Section 3.5.2). The network was trained with an evolutionary algorithm to discriminate between real calls by males and noise in the same frequency range. The evolved networks were then tested on 34 novel stimuli, including calls of related species and reconstructed calls of ancestor species. The reactions of real Túngara frog females to the same calls were also recorded, and the networks predicted well the females' reactions ($r = 0.80$; Figure 5.14).

This result is one of the first successful applications of a recurrent network to the study of animal behavior, but the authors went further. They wanted to use neural networks to reconstruct the evolutionary history of male calling and female responses in these frogs and to study how selection pressures from female recognition systems in the past have shaped calling in males. In frogs, male calling and female responses to the calls of their own species develop spontaneously without any functional experiences. Phelps and Ryan had shown earlier that the response of female frogs to heterospecific and (estimated) ancestral calls can be predicted not only by how similar these test calls are to the call of the female's species but also

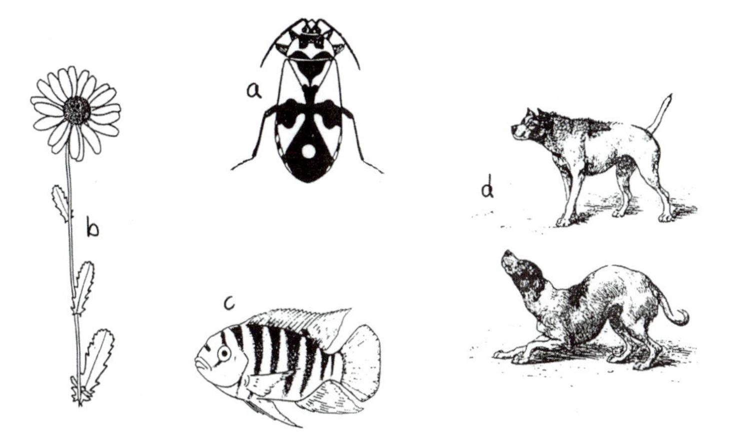

Figure 5.12 Some qualities of signal form. (a) Contrasting colors in a seed bug, *Lygaeus equestris*. (b) Radial symmetry in the Oxe-eye daisy, *Chrysanthemum leucanthemum*. (c) Spatial repetition in body coloration in the cichlid fish, *Tilapia mariae*. (d) Antithesis of threat and submissive displays in the dog. Reprinted from Enquist and Arak (1998). In *Cognitive ecology*, ed. by R. Dukas. University of Chicago Press.

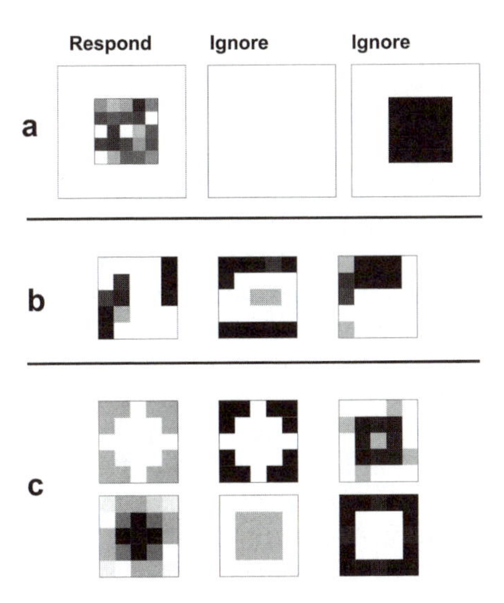

Figure 5.13 Effect of the mode of stimulus presentation on simulated signal evolution. When the receiver discriminates against both a pure black stimulus and a white background (a), signals evolve contrasting patterns of black and white (b). If signals are also presented in different orientations, the result is contrasting symmetric patterns (c). The signal starts out with random gray colors (a).

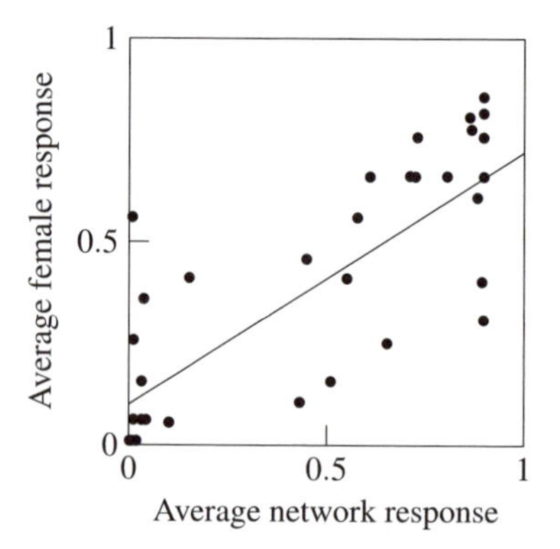

Figure 5.14 Correlation ($r = 0.80$) between responses of Túngara frog females (*P. pustolosus*) and responses of neural networks evolved to recognize the Túngara male call from noise (Section 3.5.2; Phelps 2001; Phelps & Ryan 1998). Horizontal axis: average response of 20 evolved neural networks. Vertical axis: proportion of females approaching a speaker playing the call when it was paired with another speaker playing noise. A set of 35 calls, both natural and synthetic, was used. The best-fitting line ($y = 0.62x + 0.1$) is shown.

by the phylogenetic distance of this species from the other species being tested. They suggested that the way the female frogs (or their brains) recognize male calls is influenced by the way the brain solved similar problems in its past history. This is a sensible prediction. Since responses to male calls are determined genetically, the brain may preserve traces of old responses that are no longer expressed because there are no calls to match them, at least for some evolutionary time. Over longer time, responses to old calls are likely to decay as the nervous system adapts to new situations and reuses nervous material for other purposes. The hypothesis is very difficult to test, but neural network models offer an opportunity. Phelps and Ryan had networks evolve through different histories, i.e., different sequences of calls, all ending with the call used by conspecific males today. In practice, this meant training the network to recognize the first call in the sequence and then switching to a second call while preserving network weights evolved thus far. The procedure continued until the network was trained to recognize the current male call. After training on different histories, network responses were compared with the responses of real females. The network trained with the reconstructed evolutionary history of the species responded most similarly to today's females. Although the nets with other histories evolved to recognize the present call just as well, they seem to do it in a different way that does not predict what real females do.

5.5 CONCLUSIONS

The study of behavioral evolution, together with the study of ontogeny, is the least developed area of neural network research. However, from the preceding it should be obvious that neural networks can provide insight into the evolution of behavior regarding both general issues and particular problems. Neural networks seem to fulfill the requirements we can have on models of evolving behavior mechanisms. These models also provide understanding of some of the mysteries of behavior evolution. For instance, since even very simple networks are very capable, we can grasp better how efficient behavior can evolve gradually from sloppy behavior. Neural networks also provide insight into the links among genes, nervous systems and behavior.

A number of studies have exploited neural networks to investigate the evolution of particular aspects of behavior. Even using simple networks, intriguing results have been obtained about, for instance, the occurrence of manipulation and the evolution of signal form. It is clear that many of such results cannot be obtained within either standard evolutionary game theory or using idealized behavior mechanisms such as the look-up table. Hopefully, these results are consequences of neural networks capturing essential properties of real behavior systems. This view is encouraged by the fact that neural network models are good models of many aspects of behavior, in particular, aspects crucial for evolution, such as reactions to novel stimuli (Section 3.3).

Compared with traditional approaches, neural network models also have drawbacks. One is that they are relatively complex models. This means that, although formal mathematics could, in principle, be applied to study evolution of behavior with neural network models, in practice, we need to resort to computer simulation. A simulation is always about a particular choice of parameters and does not have the expressiveness of an analytical solution. Thus, to establish general results, extensive studies are needed that simulate evolution under many different conditions. It is also important to be careful in asking questions and choosing models to ensure that results are possible to understand and have some generality. To this aim, it is advisable to start with simple models.

We want to end with a note of caution. This chapter is by necessity somewhat speculative because its subject is not well developed. We have tried to suggest potentials for neural networks in evolutionary studies of behavior, but we do not know to what extent neural networks will be a viable research strategy and for what kind of problems insights may be gained.

CHAPTER SUMMARY

- Neural network models have most of the properties needed to study evolving behavior systems, such as flexibility and efficiency.
- In evolutionary simulations, all properties of a neural network can be put under genetic control, such as network architecture, properties of network nodes (including receptors and effectors) and rules of weight change.
- Neural network models can thus show the links between genetic evolution, nervous system evolution and the evolution of behavior. Today this research is by necessity somewhat speculative because we do not understand fully how genes code for nervous systems.
- Neural network models can complement more standard approaches (e.g., ESS

analysis) when properties of behavior systems cannot be ignored.
- Evolutionary simulations with neural networks have brought insight into specific problems in evolutionary biology, such as the role of stimulus recognition in signal evolution.

FURTHER READING

Evolution of nervous system:

Butler AB, Hodos WC, 1996. *Comparative Vertebrate Neuroanatomy: Evolution and Adaptation*. New York: Wiley-Liss.

Ghysen A, 2003. The origin and evolution of the nervous system. *International Journal of Developmental Biology* 47, 555–562.

Katz PS, Harris-Warrick RM, 1999. The evolution of neuronal circuits underlying species-specific behavior. *Current Opinion in Neurobiology* 9, 628–633.

Nishikawa KC, 1997. The emergence of novel functions during brain evolution. *Bioscience* 47, 341–354.

Osorio DJ, Bacon PB, Whitington PM, 1997. The evolution of arthropod nervous systems. *American Scientist* 85, 244–252.

Studies of neural network evolution:

Nolfi S, 2003. Evolution and learning in neural networks. In Arbib (2003), 415–418.

Rolls ET, Stringer SM, 2000. On the design of neural networks in the brain by genetic evolution. *Progress in Neurobiology* 61, 557–579.

Studies of behavioral evolution using artificial neural networks:

Enquist M, Arak A, 1998. Neural representation and the evolution of signal form. In Dukas (1998a), 21–87

Enquist M, Arak A, Ghirlanda S, Wachtmeister CA, 2002. Spectacular phenomena and limits to rationality in genetic and cultural evolution. *Transactions of the Royal Society* B357, 1585–1594

Kamo M, Kubo T, Iwasa Y, 1998. Neural network for female mate preference, trained by a genetic algorithm. *Philosophical Transactions of the Royal Society* B353, 399–406

Phelps SM, Ryan MJ, 1998. Neural networks predict response biases of female Túngara frogs. *Proceedings of the Royal Society of London* B265, 279–285

Chapter Six

Conclusions

In this book we have explored the potential of neural networks to model behavior. The results suggest that neural networks can model behavior systems in all their parts, reproducing a wide range of behavioral phenomena. Figure 6.1 provides a summarizing sketch of a simple but complete model of an animal's behavior mechanism. The model includes reception and further processing of sensory input, central mechanisms of decision making and the control of muscles and other effectors. The figure shows only a few recurrent connections, but in principle connections between any nodes in the system can be included, e.g., so that sensory information affects motor control directly. Additional flexibility can be gained by including nodes with different dynamics, although we have only touched upon this possibility. By putting together these components in suitable ways, virtually any behavior map can be implemented. In addition, the model may include networks able to convey feedback signals from the environment ("reinforcement") that can be used to modify the behavior map (i.e., learn) through weight changes (Chapter 4).

In this last chapter we make an assessment of neural networks as models of behavior and compare them with other modeling approaches, in particular, cognitive models. We recall that our aim is to understand behavior, and we use neural networks because we think that they can provide better models of behavior. Of course, behavior can be used to study what occurs inside animals, but this is not our principal aim. See Staddon (2001) for a penetrating discussion of the different aims of behavior studies, with a focus on behaviorism.

6.1 ARE NEURAL NETWORKS GOOD MODELS OF BEHAVIOR?

In Chapter 1 we listed some requirements that a general model of behavior should satisfy (page 7). Below we reproduce these requirements (in italics), and we evaluate to what extent neural network models fulfill them. We recall that we consider animals in general, ranging from the simplest to the most complex. Our interest is in how behavior systems in general operate and how they can be elaborated during evolution. Humans have very special abilities, sometimes shared with our closest relatives (Tomasello 1999), that are outside our scope. Currently we do not have good neural network models of, say, human symbolic reasoning. This has sometimes generated strong criticism of the connectionist approach (Fodor & Pylyshyn 1988; Pinker & Prince 1988). Such critiques, however, are less relevant when we consider animals in general. Starting from humans would mean starting from a very special and complex case of behavior rather than from typical or simple behavior.

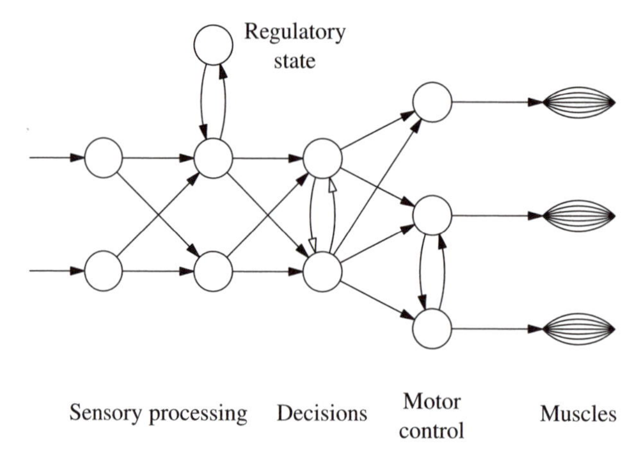

Figure 6.1 A behavior mechanism modeled as a neural network connected to receptors and muscles. The model includes reception of stimuli, processing of sensory input, a central decision making mechanism and a part that controls muscles and other effectors. An example with more sophisticated decision making is in Figure 3.27 on page 106.

6.1.1 Versatility

We observe great diversity in behavior, both between species and, on a less dramatic scale, within species. The basic structure of a general model of behavior should allow for a diversity of behavior maps to be formed.

We have shown throughout this book that neural network models seem able to implement practically any behavior system we consider. Additionally, neural networks can integrate diverse features and subsystems within a single basic architecture, allowing unified models of fields that earlier have been treated separately (e.g., reaction to stimuli, motor control, control theory, and learning).

Formally, it has been proved that a three-layer feedforward network with sufficiently many hidden nodes can approximate to any degree of accuracy any continuous map from the input to the output space (Cybenko 1989; Funahashi 1989; Hornik et al. 1989). More generally, a recurrent network can reproduce practically any *sequence* of input-output mappings (Hyötyniemi 1996; Jin et al. 1995; Šíma & Orponen 2003). These theorems guarantee that networks are general enough to encompass the diversity we observe in animal behavior, although they do not show how to build a network that implements any particular map. For this, the techniques described in Chapter 2 can be used.

6.1.2 Robustness

Regardless what explanation we seek, the mechanisms should display some robustness; otherwise, behavior would be vulnerable to any genetic or environmental disturbance. Thus small disturbances should not cause any major changes in performance.

Robustness is a landmark of neural networks, as demonstrated by many studies. We have discussed this issue in Section 3.9. Even rather small networks (few tens of nodes) can show graceful degradation of performance as nodes or connections are eliminated. Neural networks are robust also in the sense that small changes in input patterns usually result in small changes in network output (Section 3.3).

6.1.3 Learning

Learning from experience is a general feature of animal behavior. The model should allow learning, which should integrate realistically with the behavior map.

Neural network models offer a concrete interpretation of learning as changes to connection weights, corresponding to synapses in nervous systems. Finding biologically realistic learning procedures for neural networks, however, has been somewhat of a struggle (in fairness, the same is also true for other models). The essential issue is that in a plausible learning algorithm weights should change based solely on information that is in principle available at biological synapses. This may not be a major problem when the goal of learning is to organize the network based on general features of incoming input, e.g., to produce receptive fields of the Hubel-Wiesel type (Sections 2.4 and 4.7; Kohonen 2001). Neural network studies show in fact that such organization can be achieved with easily available information such as the activity of the pre- and postsynaptic nodes. However, learning about definite input-output relationships (e.g., classical and instrumental conditioning) also needs feedback from the animal's behavior and its consequences. In early algorithms such as back-propagation, such feedback was poorly integrated with the functioning of the network itself (Section 4.2). So-called reinforcement learning (Sections 2.3.4 and 4.2) represents a more biologically plausible alternative and can be materialized within the network. Recently, other promising algorithms have appeared (Section 4.2.2).

6.1.4 Ontogeny

The behavior system of an individual develops in a sequence of events in which genes, the developing individual and the environment interact. The structure of the model should allow for gradual development of a behavior map from scratch to an adult form.

This is perhaps the weak point of neural network models today, not because we know that neural networks suffer from particular shortcomings but rather because of our unsatisfactory understanding of the interactions among genes, the environment and the developing individual. Nevertheless, we saw in Chapter 4 that neural networks allow us to make interesting speculations about ontogeny, and the whole process appears at least less mysterious. For instance, since neural networks allow to model learning, they demonstrate how genetic information controlling the development of the network can be integrated with environmental effects. Moreover, the ontogeny of sensory processing abilities has been illuminated by network studies on the development of receptive fields, as recalled in the preceding point.

6.1.5 Evolution

For evolution to occur, genetic variation must exist that affects the development of behavior mechanisms. In a model, it should be possible to specify how genes control features of behavior maps and learning mechanisms. In addition, a model should allow the evolution of nervous systems from very simple forms to the complexities seen in, e.g., birds and mammals.

Realistic studies of the evolution of genetically controlled features of nervous systems suffer from the same lack of knowledge just discussed for ontogeny. However, the requirement that complex systems can evolve from simpler ones seems overwhelmingly fulfilled. Already very small networks are very capable and can be improved incrementally in a variety of ways, leading from sloppy to near-perfect behavior. In addition, we have seen that neural network studies are contributing to the understanding of particular issues about the evolution of behavior (Chapter 5).

6.2 DO WE USE TOO SIMPLE NETWORK MODELS?

Most analyses in this book rely on simple networks operating in discrete time. Nervous systems are much more complex and operate in continuous time. Such complexities are considered often in neuroethology and neuroscience, which are based on reconstructions of nervous machinery, including the detailed description of patterns of connectivity and the dynamics of neurons and synapses. The main goal of neuroscience, however, is to understand the nervous system and not behavior. We have already stated and defended our parsimonious approach in Chapter 1. Simple models have the advantage of being easier to understand and potentially more general because they do not include details that vary between species. On the other hand, there is always the possibility that models are too simple or in other ways unrealistic. The crucial test of simple neural network models of behavior is the extent to which they capture essential features of the behavior of real animals. If they do, the fact that the model is simple and neurally inspired (rather than based on some other metaphor, such as the digital computer) should be considered bonuses rather than problems. Of course, our approach would be in trouble if nervous systems turned out to operate in a dramatically different way from the models used in this book.

6.3 COMPARISONS WITH OTHER MODELS

In Chapter 1 we discussed various approaches to modeling behavior. Do neural networks offer any advantages over these other approaches? And what are the disadvantages? We start with advantages. First, the neural network approach provides a material implementation of sensory reception, processing, memory, decision making and motor control that is consistent with neurobiology and can model the whole causal chain from stimulus reception to behavior. Network models also show potential for modeling behavioral ontogeny and nature-nurture issues, although these areas are not well explored. In contrast, most modeling frameworks describe the inside of animals in abstract terms and often cover only some aspects of behavior systems. For instance, network models can combine motivational process with learning in one single structure, whereas today motivational and learning models tend to be specialized to their respective scopes. A related consideration is that neural network models benefit from two independent sources of information: observations of behavior and neurobiology. Most other approaches, on the other hand, base their modeling solely on behavioral observations (see also Section 1.3.6).

Neural network models used in this book are simpler than typical neuroethological models and thus are more useful for conceptual and intuitive understanding. This also renders them a degree of generality that more detailed neuroethological models may lack. If we want to use them as models of behavior, such generality is necessary because of the diversity of animals and behavior they need to cover. To draw a line between simplified neural networks and neuroethology is neither possible nor desirable. Our strategy is rather to always try to minimize neurobiological complexity without loosing precision at the behavioral level. The current progress in neuroethology is likely to further strengthen the network approach by correcting existing models and producing new ideas. The study of imperfect and dysfunctional behavior, as well as behavior following damage to the nervous system, is another area where neural networks show potential (Chapter 3). The reason is that when a problem is difficult, or when a system sustains damage, details of system machinery become more important and we can't just consider what function the system should perform (Section 1.1).

The insight that neural networks can provide about the evolution of behavior is perhaps one of their more interesting contributions. Neural networks make it possible to understand the evolution of behavior by offering material explanations for the evolution from simple to complex and from sloppy to perfect. Most other models cannot account for this. Behavioral ecology just assumes that optimal behavior has been produced by evolution, and cognitive, ethological, and learning models are static. For instance, consider a cognitive model that prescribes a mathematical algorithm that operates on conceptual or symbolic information. Such a model may operate fine, once established, but it gives little insight about its own evolution. Sometimes even small changes to the algorithm may ruin functionality completely.

What about drawbacks? Although network models have been criticized for excessive simplicity (in comparison with nervous systems), they are indeed more complex than most other models of behavior, and it can be difficult to understand what is going on even in simple networks. This is particularly the case when the network is just fitted to behavioral data. The contributions from different motivational factors become hidden in the weighted sum between these factors and memory. This is similar to what occurs in reality, but it is not always helpful for understanding. A model based on control theory, for instance, might describe feedback mechanisms more clearly. One worry is that neural networks introduce unnecessary complexity and that simpler models would do the job rather well. At the same time, reality is even more complex than neural networks. Thus our models will never be perfect models of behavior unless we include all relevant details. But then we have a neuroethological model, and all the advantages of simplification are lost.

Another issue is that neural networks, with their flexibility and power in producing input-output maps, perhaps invite one to play around with different ingredients without a thorough understanding of model operation. It is important to proceed carefully by asking precise questions, starting with simple models that allow us to analyze and understand what goes on. Lack of understanding may be particularly risky in evolutionary simulations, especially when complex problems or network architecture require the evolutionary process to search in very large state spaces.

Finally, we have seen in Chapters 3 and 4 that both learning theory and classical ethological theory are relatively consistent with neural network theory, whereas parts of the cognitive approach to animal behavior are more difficult to reconcile with neural network models. We have considered only a few cognitive models in this book, but they cannot be ignored because cognitive thinking dominates animal behavior today (Bekoff et al. 2002; Dukas 1998a; Gallistel 1990; Kamil 1998; Pearce 1997; Real 1994; Ristau 1991). So we need more discussion of cognitive modeling and how it relates to a neural network perspective.

6.4 NEURAL NETWORKS AND ANIMAL COGNITION

Most views of animal cognition (as well as of human and machine cognition) emphasize such things as representation of information or knowledge, computation and concepts and the manipulation of symbolic or conceptual information (Gallistel 1990; Kamil 1998; McFarland 1999). It is "cognition" in this sense we will discuss here. Other prominent concepts in texts on animal cognition are purpose, intentions, expectations and goal seeking (Gallistel 1990; Pearce 1997; and articles in Bekoff et al. 2002). Broader definitions of cognition also exist, such as "cognition refers to the mechanisms by which animals acquire, process, store and act on information from the environment" (Shettleworth 1998). However, with such a definition, all behavior is cognitive, including simple reflexes, and all models are cognitive—cognitive models would not be characterized by any specific structural element (see Chapter 1).

In cognitive science, the meaning of cognition, computation and knowledge representation has been discussed extensively (see, e.g., articles in Wilson & Keil 1999). A common topic in these discussions is the relationship between neural networks and classical cognitive models based on formal manipulation of symbols. Some have concluded that the two approaches are basically equivalent, whereas others argue that they are so different that they cannot be reconciled. A summary of this debate is offered by Ramsey (1999). Cognitive models and neural network models seem very different (see, e.g., Figure 6.4). If we look at a feedforward network, we see its architecture, memories stored as weights and patterns of node activity in response to input. The network definitely "processes information," but it hardly manipulates symbols or concepts according to formal rules, as, for instance, a computer program does. Such nonsymbolic processing is nevertheless capable of recognizing stimuli and making decisions.

If neural networks are not "cognitive," in the narrow sense defined above, how can they deal with observations of animal behavior usually explained in terms of "cognition"? There are at least two possible answer to this question that we explore in the next two sections. The first possibility is that the kind of neural network models that exist today cannot account for the most advanced modes of behavioral control observed in animals ("cognition"). If this turns out to be true, it is important to ask whether neural networks can be made "more cognitive." The second possibility is that many observations of animal cognition are indeed within the scope of existing neural network models. A feedforward network, for instance, is a very capable mechanism, and evolution does not select for concepts per se. What matters in evolution is appropriate responses, speed and the cost of the behavior mechanism.

6.4.1 Can neural networks be made "more cognitive"?

It may very well be that a mechanism with certain complex abilities requires concepts and symbolic manipulation. Can networks be developed in this direction, or are such processes outside the reach of neural networks, whether models or real systems?

Let us first consider feedforward networks. Can they form concepts and manipulate them in a symbolic fashion? Some network models do have "concept nodes" that represent the external world at an abstract level (Figure 6.2). For instance, one may assume that a particular node is active whenever an object belonging to a particular class is perceived. The activity of such nodes is then processed by the network in the usual way. In this sense, any network model can process symbolic or conceptual information. However, processing would occur according to the usual, nonsymbolic rules of network operation. Moreover, these models do not explain how stimulation from the external world is transformed into classes or concepts. A step in this direction is taken by feedforward networks based on feature detectors (Figure 6.2, left). These networks can categorize stimuli based on the presence of features or combinations of features, but the operation of feature detectors is, again, often just assumed and not implemented explicitly in the model. See Section 3.4.4 for further discussion of feature detectors.

We should also ask whether cognitive feedforward networks offer any advantages. It is common to view cognitive mechanism as superior either because they can solve more difficult problems, or because they speed up decision making, or because they are less costly. Standard feedforward networks, however, are already vary capable at forming input-output maps and are also fast and economical. It seems unlikely that introducing concept nodes could lead to faster or less costly networks. In fact, the opposite may be true, because introducing feature detectors may increase the number of nodes and layers. For instance, to recognize combinations of features, as well as single features, may require a very large number of nodes ("combinatioral explosion"; see Rolls & Deco 2002). We must also remember that a model that forms only hard (yes/no) categories would require additional mechanisms to generate smooth generalization gradients (Section 3.3.1) and to respond to novel stimuli (Section 4.4.1).

Another way feedforward neural networks may be made more cognitive is by introducing more layers. This creates some independence between sensory processing and decision making, which can be interpreted as representation of infor-

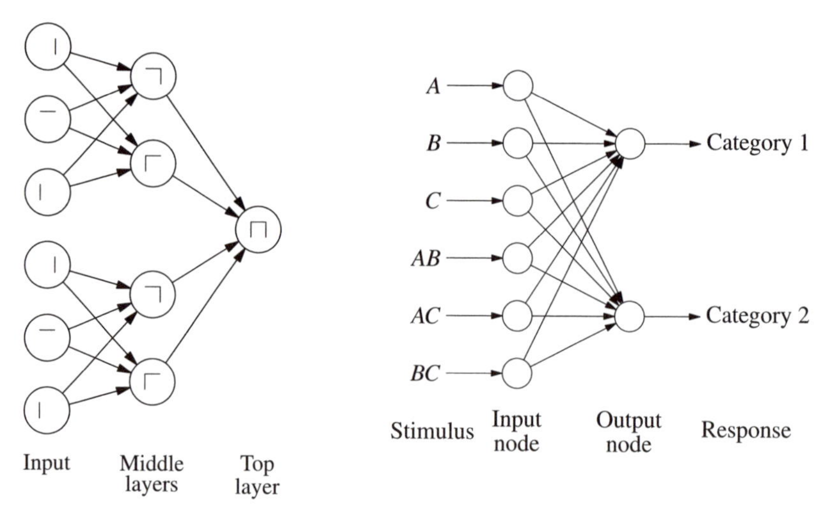

Figure 6.2 "Cognitive" neural network models. Left: Model of visual object recognition (redrawn based on Rolls & Deco 2002). The model builds on feature detecting neurons (Section 3.4.4) whose outputs are combined in successive steps to achieve object recognition insensitive to transformations such as translation over the retina. Right: A network redrawn from Gluck (1991). Each input node is sensitive to a single stimulus or to a combination of stimuli.

mation on which decisions are based (Chapter 3). This may lead to the formation of categories or concepts if thresholds are introduced in the output layer so that a number of different stimuli may be reacted to in exactly the same way (but this would impair smooth generalization abilities, as noted in the preceding paragraph). A network without thresholds may form graded categories or concepts, which may approximate hard categories. Whether multilayer networks could encompass observations of animal cognition is largely unknown (see below), but increasing the number of layers clearly has some advantages. It makes the network more potent in forming input-output maps (Section 2.2.1.3), and it allows for perceptual learning, i.e., learning that is not expressed directly in behavior (Section 4.4).

What about recurrent networks? Cognitive scientists and early ethologists have often criticized behaviorism for proposing S-R (stimulus-response) theories that ignore internal factors, although this is not true of all behaviorism (Staddon 2001). Feedforward networks are an S-R theory, although more complex than, say, a lookup table. The critique of S-R theories carries important insights because including recurrent connections or other kinds of internal states allows better control of behavior than possible within an S-R system. Indeed, recurrent networks can be sensitive to temporal structures in stimulation and can produce organized sequences of behavior (Chapter 3). This is achieved by building "regulatory states," i.e., node activities that, through the recurrent connections, are sensitive to sequences of inputs and/or outputs. So are recurrent networks more cognitive? Not necessarily (many recurrent networks cannot be interpreted as cognitive, e.g., networks for

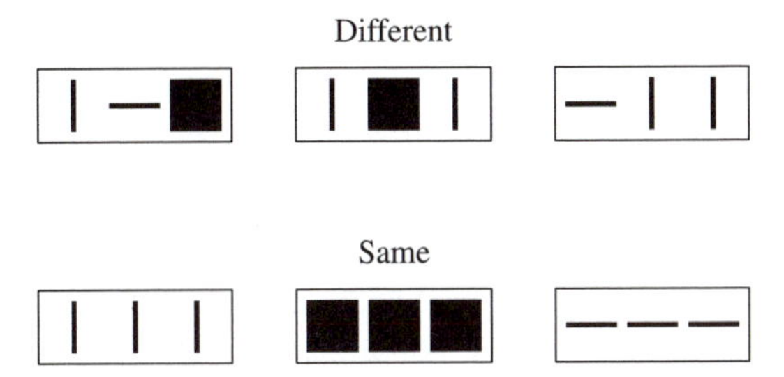

Figure 6.3 Can a simple network master this discrimination problem, involving the rather abstract notions of same and different? The first row shows three examples of "different" stimuli, and the second row shows three "same" stimuli made up of simple patterns. Wasserman (1995) has trained pigeons on a similar task, showing that the animals learn to respond differently to "same" and "different" stimuli and generalize well if shown examples of "same" and "different" composed with novel patterns.

motor control), but recurrence opens up new possibilities. For instance, regulatory states partially detached from external stimulation could develop ("thinking"). Also, regulatory states could occur in structured sequences that might be interpreted as "rules" of processing. This is an exciting yet largely unexplored area of potential development. Some recurrent networks also allow for such operations as the retrieval of a stored memory based on a fragment of information or the classification of stimuli into categories (Amit 1989; Haykin 1999). These networks are capable of learning and thus could be said to "learn concepts," although they do not seem to have been tested as models of concept formation.

6.4.2 Can neural networks account for animal cognition?

A number of empirical findings have been offered as evidence of "cognitive" abilities in animals, such as concept formation, numerical abilities and manipulation of symbolic information (Balda et al. 1998; Bekoff et al. 2002; Shettleworth 1998). Alternative accounts of these findings, however, are rarely tested, and there is the possibility that standard neural networks may account for some or most observations of animal cognition. Let us consider an example of concept formation relative to the abstract concepts of "same" and "different," as illustrated in Figure 6.3. Experiments have shown that some animal species can learn to discriminate pictures composed of identical elements from pictures composed of different elements. The animal is first trained, by instrumental conditioning, to respond to pictures of one kind only, e.g., to "same" pictures but not to "different" pictures. The animal is then tested with examples of "same" and "different" composed of picture elements that were never shown during training. If the animal continues to respond appropriately, this is taken as evidence that the discrimination was solved by using a concept of "same" and "different" rather than, say, simply by associating each training picture with the appropriate response (see Cook 2002; Delius 1994; and Wasserman 2002 for both data and details on experimental methods).

Can neural networks accomplish such behavior? In one sense, we know they can because networks can form almost any behavior map. The question is rather whether cognitive behavior emerges from either simulated evolution or biologically plausible models of learning processes. We do not know the answer to this question, but it would be premature to conclude that networks, even feedforward ones, cannot match animals' abilities on this problem. It seems important to simulate the empirical studies with various types of networks, both feedforward and recurrent, to sort out the extent to which they are capable of "cognitive" behavior. If simple networks fail, we will conclude that they cannot account for certain "cognitive" abilities, and we will need to develop new theory. But if the networks can reproduce these results, we might reconsider our ideas about what goes on inside animals, e.g., the extent to which symbolic manipulation is necessary for cognitive behavior.

We can also sketch some interesting predictions that may come from a successful network model of the "same" and "different" problem. In fact, given what we know about generalization in networks (Section 3.3), we expect that the network would not respond in exactly the same way to all novel pictures. Although this might be interpreted as a failure to develop a "same" concept, we must also recognize that animals do make mistakes (e.g., successful conceptualization may be claimed based on 80% correct responses). We have seen a similar situation relative to perceptual constancy, where not all novel views of a familiar object are recognized with the same accuracy (Section 3.4.3). In a simple network model, performance on a novel picture should depend on its similarity to known pictures, and we thus expect the network to make most mistakes on those novel pictures that are less similar to the training ones. It would be very interesting to check whether this holds for animals as well, including humans. Would it imply that no "same" concept is formed?

6.4.3 The problem of observability

If neural networks can account for many skills that seem to require formal symbolic abilities, we face another important issue regarding the *observability* of internal processes. That is, how much can we learn about internal mechanisms from only observing behavior? In animal cognition and most of animal psychology, behavior is used as the only window into the animal. Neuroethology, in contrast, gains almost all its knowledge from studying the structure and operation of the nervous systems itself. Our neural network approach falls in between.

Theoretical results from systems theory show that there are important limitations to what one can learn about a system's inside from the outside (Kalman 1968; Luenberger 1979; McFarland & Houston 1981; Polderman & Willems 1998). This is an acute problem for students of behavior. For instance, a number of studies have tried to distinguish between different types of memory models based on observations of behavior (e.g., prototype and instance theories; see Pearce 1994; Shanks 1995). Whether this is even theoretically possible is an important question. For instance, finding out the memory mechanism of a neural network would be difficult from just exploring stimulus-response patterns. Let us make a concrete example using a famous study of dead reckoning in desert ants. Gallistel (1990) offers two models for this ability. One model stores the ant's position in Cartesian coordinates, the other in polar coordinates (Figure 6.4). In the figure we have also sketched a network model for this problem. For the sake of the argument, we assume that it can reproduce the ant's behavior. Which model should we prefer? All do the job, which means that we cannot tell them apart by observing normal behavior. Perhaps we could experimentally fake the sun in impossible locations in the sky or introduce a second "sun" and see how both models and ants respond (note, however, that the cognitive models have no built-in mechanism for handling such unforeseen situations; the networks would at least do something based on the input they receive). We could also look inside the models. It is clear that the computational details are different. The network does not calculate sines and cosines but only weighted sums and transfer functions. The point of view of classical cognitive science would be that the models are identical at the computational level (i.e., they ultimately perform the same function; see below) and that the symbolic models would have to be preferred because they display the fundamental computation. Gallistel, discussing which cognitive model to prefer, suggests that the Cartesian model is more robust than the polar model when small errors are introduced. The network, on the other hand, probably could be made even more robust and have a richer set of states to describe the ant's current position relative to the home burrow. The cognitive models, on the other hand, are potentially easier to understand, and they show clearly that the information available to the ant is sufficient for dead reckoning to be logically possible.

6.4.4 Evolution

We must also ask how cognitive models and neural networks relate to evolution. Neural networks are a framework of great generality, capable of implementing practically any behavior map, and within which small changes usually have small effects. These properties may offer an understanding of how a behavior system can change and improve gradually during evolution (see Section 5.2.4). Cognitive models, on the other hand, usually focus on the end point or optimal solution to an evolutionary problem, and the structure of the model is not general but specific to that particular problem (apart from falling within the broad framework of symbolic computation). Consider again Gallistel's model of dead reckoning in the desert ant (Figure 6.4). It is difficult imagine how the model looked before it got its final form and how it evolved from less perfect mechanisms. The main problem is that the model does not explain how the ant brain actually implements the assumed computations. Marr (1982), in his influential book, *Vision*, often cited as the most complete cognitive manifesto, argued that the operation of any intelligent mechanism (including, e.g., nervous systems and computers) should be understood at three levels. The highest level is the computational level of abstract problem analysis. It is followed by the algorithmic level, which is concerned with defining a particular formal procedure to solve a given problem. The last level concerns the physical implementation of the algorithm. An important aspect of this subdivision was that issues relating to a higher level were claimed to be independent of issues at lower levels (for a similar view directly concerned with animal behavior, see

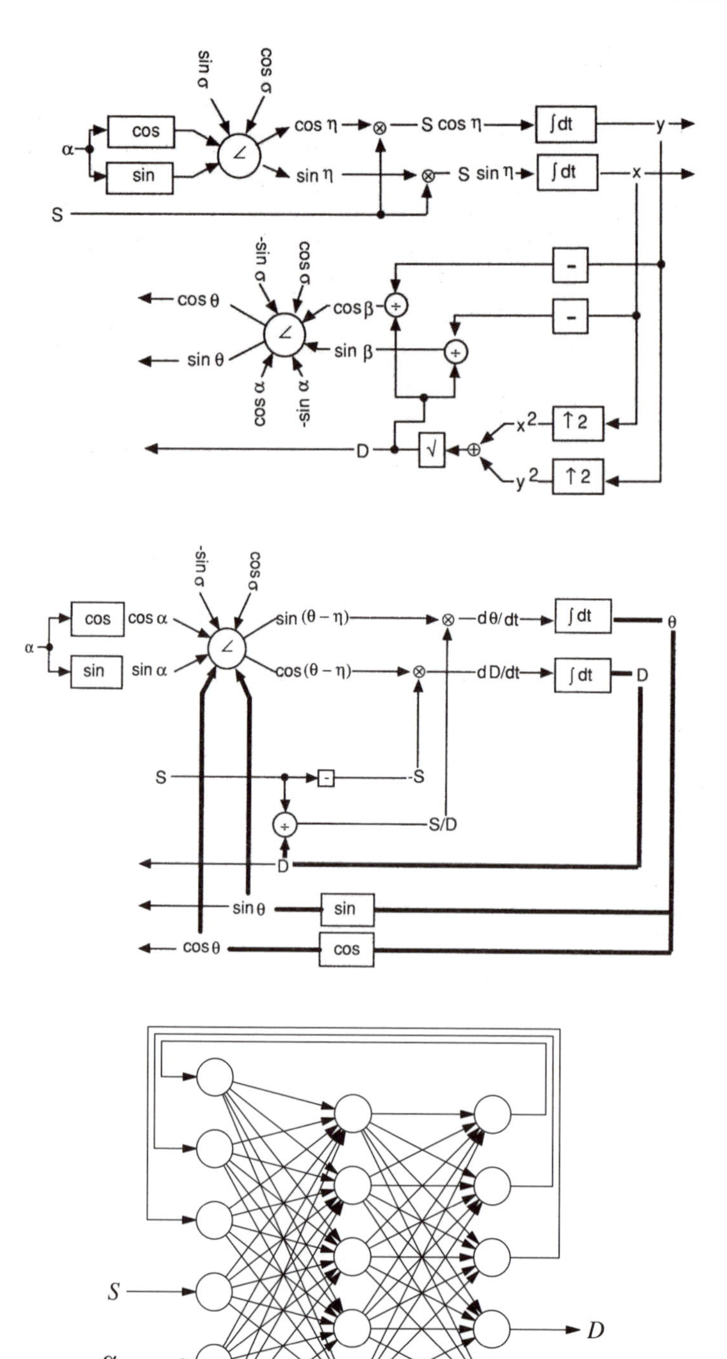
sin σ
cos σ
α
cos
sin
cos η
sin η
S cos η
S sin η
∫dt
y
x
S
-sin σ
cos σ
cos θ
sin θ
cos β
sin β
cos α
-sin α
x2
y2
↑2
D
√
cos α
sin α
sin (θ – η)
cos (θ – η)
dθ/dt
dD/dt
θ
-S
S/D
sin θ
cos θ
S
α
σ
D
θ

Gallistel 1990). This claim has given justification to ignoring the implementation level, leading to the most common style of cognitive modeling today. It follows that cognitive models typically consider just the ability to solve a problem but not, for instance, the speed and cost of the mechanism. However, all three matter in evolution. In network models, speed and cost are easily defined in term of number of processing steps and number of nodes and connections. In contrast, it is difficult to estimate the cost and speed of a cognitive model. For example, what is the time required to calculate a sine or a logarithm? We cannot answer this questions unless we know the material basis for these operations, and this invalidates Marr's claim that cognition in terms of computation or formal procedures is independent of its physical implementation (Churchland & Grush 1999).

Cognitive scientists such as Marr were inspired by computer science and artificial intelligence but there are obvious differences between how computer programs are developed and biological evolution. The development of a computer program involves all three of Marr's levels, more or less as described by Marr. First, the problem is considered from an abstract point of view, then a program is written, and lastly a physical implementation of the program is created. The evolution of animal intelligence, on the other hand, is caused by changes at the level of physical implementation (Marr's third level) or, more precisely, by changes to genes that control the development of the physical implementation. The result of evolution is also manifested at the physical level, whereas explanations in terms of algorithms (decision rules) or computations, when appropriate, are interpretations of the outcome of evolution.

It follows from our discussion that to evaluate constraints and possibilities relative to the evolution of animal intelligence, we must consider the level of physical implementation. For instance, whether a problem is difficult or not depends on what is possible at the physical level, and whether the problem is complex at an abstract or algorithmic level may not always matter. Thus, to understand the evolution of cognitive abilities, we need to know how they relate to changes in neural architecture and functionality.

Figure 6.4 Top and middle: Two models of navigation in desert ants. Both models compute the bearing θ to the nest (more precisely, the sine and cosine of θ) but the first model uses Cartesian coordinates as state variables (distance from the nest along x and y axes), whereas the second model uses polar coordinates (linear distance and angle to the nest). Both models receive as input the ant's speed S, the angle of the lubber line relative to the sun (solar heading α) and the azimuth of the sun as function of day (σ), computed by an internal function. Reprinted from Gallistel (1990) by permission of C. R. Gallistel. Bottom: Sketch of a neural network that potentially could produce the same input-output relationship as the preceding models. Note that a complete network model would not use S and α as inputs but rather would extract such information from sensory input, e.g., from visual and proprioceptive inputs. Information about solar azimuth σ may be produced from within the network, based on an internal clock. Likewise, network output would not be an abstract representation of the ant's position but an actual behavior sequence that would bring the ant back to its nest.

6.5 FINAL WORDS

Despite foreshadowing neural networks at least a few times (Section 1.4.3), students of animal behavior have for too long ignored their potentials as models of behavior. Our analysis suggests that neural networks can embody all elements of a behavior system and provide understanding both at the behavioral and the physiological levels. Although we focused on simple networks, these were shown to be powerful enough to encompass a wide array of behavioral phenomena. Complexity can be added in case of need: we do not endorse simplicity when misleading or wrong. Neural networks can also provide a common ground for discussion between different subjects and research programs. We also think, as it should be clear from the preceding section, that the relationship between neural networks and cognitive models needs to be explored much further, including possibly a reevaluation of what observations of animal cognition mean. Hopefully, these two modeling styles can be reconciled and together produce a better understanding of biological intelligence and its evolution.

Perhaps we are too enthusiastic about neural networks, but we hope to have convinced the reader that these models should be investigated further and may represent a significant leap in our understanding of behavior, incorporating current progress in neuroethology without drowning in all the details of neurobiology.

FURTHER READING

Balda RP, Pepperberg IM, Kamil AC, editors, 1998. *Animal Cognition in Nature*. London: Academic Press.

Gallistel CR, 1990. *The Organization of Learning*. Cambridge, MA: MIT Press.

Shettleworth SJ, 1998. *Cognition, Evolution, and Behavior*. New York: Oxford University Press.

Staddon JER, 2001. *The New Behaviorism: Mind, Mechanism and Society*. Hove, East Sussex: Psychology Press.

Wilson RA, Keil FC, editors, 1999. *The MIT Encyclopedia of the Cognitive Sciences*. Cambridge, MA: MIT Press.

Bibliography

Abrams PA, 2000. The evolution of predator-prey interactions: Theory and evidence. *Annual Review of Ecology and Systematics* 31, 79–105.

Ackley D, Hinton GE, Sejnowski TJ, 1985. A learning algorithm for Boltzmann machines. *Cognitive Science* 9, 147–169.

Adrian ED, 1931. Potential changes in the isolated nervous system of *Dytiscus marginalis*. *Journal of Physiology* 72, 132–151.

Alcock J, Sherman P, 1994. The utility of the proximate-ultimate dichotomy in ethology. *Ethology* 96, 58–62.

Alerstam T, 1993. *Bird Migration*. Cambridge, England: Cambridge University Press.

Amari S, Arbib MA, 1977. Competition and cooperation in neural nets. In J Metzler, editor, *Systems neuroscience*, 119–165. New York: Academic Press.

Amirikian B, Georgopoulos AP, 2003. Motor cortex: Coding and decoding of directional operations. In Arbib (2003), 690–696.

Amit D, 1989. *Modeling Brain Function*. Cambridge, England: Cambridge University Press.

Andersson M, 1994. *Sexual Selection*. Princeton, NJ: Princeton University Press.

Antonini A, Stryker MP, 1993. Rapid remodeling of axonal arbors in the visual cortex. *Science* 260, 1819–1821.

Arak A, Enquist M, 1995. Conflict, receiver bias and the evolution of signal form. *Philosophical Transactions of the Royal Society* B349, 337–344.

Arbib MA, editor, 1995. *The Handbook of Brain Theory and Neural Networks*. Cambridge, MA: MIT Press.

Arbib MA, editor, 2003. *The Handbook of Brain Theory and Neural Networks*. Cambridge, MA: MIT Press, 2 edition.

Arbib MA, Erdi P, Szentagothai J, 1998. *Neural Organization: Structure, Function, and Dynamics*. Cambridge, MA: MIT Press.

Arnqvist G, Rowe L, 2002. Antagonistic coevolution between the sexes in a group of insects. *Nature* 415, 787–789.

Atkinson RR, Estes WK, 1963. Stimulus sampling theory. In RD Luce, RB Bush, editors, *Handbook of Mathematical Psychology*. New York: Wiley.

Attneave F, 1957. Transfer of experience with a class-schema to identification learning of patterns and shapes. *Journal of Experimental Psychology* 54, 81–88.

Aubin JP, 1995. Learning as adaptive control of synaptic matrices. In Arbib (1995), 527–530.

Baerends GP, 1971. The ethological analysis of fish behavior. In WS Hoar, DJ Randall, editors, *Fish Physiology*, volume 6, 279–370. London: Academic Press.

Baerends GP, 1976. The functional organization of behaviour. *Animal Behaviour* 24, 726–738.

Baerends GP, 1982. The herring gull and its world. II. The responsiveness to egg features. General discussion. *Behaviour* 82, 276–411.

Baerends GP, Baerends-van Roon JM, 1950. *An introduction to the study of cichlid fishes*. Leiden: Brill.

Baerends GP, Brouwer R, Waterbolk HT, 1955. Ethological studies on *Lebistes reticulatus* (Peters): An analysis of the male courtship pattern. *Behaviour* 8, 249–334.

Baerends GP, Drent RH, 1982. The herring gull and its world. II. The responsiveness to egg features. *Behaviour* 82, 1–416.

Baerends GP, Kruijt JP, 1973. Stimulus selection. In Hinde & Stevenson-Hinde (1973).

Baker JR, 1938. The evolution of breeding seasons. In GR de Beer, editor, *Evolution: Essays on Aspects of Evolutionary Biology*. Oxford, England: Oxford University Press.

Balda RP, Pepperberg IM, Kamil AC, editors, 1998. *Animal Cognition in Nature*. London: Academic Press.

Balkenius C, Morén J, 1998. Computational models of classical conditioning: A comparative study. In Pfeiffer et al. (1998).

Barto AJ, 1995. Reinforcement learning. In Arbib (1995), 804–809.

Barto AJ, 2003. Reinforcement learning. In Arbib (2003), 963–968.

Basil JA, Kamil AC, Balda RP, Fite KV, 1996. Differences in hippocampal volume among food storing corvids. *Brain, Behavior and Evolution* 7, 156–164.

Basolo A, 1990. Female preference for male sword length in the green swordtail, *Xiphophorus helleri* (Pisces: Poeciliidae). *Animal Behaviour* 40, 332–338.

Bateson P, Horn G, 1994. Imprinting and recognition memory: A neural net model. *Animal Behaviour* 48, 695–715.

Baylor DA, Nunn BJ, Schnapf JL, 1987. Spectral sensitivity of cones of the monkey, *Macaca fascicularis*. *Journal of Physiology* 390, 145–160.

Beer RD, Gallagher JC, 1992. Evolving dynamical neural networks for adaptive behavior. *Adaptive Behavior* 1, 91–122.

Beer RD, Chiel HJ, Quinn RD, Ritzmann RE, 1998. Biorobotic approaches to the study of motor systems. *Current Opinion in Neurobiology* 8, 777–782.

Bekoff M, Allen C, Burghardt G, editors, 2002. *The Cognitive Animal*. Cambridge, MA: MIT Press.

Benda J, Herz AVM, 2003. A universal model for spike-frequency adaptation. *Neural Computation* 15, 2523–2564.

Berridge KC, 1994. The development of action patterns. In Hogan & Bolhuis (1994), 147–180.

Bienenstock EL, Cooper LN, Munro PW, 1982. Theory for the development of neuron sensitivity: Orientation specificity and binocular interaction in visual cortex. *Journal of Neuroscience* 2, 32–48.

Binmore K, 1987. Modeling rational players. I. *Economics and Philosophy* 3, 179–214.

Blass E, editor, 2001. *Developmental Psychobiology*, volume 13 of *Handbook of Behavioral Neurobiology*. New York: Kluwer.

Bloomfield TM, 1967. A peak shift on a line-tilt continuum. *Journal of the Experimental Analysis of Behavior* 10, 361–366.

Blough DS, 1975. Steady state data and a quantitative model of operant generalization and discrimination. *Journal of Experimental Psychology: Animal Behavior Processes* 104, 3–21.

Bolhuis JJ, 1991. Mechanisms of avian imprinting: A review. *Biological Review* 66, 303–345.

Bolhuis JJ, Hogan JA, editors, 1999. *The Development of Animal Behaviour: A Reader*. Malden, MA: Blackwell.

Bolles RC, 1975. *Theory of Motivation*. New York: Harper and Row, 2 edition.

Braitenberg V, Schüz A, 1998. *Cortex: Statistics and Geometry of Neuronal Connectivity*. Berlin: Springer-Verlag, 2 edition.

Breland K, Breland M, 1961. The misbehavior of organisms. *American Psychologist* 61, 681–684.

Brodfuehrer PD, Debski EA, O'Gara BA, Friesen WO, 1995. Neuronal control of leech swimming. *Journal of Neurobiology* 27, 403–418.

Brown JL, 1975. *The Evolution of Behaviour*. New York: Norton.

Brush FR, Bush RR, Jenkins WO, John WF, Whiting JWM, 1952. Stimulus generalization after extinction and punishment: An experimental study of displacement. *Journal of Abnormal and Social Psychology* 47, 633–640.

Bullock S, Cliff D, 1997. The role of hidden preferences in the artificial coevolution of symmetrical signals. *Proceedings of the Royal Society of London* B264, 505–511.

Bullock TH, 1961. The origins of patterned nervous discharge. *Behaviour* 17, 48–59.

Buonomano DV, Karmarkar UR, 2002. How do we tell time? *Neuroscientist* 8, 42–51.

Burrows M, 1979. Synaptic potentials affect the release of transmitter from locust non-spiking neurons. *Science* 204, 81–83.

Bush P, Sejnowski TJ, 1991. Simulation of a reconstructed Purkinje cell based on simplified channel kinetics. *Neural Computation* 3, 321–332.

Bush RR, Mosteller F, 1951. A model for stimulus generalization and discrimination. *Psychological Review* 58, 413–423.

Butler AB, Hodos WC, 1996. *Comparative Vertebrate Neuroanatomy: Evolution and Adaptation*. New York: Wiley-Liss.

Cacioppo JT, Tassinary LG, Berntson G, editors, 2000. *Handbook of Psychophysiology*. Cambridge, England: Cambridge University Press, 2 edition.

Calabretta R, Di Ferdinando A, Parisi D, 2004. Ecological neural networks for object recognition and generalization. *Neural Processing Letters* 19, 37–48.

Catchpole C, Slater P, 1995. *Bird Song: Biological Themes and Variations*. Cambridge, England: Cambridge University Press.

Cervantes-Péres F, 2003. Visuomotor coordination in frog and toad. In Arbib (2003), 1219–1224.

Chapman T, Partridge L, 1996. Sexual conflict as fuel for evolution. *Nature* 381, 189–190.

Cheng K, Spetch ML, Johnson M, 1997. Spatial peak shift and generalization in pigeons. *Journal of Experimental Psychology: Animal Behavior Processes* 23, 469–481.

Churchland PM, 1995. *The Engine of Reason, the Seat of the Soul: A Philosophical Journey into the Brain*. Cambridge, MA: MIT Press.

Churchland PS, Grush R, 1999. Computation and the brain. In Wilson & Keil (1999), 155–158.

Churchland PS, Sejnowski T, 1992. *The Computational Brain*. Cambridge, MA: MIT Press.

Clayton KN, 1969. Reward and reinforcement in selective learning: Considerations with respect to a mathematical model of learning. In JT Tapp, editor, *Reinforcement and Behavior*, 96–119. New York: Academic Press.

Cliff D, 2003. Neuroethology, computational. In Arbib (2003), 737–741.

Cobas A, Arbib MA, 1992. Prey-catching and predator-avoidance in frog and toad: Defining the schemas. *Journal of Theoretical Biology* 157, 271–304.

Cohen AH, Boothe DL, 2003. Sensorimotor interactions and central pattern generators. In Arbib (2003), 1017–1020.

Cohen MA, Grossberg S, 1983. Absolute stability of global pattem formation and parallel memory storage by competitive neural networks. *IEEE Transactions on Systems, Man and Cybernetics* 13, 815–826.

Cook RG, 2002. Same-different concept formation in pigeons. In Bekoff et al. (2002), 229–237.

Coren S, Ward LM, Enns JT, 1999. *Sensation and Perception*. Fort Worth: Hartcourt Brace, 5 edition.

Cott H, 1940. *Adaptive Colouration in Animals*. London: Methuen.

Crevier D, 1993. *AI: The Tumultuous History of the Search for Artificial Intelligence*. New York: Basic Books.

Cybenko G, 1989. Approximation by superpositions of a sigmoidal function. *Mathematics of Control, Signals, and Systems* 2, 303–314.

Dane B, Wakott C, Drury WH, 1959. The form and duration of the display actions of the goldeneye (*Bucephala clangula*). *Behaviour* 14, 265–281.

Darwin C, 1872. *The Expression of the Emotions in Man and Animals*. London: Murray.

Davies N, 1992. *Dunnock behaviour and social evolution*, volume 1. Oxford, England: Oxford University Press.

Davis WJ, 1979. Behavioural hierarchies. *Trends in Neuroscience* 2, 5–7.

Dawkins M, Dawkins R, 1974. Some descriptive and explanatory stochastic models of decision making. In McFarland (1974a).

Dawkins R, 1982. *The Extended Phenotype*. New York: Oxford University Press.

Dawkins R, Krebs J, 1979. Arms races between and within species. *Proceedings of the Royal Society of London* B205, 489–511.

Dayan P, Abbott LF, 2001. *Theoretical Neuroscience: Computational and Mathematical Modeling of Neural Systems*. Cambridge, MA: MIT Press.

Dean J, Cruse H, 2003. Motor pattern generation. In Arbib (2003), 696–701.

De Felipe J, Jones EG, editors, 1989. *Cajal on the Cerebral Cortex: An Annotated Translation of the Complete Writings*. Oxford, England: Oxford University Press.

Delius J, 1994. Comparative cognition of identity. In P Bertelson, P Eelen, G Ydewalle, editors, *International Perspectives on Psychological Science*, volume 1, 25–39. Hillsdale, NJ: Lawrence Erlbaum Associates.

DeVoogd TJ, 1991. Endocrine modulation of the development and adult function of the avian song system. *Psychoneuroendocrinology* 16, 41–66.

DeVoogd TJ, Krebs JR, Healy SD, Purvis A, 1993. Relations between song repertoire size and the volume of brain nuclei related to song: Comparative evolutionary analyses amongst oscine birds. *Proceedings of the Royal Society of London* B254, 75–82.

Dewsbury DA, 1992. On the problems studied in ethology, comparative psychology, and animal behavior. *Ethology* 92, 89–107.

Dewsbury DA, 1999. The proximate and the ultimate: Past, present, and future. *Behavioural Processes* 46, 189–199.

Dickinson A, 1980. *Contemporary Animal Learning Theory*. Cambridge, England: Cambridge University Press.

Dickinson A, 1989. Expectancy theory in animal conditioning. In Klein & Mowrer (1989).

Dickinson PS, 2003. Neuromodulation in invertebrate nervous systems. In Arbib (2003), 757–761.

Didday RL, 1976. A model of visuomotor mechanisms in the frog optic tectum. *Mathematical Biosciences* 30, 169–180.

Dieckmann U, Law R, 1996. The dynamical theory of coevolution: A derivation from stochastic ecological processes. *Journal of Mathematical Biology* 34, 579–612.

Dieckmann U, Marrow P, Law R, 1995. Evolutionary cycling in predator-prey interactions: Population dynamics and the Red Queen. *Journal of Theoretical Biology* 176, 91–102.

Dill M, Edelman S, 2001. Imperfect invariance to object translation in the discrimination of complex shapes. *Perception* 30, 707–724.

Dill M, Fahle M, 1998. Limited translation invariance of human visual pattern recognition. *Perception and Psychophysics* 60, 65–81.

Dill M, Wolf R, Heisenberg M, 1993. Visual pattern recognition in *Drosophila* involves retinotopic matching. *Nature* 365, 751–753.

Dittrich W, Gilbert F, Green P, McGregor P, Grewcock D, 1993. Imperfect mimicry: A pigeon's perspective. *Proceedings of the Royal Society of London* B251, 195–200.

Domjan M, 1980. Ingestional aversion learning: Unique and general processes. *Advances in the Study of Behavior* 11, 276–336.

Doty RW, 1951. Influence of stimulus pattern on reflex deglutition. *American Journal of Physiology* 166, 142–158.

Doty RW, Bosma JF, 1956. An electromyographic analysis of reflex deglutition. *Journal of Physiology* 19, 44–60.

Dougherty DM, Lewis P, 1991. Stimulus generalization, discrimination learning, and peak shift in horses. *Journal of the Experimental Analysis of Behavior* 56, 97–104.

Doya K, 2003. Recurrent networks: Learning algorithms. In Arbib (2003), 955–960.

Dudai Y, 1989. *The Neurobiology of Memory: Concepts, Findings, Trends*. Oxford, England: Oxford University Press.

Dukas R, editor, 1998a. *Cognitive Ecology*. Chicago: University of Chicago Press.

Dukas R, 1998b. Constraints on information procesing and their effects on evolution. In Dukas (1998a), 89–127.

Eibl-Eibesfeldt I, 1975. *Ethology: The Biology of Behavior*. New York: Holt, Rinehart & Winston.

Elman JL, 1990. Finding structure in time. *Cognitive Science* 14, 179–211.

Elman JL, Bates EE, Johnson MH, Karmiloff-Smith A, Parisi D, Plunkett K, 1996. *Rethinking Innateness: A Connectionist Perspective on Development*. Cambridge, MA: MIT Press.

Enquist M, Arak A, 1993. Selection of exaggerated male traits by female aesthetic senses. *Nature* 361, 446–448.

Enquist M, Arak A, 1994. Symmetry, beauty and evolution. *Nature* 372, 169–172.

Enquist M, Arak A, 1998. Neural representation and the evolution of signal form. In Dukas (1998a), 21–87.

Enquist M, Arak A, Ghirlanda S, Wachtmeister CA, 2002. Spectacular phenomena and limits to rationality in genetic and cultural evolution. *Transactions of the Royal Society* B357, 1585–1594.

Enquist M, Johnstone R, 1997. Generalization and the evolution of symmetry preferences. *Proceedings of the Royal Society of London* B264, 1345–1348.

Enquist M, Leimar O, 1993. The evolution of cooperation in mobile organisms. *Animal Behaviour* 45, 747–757.

Eshel I, 1983. Evolutionary and continuous stability. *Journal of Theoretical Biology* 103, 99–111.

Eshel I, 1996. On the changing concept of evolutionary population stability as a reflection of a changing point of view in the quantitative theory of evolution. *Journal of Mathematical Biology* 34, 485–510.

Ewert JP, 1980. *Neuroethology*. Berlin: Springer-Verlag.

Ewert JP, 1985. Concepts in vertebrate neuroethology. *Animal Behaviour* 33, 1–29.

Ewert JP, Capranica RR, Ingle DJ, editors, 1983. *Advances in Vertebrate Neuroethology*. New York: Plenum Press.

Fay R, 1970. Auditory frequency generalization in the goldfish (*Carassius auratus*). *Journal of the Experimental Analysis of Behavior* 14, 353–360.

Feldman JM, 1975. Blocking as a function of added cue intensity. *Animal Learning & Behavior* 3, 98–102.

Fentress JC, editor, 1976. *Simpler Networks and Behavior*. Sunderland, MA: Sinauer Associates.

Field DJ, 1995. Visual coding, redundancy, and feature detection. In Arbib (1995), 1012–1016.

Flash T, Sejnowski TJ, 2001. Computational approaches to control. *Current opinion in neurobiology* 11, 655–662.

Fodor J, Pylyshyn Z, 1988. Connectionism and cognitive architecture: A critical analysis. *Cognition* 28, 3–71.

Földiák P, 1991. Learning invariance from transformation sequences. *Neural Computation* 3, 194–200.

Forger NG, 2001. Development of sex differences in the nervous system. In Blass (2001), 143–198.

Forti S, Menini A, Rispoli G, Torre V, 1989. Kinetics of phototransduction in retinal rods of the newt, *Triturus cristatus*. *Journal of Physiology* 419, 265–295.

French R, 1999. Catastrophic forgetting in connectionist networks: Causes, consequences and solutions. *Trends in Cognitive Sciences* 3, 128–135.

Fudenberg D, Tirole J, 1992. *Game Theory*. Cambridge, MA: MIT Press.

Funahashi K, 1989. On the approximate realization of continuous mappings by neural networks. *Neural Networks* 2, 183–192.

Futuyma DJ, 1998. *Evolutionary Biology*. Sunderland, MA: Sinauer Associates.

Galizio M, 1985. Human peak shift: Analysis of the effects of three-stimulus discrimination training. *Learning and Motivation* 16, 478–494.

Galli L, Maffei L, 1988. Spontaneous impulse activity of rat retinal ganglion cells in prenatal life. *Science* 242, 90–91.

Gallistel CR, 1980. *The Organization of Action*. Hillsdale, NJ: Lawrence Erlbaum Associates.

Gallistel CR, 1990. *The Organization of Learning*. Cambridge, MA: MIT Press.

Ganong WF, 2001. *Review of Medical Physiology*. Upper Saddle River, NJ: Prentice-Hall International, 15 edition.

Garcia J, Ervin FA, Koelling RA, 1966. Learning with prolonged delay of reinforcement. *Psychonomic Science* 5, 121–122.

Garcia J, Koelling RA, 1966. Relation of cue to consequence in avoidance learning. *Psychonomic Science* 4, 123–124.

Gavrilets S, Arnqvist G, Friberg U, 2001. The evolution of female mate choice by sexual conflict. *Proceedings of the Royal Society of London* B268, 531–539.

Gazzaniga MS, 2000. *The New Cognitive Neurosciences*. Cambridge, MA: MIT Press.

Gegenfurter KR, Sharpe LT, 1999. *Color Vision*. Cambridge, UK: Cambridge University Press.

Geritz SAH, Kisdi É, Meszéna G, Metz JAJ, 1998. Evolutionarily singular strategies and the adaptive growth and branching of the evolutionary tree. *Evolutionay Ecology* 12, 35–57.

Ghirlanda S, 2002. Intensity generalization: Physiology and modelling of a neglected topic. *Journal of Theoretical Biology* 214, 389–404.

Ghirlanda S, 2005. Retrospective revaluation as simple associative learning. *Journal of Experimental Psychology: Animal Behavior Processes* 31, 107–111.

Ghirlanda S, Enquist M, 1998. Artificial neural networks as models of stimulus control. *Animal Behaviour* 56, 1383–1389.

Ghirlanda S, Enquist M, 1999. The geometry of stimulus control. *Animal Behaviour* 58, 695–706.

Ghirlanda S, Enquist M, 2003. A century of generalization. *Animal Behaviour* 66, 15–36.

Ghysen A, 2003. The origin and evolution of the nervous system. *International Journal of Developmental Biology* 47, 555–562.

Gibbs ME, 1991. Behavioral and pharmacological unraveling of memory formation. *Neurochemical Research* 16, 715–726.

Gibbs ME, Ng KT, 1977. Psychobiology of memory: Towards a model of memory formation. *Biobehavioral Reviews* 1, 113–136.

Gibson EJ, 1969. *Principles of Perceptual Learning and Development*. New York: Appleton-Century-Crofts.

Gibson EJ, Walk RD, 1956. The effect of prolonged exposure to visually presented patterns on learning to dsciriminate between them. *Journal of Comparative and Physiological Psychology* 49, 239–242.

Gibson JR, Connors BW, 2003. Neocortex: Chemical and electrical synapses. In Arbib (2003), 725–729.

Gluck MA, 1991. Stimulus generalisation and representation in adaptive network models of category learning. *Psychological Science* 2, 50–55.

Gluck MA, Thompson RF, 1987. Modeling the neural substrate of associative learning and memory: A computational approach. *Psychological Review* 94, 176–191.

Golden RM, 1996. *Mathematical Methods for Neural Network Analysis and Design*. Cambridge, MA: MIT Press.

Goldsmith TH, 1990. Optimization, constraint and history in the evolution of the eyes. *Quarterly Review of Biology* 65, 281–322.

Grafen A, 1991. Modelling in behavioural ecology. In Krebs & Davies (1991), 5–31.

Grafen A, 1999. Formal Darwinism, the individual-as-a-maximizing-agent analogy and bet-hedging. *Proceedings of the Royal Society of London* B266, 799–803.

Graham LJ, Kado RT, 2003. Biophysical mosaic of the neuron. In Arbib (2003), 170–175.

Graves JA, Whiten A, 1980. Adoption of strange chicks by herring gulls, *Larus argentatus* L. *Zeitschrift für Tierpsychologie* 54, 267–278.

Grice GR, Saltz E, 1950. The generalization of an instrumental response to stimuli varying in the size dimension. *Journal of Experimental Psychology* 40, 702–708.

Grillner S, 1996. Neural networks for vertebrate locomotion. *Scientific American* 274, 64–69.

Grillner S, Deliagina T, Ekeberg Ö, Manira AE, Hill R, Lansner A, Orlovsky G, Wallén P, 1995. Neural networks that coordinate locomotion and body orientation in lamprey. *Trends in Neuroscience* 18, 270–279.

Guthrie ER, 1935. *The Psychology of Learning.* New York: Harper.

Hall G, 2001. Perceptual learning. In RR Mowrer, SB Klein, editors, *Contemporary Learning Theory*, 367–407. Mahwah, NJ: Lawrence Erlbaum Associates.

Halliday TR, Slater PJB, 1983. *Genes, Development and Learning*, volume 3 of *Animal Behaviour*. Oxford, England: Blackwell.

Hammerstein P, 1996. Darwinian adaptation, population gentics and the streetcar theory of evolution. *Journal of Mathematical Biology* 34, 511–532.

Hanson H, 1959. Effects of discrimination training on stimulus generalization. *Journal of Experimental Psychology* 58, 321–333.

Harris-Warrick RM, Marder E, 1991. Modulation of neural networks for behavior. *Annual Review of Neuroscience* 14, 39–57.

Harrison RR, Koch C, 2000. A silicon implementation of the fly's optomotor control system. *Neural Computation* 12, 2291–2304.

Hartline DK, 1979. Pattern generation in the lobster (*Palinurus*) stomatogastric ganglion. II. Pyloric network stimulation. *Biological Cybernetics* 33, 223–236.

Hartline DK, Gassie DV, 1979. Pattern generation in the lobster (*Palinurus*) stomatogastric ganglion. I. Pyloric neuron kinetics and synaptic interactions. *Biological Cybernetics* 33, 209–222.

Hasselmo ME, Wyble BP, Fransen E, 2003. Neuromodulation in mammalian nervous systems. In Arbib (2003), 761–765.

Hassenstein B, Reichardt W, 1959. Wie sehen Insekten Bewegungen? *Umschau* 10, 302–305.

Hayes CM, Galef BG Jr, 1996. *Social Learning in Animals: The Roots of Culture.* London: Academic Press.

Hayes RD, Byrne JH, Baxter DA, 2003. Neurosimulation: Tools and resources. In Arbib (2003), 776–780.

Haykin S, 1999. *Neural Networks: A Comprehensive Foundation.* New York: Macmillan, 2 edition.

Hebb DO, 1949. *The Organization of Behaviour*. London: Wiley.

Hebb DO, 1966. *A Textbook of Psychology*. Philadelphia: Saunders, 2 edition.

Hedgecock EM, Russell RL, 1975. Normal and mutant thermotaxis in the nematode, *Caenorhabditis elegans*. *Proceedings of the National Academy of Science of the U.S.A.* 72, 4061–4065.

Heiligenberg W, 1974. A stochastic analysis of fish behavior. In McFarland (1974a), 87–118.

Hertz J, Krogh A, Palmer R, 1991. *Introduction to the Theory of Neural Computation*. Reading, MA: Addison-Wesley.

Hill AAV, Van Hooser SD, Calabrese RL, 2003. Half-center oscillators underlying rhythmic movements. In Arbib (2003), 507–511.

Hinde RA, 1970. *Animal Behaviour*. Tokyo: McGraw-Hill Kogakusha, 2 edition.

Hinde RA, Stevenson-Hinde J, editors, 1973. *Constraints on Learning*. New York: Academic Press.

Hinton GE, 1989. Connectionist learning procedures. *Artificial Intelligence* 40, 185–234.

Hogan JA, 1994a. Structure and development of behavior systems. *Psychonomic Bulletin & Review* 1, 439–450.

Hogan JA, 1994b. The concept of cause in the study of behavior. In Hogan & Bolhuis (1994).

Hogan JA, 1997. Energy models of motivation: A reconsideration. *Applied Animal Behaviour Science* 53, 89–105.

Hogan JA, 1998. Motivation. In G Greenberg, M Haraway, editors, *Comparative Psychology: A Handbook*, 164–175. New York: Garland.

Hogan JA, 2001. Development of behavior systems. In Blass (2001), 229–279.

Hogan JA, Bolhuis JJ, editors, 1994. *Causal Mechanisms of Behavioural Development*. Cambridge, England: Cambridge University Press.

Hogan JA, van Boxel F, 1993. Causal factors controlling dustbathing in Burmese red junglefowl: Some results and a model. *Animal Behaviour* 46, 627–635.

Holland JH, 1975. *Adaptation in Natural and Artificial Systems*. Ann Arbor, MI: University of Ann Arbor Press.

Hölldobler B, Wilson EO, 1990. *The Ants*. Berlin: Springer-Verlag.

Holmgren N, Enquist M, 1999. Dynamics of mimicry evolution. *Biological Journal of the Linnéan Society* 66, 145–158.

Holmgren N, Getz W, 2000. Perceptual constraints and the evolution of host-plant selection in insects: Generalists versus specialists. *Evolutionary Ecology Research* 2, 81–106.

Hooper SL, 2003. Crustacean stomatogastric system. In Arbib (2003), 304–308.

Hopfield JJ, 1982. Neural networks and physical systems with emergent collective computational abilities. *Proceedings of the National Academy of Science of the U.S.A.* 79, 2554–2558.

Horn G, 1967. Neuronal mechanisms for habituation. *Nature* 215, 707–711.

Horn G, 1985. *Memory, Imprinting and the Brain.* Oxford, England: Clarendon Press.

Hornik K, Stinchcombe M, White H, 1989. Multilayer feedforward networks are universal approximators. *Neural Networks* 2, 359–366.

Houston AI, McNamara JM, 1999. *Models of Adaptive Behaviour.* Cambridge, England: Cambridge University Press.

Hsiung CY, Mao GY, 1998. *Linear Algebra.* Singapore: World Scientific.

Hubel DH, 1988. *Eye, Brain, and Vision.* New York: American Scientific Library.

Hubel NM, Wiesel NM, 1962. Receptive fields, binocular interaction and functional architecture in the cat's visual cortex. *Journal of Physiology* 160, 106–154.

Huber L, Lenz R, 1996. Categorization of prototypical stimulus classes by pigeons. *Quarterly Journal of Experimental Psychology* 49B, 111–133.

Huff RC, Sherman JE, Cohn M, 1975. Some effects of response-independent reinforcement in auditory generalization gradients. *Journal of the Experimental Analysis of Behavior* 23, 81–86.

Hull CL, 1943. *Principles of Behaviour.* New York: Appleton-Century-Crofts.

Hull CL, 1949. Stimulus intensity dynamisn (V) and stimulus generalization. *Psychological Review* 56, 67–76.

Hyötyniemi H, 1996. Turing machines are recurrent neural networks. In J Alander, T Honkela, M Jakobsson, editors, *STeP'96: Genes, Nets and Symbols*, 13–24. Helsinki: Finnish Artificial Intelligence Society.

Ijspeert AJ, 2001. A connectionist central pattern generator for the aquatic and terrestrial gaits of a simulated salamander. *Biological Cybernetics* 84, 331–348.

Ijspeert AJ, 2003. Locomotion, vertebrate. In Arbib (2003), 649–654.

Ijspeert AJ, Hallam J, Willshaw D, 1998. From lampreys to salamanders: Evolving neural controllers for swimming and walking. In Pfeiffer et al. (1998).

Ijspeert AJ, Kodjabachian J, 1999. Evolution and development of a central pattern generator in the lamprey. *Artificial Life* 5, 247–269.

Immelmann K, 1972. The influence of early experience upon the development of social behaviour in estrildine finches. *Proceedings of the XVth Ornithological Congress, Den Haag 1970*, 316–338.

Intrator N, Cooper LN, 1995. BCM theory of visual cortical plasticity. In Arbib (1995), 153–157.

Jansson L, Forkman B, Enquist M, 2002. Experimental evidence of receiver bias for symmetry. *Animal Behaviour* 63, 617–621.

Jenkins HM, Harrison RH, 1960. Effect of discrimination training on auditory generalization. *Journal of Experimental Psychology* 59, 246–253.

Jin L, Nikiforuk P, Gupta M, 1995. Approximation of discrete time state space trajectories using dynamic recurrent networks. *IEEE Transactions on Automatic Control* 40, 1266–1270.

Johannesson M, 1997. Modelling asymmetric similarity with prominence. *British Journal of Mathematical and Statistical Psychology* 53, 121–139.

Johnston WA, Dark VJ, 1986. Selective attention. *Annual Review of Psychology* 37, 43–75.

Johnstone R, 1994. Female preferences for symmetrical males as a by-product of selection for mate recognition. *Nature* 372, 172–175.

Jordan M, 1986. Attractor dynamics and parallelism in a connectionist sequential machine. In *Proceedings of the Eighth Annual Conference of the Cognitive Science Society*, 531–546. Hillsdale, NJ: Lawrence Erlbaum Associates.

Kahn JI, Foster DH, 1985. Visual comparison of rotated and reflected random-dot patterns as a function of their positional symmetry and separation in the field. *Quarterly Journal of Experimental Psychology* 33A, 155–166.

Kalish H, 1969a. Alternative explanations. In Marx (1969), 276–297.

Kalish H, 1969b. Generalization as a fundamental phenomenon. In Marx (1969), 259–275.

Kalish H, 1969c. Generalization as an epiphenomenon. In Marx (1969), 205–297.

Kalish H, 1969d. Methods and theory. In Marx (1969), 207–218.

Kalman RE, 1960. A new approach to linear filtering and prediction problems. *Transactions of the ASME—Journal of Basic Engineering* 82D, 35–45.

Kalman RE, 1968. New development in systems theory relevant to biology. In MC Mesarovic, editor, *Systems Theory and Biology*, 222–232. Berlin: Springer.

Kamil A, 1998. On the proper definition of cognitive ethology. In Balda et al. (1998), 1–28.

Kamin LJ, 1968. "Attention-like" processes in classical conditioning. In MR Jones, editor, *Miami Symposium on the Prediction of Behavior: Aversive Stimulation*, 9–31. Miami, FL: University of Miami Press.

Kamin LJ, 1969. Predictability, surprise, attention, and conditioning. In BA Campbell, MR Church, editors, *Punishment and Aversive Behavior*, 279–296. New York: Appleton-Century-Crofts.

Kamin LJ, Schaub RE, 1963. Effects of conditioned stimulus intensity on the conditioned emotional response. *Journal of Comparative and Physiological Psychology* 56, 502–507.

Kamo M, Ghirlanda S, Enquist M, 2002. The evolution of signal form: Effects of learned vs. inherited recognition. *Proceedings of the Royal Society of London* B269, 1765–1771.

Kamo M, Kubo T, Iwasa Y, 1998. Neural network for female mate preference, trained by a genetic algorithm. *Philosophical Transactions of the Royal Society* B353, 399–406.

Kandel E, Schwartz J, Jessell T, 2000. *Principles of Neural Science*. London: Prentice-Hall, 4 edition.

Katz PS, Harris-Warrick RM, 1999. The evolution of neuronal circuits underlying species-specific behavior. *Current Opinion in Neurobiology* 9, 628–633.

Kehoe EJ, 1988. A layered model of associative learning: Learning-to-learn and configuration. *Psychological Review* 95, 411–433.

Kennedy JS, 1983. Zigzagging and casting as a programmed response to windborne odour: A review. *Physiological Entomology* 8, 109–120.

Kiehn O, Hounsgaard J, Sillar KT, 1997. Basic building blocks of vertebrate spinal central pattern generators. In PSQ Stein, S Grillner, A Selverston, DG Stuart, editors, *Neurons, Networks and Motor Behavior*, 47–59. Cambridge, MA: MIT Press.

Klein SB, 2002. *Learning: Principles and Applications*. New York: McGraw-Hill, 4 edition.

Klein SB, Mowrer RR, editors, 1989. *Contemporary Learning Theories: Pavlovian Conditioning and the State of Traditional Learning Theory*. Hillsdale, NJ: Lawrence Erlbaum Associates.

Kohonen T, 1982. Self-organized formation of topologically correct feature maps. *Biological Cybernetics* 43, 59–69.

Kohonen T, 1984. *Self-Organization and Associative Memory*. Berlin: Springer-Verlag.

Kohonen T, 2001. *Self-Organizing Maps*. Berlin: Springer-Verlag, 3 edition.

Kojima S, Kobayashi S, Yamanaka M, Sadamoto H, Nakamura H, Fujito Y, Kaway R, Skakibara M, Ito E, 1998. Sensory preconditioning for feeding response in the pond snail, *Lymnaea stagnalis*. *Brain Research* 808, 113–115.

Krakauer D, Johnstone R, 1995. The evolution of exploitation and honesty in animal communication: A model using artificial neural networks. *Philosophical Transactions of the Royal Society* B348, 355–361.

Krantz DH, Tversky A, 1971. Conjoint-measurement analysis of composition rules in psychology. *Psychological Review* 78, 151–169.

Krebs JR, Davies NB, editors, 1977. *Behavioural Ecology. An Evolutionary Approach*. London: Blackwell.

Krebs JR, Davies NB, 1987. *An Introduction to Behavioural Ecology*. London: Blackwell, 2 edition.

Krebs JR, Davies NB, editors, 1991. *Behavioural Ecology. An Evolutionary Approach*. London: Blackwell, 3 edition.

Krebs JR, Sherry DF, Healy SD, Perry VH, Vaccarino AL, 1989. Hippocampal specialization of food-storing birds. *Proceedings of the National Academy of Science of the U.S.A.* 86, 1388–1392.

Kruit JP, 1964. Ontogeny of social behaviour in the Burmese red jungle fowl (*Gallus gallus spadiceus*). *Behaviour Supplement* 12, 1–201.

Larsen BH, Hogan JA, Vestergaard KS, 2000. Development of dustbathing behavior sequences in the domestic fowl: The significance of functional experience. *Developmental Psychobiology* 36, 5–12.

Laughlin SB, 1981. Neural principles in the peripheral system of invertebrates. In H Autrun, editor, *Comparative Physiology and Evolution of Vision in Invertebrates: Invertebrate Visual Centers and Behavior*, 133–280. Berlin: Springer-Verlag.

Leahey TH, 2004. *A History of Psychology*. Englewood Cliffs, NJ: Prentice-Hall, 6 edition.

Lehrman DS, Hinde RA, Shaw E, editors, 1965. *Advances in the Study of Behaviour*. New York: Academic Press.

Leimar O, in press. Multidimensional convergence stability and canonical adaptive dynamics. In U Dieckmann, JAJ Metz, editors, *Elements of Adaptive Dynamics*. Cambridge, England: Cambridge University Press.

Leong CY, 1969. The quantitative effect of releasers on the attack readiness of the fish *Haplochromis burtoni* (Cichlidae, Pisces). *Zeithschrift für vergleichende Physiologie* 65, 29–50.

Lettvin JY, Maturana H, McCulloch WS, Pitts W, 1959. What the frog's eye tells the frog's brain. *Proceedings of the Institute of Radio Engineers* 47, 1940–1951.

Linsker R, 1986. From basic network principles to neural architecture. *Proceedings of the National Academy of Sciences of the U.S.A.* 83, 7508–7512, 8390–8394, 8779–8783.

Lorenz KZ, 1935. Der Kumpan in der Umwelt des Vogel. *Journal of Ornithology* 83, 137–413.

Lorenz KZ, 1937. Ueber die Bildung des Iinstinktsbegriff. *Naturwissenschaft* 25, 289–300, 307–318, 324–331.

Lorenz KZ, 1941. Vergleichende Bewegungsstudien bei Anatiden. *Journal für Ornithologie* 89, 194–294.

Lorenz KZ, 1981. *The Foundations of Ethology*. New York: Springer-Verlag.

Lorenz KZ, Tinbergen N, 1938. Taxis and Instinkthandlung in der eirollbewegung der graugans. *Zeitschrift für Tierpsychologie* 2, 1–29.

Lubow RE, 1989. *Latent Inhibition and Conditioned Attention Theory*. Cambridge, England: Cambridge University Press.

Lubow RE, Rifkin B, Alek M, 1976. The context effect: The relationship between stimulus preexposure and environmental preexposure determines subsequent learning. *Journal of Experimental Psychology: Animal Behavior Processes* 2, 38–47.

Ludlow AR, 1980. The evolution and simulation of a decision maker. In Toates & Halliday (1980), 273–296.

Ludlow AT, 1976. The behaviour of a model animal. *Behaviour* 58, 172.

Luenberger DG, 1979. *Introduction to Dynamic Systems: Theory, Models and Applications*. New York: Wiley.

Lytton WW, Sejnowski TJ, 1991. Simulations of cortical pyramidal neurons synchronized by inhibitory interneurons. *Journal of Neurophisiology* 66, 1059–1079.

Mackintosh NJ, 1974. *The Psychology of Animal Learning*. London: Academic Press.

Mackintosh NJ, 1975. A theory of attention: Variations in the associability of stimuli with reinforcement. *Psychological Review* 82, 276–298.

Mackintosh NJ, 1976. Overshadowing and stimulus intensity. *Animal Learning & Behavior* 4, 186–192.

Mackintosh NJ, 1983. General Principles of Learning. In Halliday & Slater (1983), 149–177.

Mackintosh NJ, editor, 1994. *Animal Learning and Cognition.* New York: Academic Press.

Mackintosh NJ, 1995. Categorization by people and pigeons: The twenty-second Bartlett memorial lecture. *Quarterly Journal of Experimental Psychology* 48B, 193–214.

Mackintosh NJ, 2000. Abstraction and discrimination. In C Heyes, L Huber, editors, *The evolution of cognition*, 123–142. Cambridge, MA: MIT Press.

Magnus D, 1958. Experimentelle untersuchung zur bionomie und ethologie des kaisermantels *Argynnis paphia* L. (Lep. Nymph.). I. Uber optische auslöser von anfliegereaktio ind ihre bedeutung fur das sichfinden der geschlechter. *Zeitschrift für Tierpsychologie* 15, 397–426.

Marder E, Calabrese R, 1996. Principles of rhythmic motor pattern generation. *Physiological Reviews* 76, 687–717.

Mareschal D, Shultz T, Mareschal D, 1996. Generative connectionist networks and constructivist cognitive development. *Cognitive Development* 11, 571–603.

Marr D, 1982. *Vision.* San Francisco: Freeman.

Marsh G, 1972. Prediction of the peak shift in pigeons from gradients of excitation and inhibition. *Journal of Comparative and Physiological Psychology* 81, 262–266.

Martin-Soelch C, Leenders KL, Chevalley AF, Missimer J, Kúnig G, Magyar S, Mino A, Schultz W, 2001. Reward mechanisms in the brain and their role in dependence: Evidence from neurophysiological and neuroimaging studies. *Brain Research Reviews* 36, 139–149.

Marx M, editor, 1969. *Learning: Processes.* London: MacMillan.

Mauk MD, Donegan NH, 1997. A model of pavlovian eyelid conditioning based on the synaptic organization of the cerebellum. *Learning & Memory* 4, 130–158.

Maynard Smith J, 1978. Optimization theory in evolution. *Annual Review of Ecology and Systematics* 9, 31–56.

Maynard Smith J, 1982. *Evolution and the Theory of Games.* Cambridge, England: Cambridge University Press.

Maynard Smith J, 1987. When learning guides evolution. *Nature* 329, 761–762.

Maynard Smith J, Burian S, Kauffman P, Alberch J, Campbell B, Goodwin R, Lande D, 1985. Developmental constraints and evolution. *Quarterly Review of Biology* 60, 265–287.

Maynard Smith J, Price GR, 1973. The logic of animal conflict. *Nature* 246, 15–18.

Maynard Smith J, Szathmáry E, 1995. *The Major Transitions in Evolution*. San Francisco: Freeman.

Mayr E, 1961. Cause and effect in biology. *Science* 134, 1501–1506.

McClelland JL, 1999. Cognitive modeling, connectionist. In Wilson & Keil (1999).

McClelland JL, McNaughton BL, O'Reilly RC, 1995. Why there are complementary learning systems in the hippocampus and neocortex: Insights from the successes and failures of connectionist models of learning and memory. *Psychological Review* 102, 419–457.

McClelland JL, Rumelhart DE, 1985. Distributed memory and the representation of general and specific information. *Journal of Experimental Psychology: General* 114, 159–188.

McClelland JL, Rumelhart DE, editors, 1986. *Parallel Distributed Processing: Explorations in the Microstructure of Cognition*, volume 2. Cambridge, MA: MIT Press.

McCloskey M, Cohen N, 1989. Catastrophic interference in connectionist networks: The sequential learning problem. In GH Bower, editor, *The Psychology of Learning and Motivation*, 109–164. London: Academic Press.

McCulloch WS, Pitts W, 1943. A logical calculus of the ideas immanent in nervous activity. *Bulletin of Mathematical Biophysics* 5, 115–133.

McFarland DJ, 1971. *Feedback mechanisms in animal behaviour*. London: Academic Press.

McFarland DJ, editor, 1974a. *Motivational control systems analysis*. London: Academic Press.

McFarland DJ, 1974b. Time-sharing as a behavioral phenomenon. In DS Lehrman, JS Rosenblatt, RA Hinde, E Shaw, editors, *Advances in the Study of Behavior*, volume 4, 201–225. London: Academic Press.

McFarland DJ, editor, 1987. *The Oxford Companion to Animal Behaviour*. Oxford, England: Oxford University Press.

McFarland DJ, 1989. *Problems of Animal Behaviour*. Harlow, England: Longman.

McFarland DJ, 1999. *Animal Behaviour: Psychobiology, Ethology and Evolution*. Harlow, England: Longman, 3 edition.

McFarland DJ, Baher E, 1968. Factors affecting feather posture in the barbary dove. *Animal Behaviour* 16, 171–177.

McFarland DJ, Bösser T, 1993. *Intelligent Behavior in Animals and Robots.* Cambridge, MA: MIT Press.

McFarland DJ, Houston A, 1981. *Quantitative Ethology. The State Space Approach.* London: Pitman Books.

McFarland DJ, Sibly RM, 1975. The behavioural final common path. *Philosophical Transactions of the Royal Society of London* B270, 265–293.

McLaren IPL, Mackintosh NJ, 2000. An elemental model of associative learning: I. Latent inhibition and perceptual learning. *Animal Learning & Behavior* 28, 211–246.

McNeill AR, 1982. *Optima for Animals.* London: Edward Arnold.

Medina JF, Garcia KS, Nores WL, Taylor NM, Mauk MD, 2000. Timing mechanisms in the cerebellum: Testing predictions of a large scale computer simulation. *Journal of Neuroscience* 20, 5516–5525.

Medina JF, Mauk MD, 2000. Computer simulation of cerebellar information processing. *Nature Neuroscience* 3, 1205–1211.

Mednick SA, Freedman JL, 1960. Stimulus generalization. *Psychological Bulletin* 57, 169–199.

Menini A, Picco C, Firestein S, 1995. Quantal-like current fluctuations induced by odorants in olfactory receptor cells. *Nature* 373, 435–437.

Merilaita S, 2003. Visual background complexity facilitates the evolution of camouflage. *Evolution* 57, 1248–1254.

Metz HAJ, 1977. State space models of animal behaviour. *Annals of System Research* 6, 65–109.

Mineka S, Cook M, 1988. Social learning and the acquisition of snake fear in monkeys. In TR Zentall, BG Galef Jr, editors, *Social Learning: Psychological and biological perspectives.* Hillsdale, NJ: Lawrence Erlbaum Associates.

Minsky ML, 1969. *Computation: Finite and infinite machines.* Cambridge, MA: MIT Press.

Minsky ML, Papert SA, 1969. *Perceptrons.* Cambridge, MA: MIT Press.

Mitchell M, 1996. *An Introduction to Genetic Algorithms.* Cambridge, MA: MIT Press.

Mittelstaedt H, 1957. Prey capture in mantids. In BT Scheer, editor, *Recent Advances in Invertebrate Physiology: A Symposium*, 1–50. Eugene, OR: University of Oregon Publications.

Morris RGM, Kandel ER, Squire LR, 1988. The neuroscience of learning and memory: Cells, neural circuits and behavior. *Trends in Neuroscience* 11, 125–179.

Mostofsky DI, editor, 1965. *Stimulus Generalization.* Stanford, CA: Stanford University Press.

Mowrer RR, Klein SB, editors, 2001. *Handbook of Contemporary Learning Theories.* Hillsdale, NJ: Lawrence Erlbaum Associates.

Mulloney B, Selverston AI, 1974. Organization of the stomatogastric ganglion in the spiny lobster. *Journal of Comparative Physiology* 91, 1–78.

Müller D, Gerber B, Hellstern F, Hammer M, Menzel R, 2000. Sensory preconditioning in honeybees. *Journal of Experimental Biology* 203, 1351–1364.

Neisser U, 1967. *Cognitive Psychology.* New York: Appleton-Century-Crofts.

Nieuwenhuys R, Ten Donkelaar H, Nicholson C, 1998. *The Central Nervous System of Vertebrates.* New York: Springer.

Nishikawa KC, 1997. The emergence of novel functions during brain evolution. *Bioscience* 47, 341–354.

Nishikawa KC, 2002. Evolutionary convergence in nervous systems: Insights from comperative phylogenetic studies. *Brain, Behavior and Evolution* 59, 240–249.

Nolfi S, 2002. Power and limits of reactive agents. *Neurocomputing* 42, 119–145.

Nolfi S, 2003. Evolution and learning in neural networks. In Arbib (2003), 415–418.

Nolfi S, Floreano D, editors, 2000. *Evolutionary Robotics.* Cambridge, MA: MIT Press.

Nolfi S, Miglino O, Parisi D, 1994. Phenotypic plasticity in evolving neural networks. In JD Nicoud, P Gaussier, editors, *Proceedings of the Conference from Perception to Action.* Los Alamitos, CA: IEEE Computer Society Press.

Nordeen KW, Nordeen EJ, 1988. Projection neurons within a vocal motor pathway are born during song learning m zebra finches. *Nature* 334, 149–151.

Nosofsky RM, 1986. Attention, similarity, and the identification-categorization relationship. *Journal of Experimental Psychology: General* 115, 39–57.

Nosofsky RM, 1987. Attention and learning processes in the identification and categorization of integral stimuli. *Journal of Experimental Psychology* 13, 87–108.

Nosofsky RM, 1991. Stimulus bias, asymmetric similarity, and classification. *Cognitive Psychology* 23, 91–140.

Nottebohm F, Stokes TM, Leonard CM, 1976. Central control of song in the canary. *Journal of Comparative Neurology* 165, 457–486.

Ohinata S, 1978. Postdiscrimination shift of the goldfish (*Carassius auratus*) on a visual wavelength continuum. *Annual of Animal Psychology* 28, 113–122.

Oja E, 1982. A simplified neuron model as a principal component analyzer. *Journal of Mathematical Biology* 15, 267–273.

O'Reilly RC, 1996. Biologically plausible error-driven learning using local activation differences: The generalized recirculation algorithm. *Neural Computation* 8, 895–938.

O'Reilly RC, Johnson M, 1994. Object recognition and sensitive periods: A computational analysis of visual imprinting. *Neural Computation* 6, 357–389.

O'Reilly RC, Munakata Y, 2000. *Computational Explorations in Cognitive Neuroscience*. Cambridge, MA: MIT Press.

Osorio DJ, Bacon PB, Whitington PM, 1997. The evolution of arthropod nervous systems. *American Scientist* 85, 244–252.

Parker GA, 1979. Sexual selection and sexual conflict. In M Blum, N Blum, editors, *Sexual Selection and Reproductive Competition in Insects*, chapter 4, 123–166. New York: Academic Press.

Parker GA, Maynard Smith J, 1990. Optimality theory in evolutionary biology. *Nature* 348, 27–33.

Parsons J, 1971. Cannibalism in herring gulls. *British Birds* 64, 528–537.

Pavlov IP, 1927. *Conditioned Reflexes*. Oxford, England: Oxford University Press.

Pearce JM, 1987. A model for stimulus generalization in Pavlovian conditioning. *Psychological Review* 94, 61–73.

Pearce JM, 1994. Similarity and discrimination: A selective review and a connectionist model. *Psychological Review* 101, 587–607.

Pearce JM, 1997. *Animal Learning and Cognition*. Hove, East Sussex: Psychology Press, 2 edition.

Pearce JM, Hall G, 1980. A model for Pavlovian learning: Variations in the effectiveness of conditioned but not of unconditioned stimuli. *Psychological Review* 87, 532–552.

Pennartz CMA, 1996. The ascending neuromodulatory system in learning by reinforcement: Comparing computational conjectures with empirical findings. *Brain Research Reviews* 21, 219–245.

Pfeiffer R, Blumberg B, Meyer JA, Wilson SW, editors, 1998. *From Animals to Animats 5. Proceedings of the Fifth Conference on the Simulation of Adaptive Behavior*. Cambridge, MA: MIT Press.

Phelps SM, 2001. History's lessons. In Ryan (2001), 167–180.

Phelps SM, Ryan MJ, 1998. Neural networks predict response biases of female Túngara frogs. *Proceedings of the Royal Society of London* B265, 279–285.

Phelps SM, Ryan MJ, 2000. History influences signal recognition: Neural network models of túngara frogs. *Proceedings of the Royal Society of London B* 267, 1633–1699.

Pinker S, Prince A, 1988. On language and connectionism: Analysis of a parallel distributed processing model of language acquisition. *Cognition* 28, 73–193.

Pinker S, Ullman MT, 2002. The past-tense debate. *Trends in Cognitive Sciences* 6, 456–463.

Planta C, Conradt J, Jencik A, Verschure P, 2002. A neural model of the fly visual system applied to navigational tasks. In JR Dorronsoro, editor, *Artificial Neural Networks: ICANN 2002*, Lecture Notes in Computer Sciences, 1268–1274. Berlin: Springer-Verlag.

Plaut D, Shallice T, 1994. *Connectionist Modelling in Cognitive Neuropsychology. A Case Study*. Hove, East Sussex: Lawrence Erlbaum Associates.

Polderman JW, Willems JC, 1998. *Introduction to Mathematical Systems Theory: A Behavioral Approach*. Berlin: Springer.

Prentice SD, Patla AE, Stacey DA, 1998. Simple artificial neural network models can generate basic muscle activity patterns for human locomotion at different speeds. *Experimental Brain Research* 123, 474–480.

Quartz SR, Sejnowski TJ, 1997. The neural basis of cognitive development: A constructivist manifesto. *Behavioral and Brain Sciences* 20, 537–596.

Raben MW, 1949. The white rats discrimination of differences in intensity of illumination measured by a running response. *Journal of Comparative and Physiological Psychology* 42, 254–272.

Ramsey W, 1999. Connectionism, philosophical issues. In Wilson & Keil (1999).

Razran G, 1949. Stimulus generalisation of conditioned responses. *Psychological Bulletin* 46, 337–365.

Real LA, 1994. *Behavioral Mechanisms in Evolutionary Ecology*. Chicago: University of Chicago Press.

Redondo T, Tortosa F, Arias de Reyna L, 1995. Nest switching and alloparental care in colonial white stork. *Animal Behaviour* 49, 1097–1110.

Rescorla RA, Wagner AR, 1972. A theory of Pavlovian conditioning: Variations in the effectiveness of reinforcement and nonreinforcement. In AH Black, WF Prokasy, editors, *Classical Conditioning: Current Research and Theory*. New York: Appleton-Century-Crofts.

Rice W, 1996. Sexually antagonistic male adaptation triggered by experimantal arrest of femal evolution. *Nature* 381, 232–234.

Riddle DLea, 1997. *C. elegans II*. New York: Cold Spring Harbor Laboratory Press.

Riedman M, 1982. The evolution of alloparental care and adoption in mammals and birds. *Quarterly Review of Biology* 57, 405–435.

Ristau CA, editor, 1991. *Cognive Ethology. The Minds of Other Animals. Essays in Honor of Donald R. Griffin*. Hillsdale, NJ: Laurence Erlbaum Associates.

Ritzman RE, 1993. The neural organization of cockroach escape and its role in context-dependent orientation. In RD Beer, RE Ritzmann, T McKenna, editors, *Biological Neural Networks in Invertebrate Neuroethology and Robotics*, 113–137. New York: Academic Press.

Rizley RC, Rescorla RA, 1972. Associations in second-order conditioning and sensory preconditioning. *Journal of Comparative and Physiological Psychology* 81, 1–11.

Rochester N, Holland J, Haibt L, Duda W, 1956. Tests on a cell assembly theory of the action of the brain, using a large scale digital computer. *IRE Transactions in Information Theory* 2, 89–93.

Rolls ET, Deco G, 2002. *The Computational Neuroscience of Vision*. Oxford, England: Oxford University Press.

Rolls ET, Stringer SM, 2000. On the design of neural networks in the brain by genetic evolution. *Progress in Neurobiology* 61, 557–579.

Roper TJ, 1983. Learning as a biological phenomenon. In Halliday & Slater (1983), 178–212.

Rose JE, Hind JE, Anderson DJ, Brugge JF, 1971. Some effects of stimulus intensity on response of auditory nerve fibers of the squirrel monkey. *Journal of Neurophysiology* 34, 685–699.

Rosenblatt F, 1958. The perceptron: A probabilistic model for information storage and orgranization in the brain. *Psychological Review* 65, 386–408.

Rosenblatt F, 1962. *Principles of Neurodynamics*. New York: Spartan Books.

Rosenzweig MR, Bennett EL, 1996. Psychobiology of plasticity: Effects of training and experience on brain and behavior. *Behavioural Brain Research* 78, 57–65.

Rowell CHF, 1961. Displacement grooming in the Chaffinch. *Animal Behaviour* 9, 38–63.

Rumelhart DE, Hinton GE, Williams RJ, 1986. Learning internal representation by back-propagation of errors. *Nature* 323, 533–536.

Rumelhart DE, McClelland JL, 1986a. On learning the past tenses of English verbs. In McClelland & Rumelhart (1986), 216–271.

Rumelhart DE, McClelland JL, editors, 1986b. *Parallel Distributed Processing: Explorations in the Microstructure of Cognition*, volume 1. Cambridge, MA: MIT Press.

Russell PA, 1973. Relationships between exploratory behaviour and fear: A review. *British Journal of Psychology* 64, 417–433.

Ryan M, 1990. Sexual selection sensory systems and sensory exploitation. *Oxford Surveys in Evolutionary Biology* 7, 157–195.

Ryan M, editor, 2001. *Anuran Communication*. Washington: Smithsonian Institution Press.

Scavio MJ, Gormezano I, 1974. CS intensity effects on rabbit nictitating membrane, conditioning, extinction and generalization. *Pavlovian Journal of Biological Science* 9, 25–34.

Scavio MJ, Thompson RF, 1979. Extinction and reacquisition performance alternations of the conditioned nictitating membrane response. *Bulletin of the Psychonomic Society* 13, 57–60.

Schmajuk NA, 1997. *Animal Learning and Cognition. A Neural Network Approach*. Cambridge, England: Cambridge University Press.

Schultz W, Dayan P, Read Montague P, 1997. A neural substrate of prediction and reward. *Science* 275, 1593–1599.

Schütz A, 2003. Neuroanatomy in a computational perspective. In Arbib (2003), 733–737.

Schyns P, Goldstone R, Thilbaut JP, 1998. The development of features in object concepts. *Behavioral and Brain Sciences* 21, 1–54.

Seitz A, 1940–1941. Die Paarbildung bei einigen Cichliden: I. *Zeitschrift für Tierpsychologie* 4, 40–84.

Seitz A, 1943. Die Paarbildung bei einigen Cichliden: II. *Zeitschrift für Tierpsychologie* 5, 74–101.

Shanks DS, 1995. *The Psychology of Associative Learning*. Cambridge, England: Cambridge University Press.

Shepard RN, 1987. Toward a universal law of generalization for psychological science. *Science* 237, 1317–1323.

Shettleworth SJ, 1975. Reinforcement and the organisation of behavior in golden hamsters: Hunger, environment and food reinforcement. *Journal of Experimental Psychology: Animal Behavior Processes* 1, 56–87.

Shettleworth SJ, 1978. Reinforcement and the organisation of behavior in golden hamsters: Sunflower seed and nest paper reinforcers. *Animal Learning and Behavior* 6, 352–362.

Shettleworth SJ, 1998. *Cognition, Evolution, and Behavior.* New York: Oxford University Press.

Shin HJ, Nosofsky RM, 1992. Similarity-scaling studies of dot-pattern classification and recognition. *Journal of Experimental Psychology: General* 121, 137–159.

Shouval HZ, Perrone MP, 1995. Post-Hebbian learning rules. In Arbib (1995), 745–748.

Šíma J, Orponen P, 2003. General-purpose computation with neural networks: A survey of complexity theoretic results. *Neural Computation* 15, 2727–2778.

Simmons PJ, Young D, 1999. *Nerve Cells and Animal Behaviour.* Cambridge, England: Cambridge University Press, 2 edition.

Skinner BF, 1938. *The Behavior of Organisms.* Acton, MA: Copley Publishing Group.

Skinner BF, 1985. Cognitive science and behaviorism. *British Journal of Psychology* 76, 291–301.

Slater PJB, 2003. Fifty years of bird song research: A case study in animal behaviour. *Animal Behaviour* 65, 633–639.

Slatkin M, Maynard Smith J, 1979. Models of coevolution. *Quaterly Review of Biology* 54, 233–263.

Sluckin W, 1972. *Imprinting and Early Learning*. London: Methuen, 2 edition.

Smith JC, Roll DL, 1967. Trace conditioning with X-rays as the aversive stimulus. *Psychonomic Science* 9, 11–12.

Smith W, 1977. *The Behaviour of Communicating: An Ethological Approach*, volume 1. Cambridge, MA: Harvard University Press.

Spence K, 1937. The differential response in animals to stimuli varying within a single dimension. *Phychological Review* 44, 430–444.

Squire LR, editor, 1992. *Encyclopedia of Learning and Memory.* New York: Macmillan.

Staddon JER, 2001. *The New Behaviorism: Mind, Mechanism and Society*. Hove, East Sussex: Psychology Press.

Staddon JER, Reid AK, 1990. On the dynamics of generalization. *Psychological Review* 97, 576–578.

Stephens DW, Krebs JR, 1986. *Foraging Theory*. Princeton, NJ: Princeton University Press.

Sutherland NS, 1959. Stimulus analysing mechanisms. In *Proceedings of a Symposium on the Mechanization of Thought Processes*, volume 2, 575–609. London: Her Majesty's Stationery Office.

Sutherland NS, 1964. The learning of discrimination by animals. *Endeavour* 23, 148–152.

Sutherland NS, Mackintosh NJ, 1971. *Mechanisms of Animal Discrimination Learning*. London: Academic Press.

Sutton RS, Barto AG, 1981. Toward a modern theory of adaptive networks: Expectation and prediction. *Psychological Review* 88, 135–140.

Sutton RS, Barto AG, 1998. *Reinforcement Learning*. Cambridge, MA: MIT Press.

Tapper DN, Halpern BP, 1968. Taste stimuli: A behavioral categorization. *Science* 161, 708–709.

Taylor B, Lukowiak K, 2000. The respiratory central pattern generator of *Lymnaea*: A model, measured and malleable. *Respiration Physiology* 122, 197–207.

Thomas JH, Lockery S, 1999. Neurobiology. In IA Hope, editor, *C. elegans: A Practical Approach*. Oxford, England: Oxford University Press.

Thompson RF, 1965. The neural basis of stimulus generalization. In Mostofsky (1965), 154–178.

Thorpe WH, 1958. The learning of song patterns by birds, with especial reference to the song of the chaffinch, *Fringilla coelebs*. *Ibis* 100, 535–570.

Tinbergen N, 1948. Wat prikkelt een scholester tot broeden? *De Levande Natur* 51, 65–69.

Tinbergen N, 1951. *The Study of Instinct*. New York: Oxford University Press.

Tinbergen N, 1952. "Derived activities": Their causation, biological significance, origin, and emancipation during evolution. *Quarterly Review of Biology* 27, 1–32.

Tinbergen N, 1963. Some recent studies of the evolution of sexual behavior. In F Beach, editor, *Sex and Behavior*. New York: Wiley.

Tinbergen N, Meeuse BJD, Boerema LK, Varossieau WW, 1942. Die Balz des Samtfalters, Eumenis (= Satyrus) semele (L.). *Zeitschrift für Tierpsychologie* 5, 182–226.

Toates FM, 1980. *Animal Behaviour: A Systems Approach.* New York: Wiley.

Toates FM, 1986. *Motivational Systems.* Cambridge, England: Cambridge University Press.

Toates FM, editor, 1998. *Control of Behaviour*, volume 5 of *Biology: Brain and Behaviour.* Berlin: Springer.

Toates FM, 2001. *Biological Psychology: An Integrative Approach.* Englewood Cliffs, NJ: Prentice-Hall.

Toates FM, Halliday TR, 1980. *Analysis of Motivational Processes.* London: Academic Press.

Toledo-Rodriguez M, Gupta A, Wang Y, Zhi Wu C, Markram H, 2003. Neocortex: Basic neuron types. In Arbib (2003), 719–725.

Tolman EC, 1949. There is more than one kind of learning. *Psychological Review* 56, 144–155.

Tomasello M, 1999. *The Cultural Origins of Human Cognition.* Cambridge, MA: Harvard University Press.

Torre V, Ashmore JF, Lamb TD, Menini A, 1995. Transduction and adaptation in sensory receptor cells. *Journal of Neuroscience* 15, 7757–7768.

Trivers RL, 1972. Parental investment and sexual selection. In *Sexual Selection and the Descent of Man, 1871–1971.* London: Heinemann.

van Iersel JJA, Bol ACA, 1958. Preening of two tern species: A study of displacement activities. *Behaviour* 13, 1–88.

Vestergaard KS, Hogan JA, Kruijt JP, 1990. The development of a behavior system: Dustbathing in the Burmese red jungle fowl: I. The influence of the rearing environment on the organization of dustbathing. *Behaviour* 112, 99–116.

Von Holst E, 1937. Vom Wesen der Ordnung im Zentralnervsystem. *Naturwissenschaft* 25, 625–631, 641–647.

von Holst E, 1973. *The Behavioural Physiology of Animals and Man*, volume 1. Coral Gables, FL: University of Miami Press. Collected papers 1937–1963, translated by Robert Martin.

Wachtmeister CA, Enquist M, 2000. The evolution of courtship rituals in monogamous species. *Behavioral Ecology* 11, 405–410.

Waelti P, Dickinson A, Schultz W, 2001. Dopamine responses comply with the basic assumptions of formal learning theory. *Nature* 412, 43–48.

Wagner AR, Brandon SE, 1989. Evolution of a structured connectionist model of Pavlovian conditioning (AESOP). In Klein & Mowrer (1989), 149–189.

Wagner AR, Brandon SE, 2001. A componential theory of associative learning. In Mowrer & Klein (2001), 23–64.

Walsh V, Kulikowski J, 1998. *Perceptual Constancy*. Cambridge, England: Cambridge University Press.

Wang H, 1995. On "computalism" and physicalism: Some subproblems. In J Cornwell, editor, *Nature's Imagination: The Frontiers of Scientific Vision*. Oxford, England: Oxford University Press.

Wang XJ, Rinzel J, 2003. Oscillatory and bursting properties of neurons. In Arbib (2003), 835–840.

Wang Y, Smallwood PM, Cowan M, Blesh D, Lawler A, Nathans J, 1999. Mutually exclusive expression of human red and green visual pigment-reporter transgenes occurs at high frequency in murine cone photoreceptors. *Proceedings of the National Academy of Science of the U.S.A.* 96, 5251–5256.

Warren RM, 1999. *Auditory Perception: A New Analysis and Synthesis*. Cambridge, England: Cambridge University Press, 2 edition.

Wasserman EA, 1995. The conceptual abilities of pigeons. *American Scientist* 83, 246–255.

Wasserman EA, 2002. General signs. In Bekoff et al. (2002), 175–182.

Webb B, 2001. Can robots make good models of biological behaviour? *Behavioral and Brain Sciences* 24, 1033–1050.

Wehner R, Flatt I, 1972. The visual orientation of desert ants, *Cataglyphis bicolor*, by means of territorial cues. In R Wehner, editor, *Information Processing in the Visual System of Arthropods*, 295–302. New York: Springer.

Wehner R, Srinivasan MV, 1981. Searching behavior of desert ants, genus *Cataglyphis* (Formicidae, Hymenoptera). *Journal of Comparative Physiology* 142, 315–338.

Weiss P, 1941. Self-differentiation of the basic patterns of coordination. *Comparative Psychology Monographs* 17, 1–96.

Whatson T, Stirling V, editors, 1998. *Development and Flexibility*, volume 4 of *Biology: Brain and Behaviour*. Berlin: Springer.

Widrow B, Gupta NK, Maitra S, 1973. Punish/reward learning with a critic in adaptive threshold systems. *IEEE Transactions on Systems, Man and Cybernetics* 5, 455–465.

Widrow B, Hoff ME Jr, 1960. Adaptive switching circuits. In *IRE WESCON Convention Record*, volume 4, 96–104. New York: IRE.

Widrow B, Stearns SD, 1985. *Adaptive Signal Processing*. Englewood Cliffs, NJ: Prentice-Hall.

Wiener N, 1949. *The Interpolation, Extrapolation and Smoothing of Stationary Time Series*. New York: Wiley.

Wiepkema PR, 1961. An ethological analysis of the reproductive behaviour of the bitterling. *Archives Neérlandaises de Zoologie* 14, 103–199.

Wilson RA, Keil FC, editors, 1999. *The MIT Encyclopedia of the Cognitive Sciences*. Cambridge, MA: MIT Press.

Wine JJ, Krasne JB, 1982. The cellular organization of crayfish escape behavior. In ED Bliss, editor, *The Biology of Crustacea*, 241–292. New York: Academic Press.

Wolpert L, Beddington R, Brockes J, Jessel T, Lawrence P, Meyerowitz E, 1998. *Principles of Development*. Oxford, England: Oxford University Press.

Zeigler HP, Bischof HJ, editors, 1993. *Vision, Brain and Behavior in Birds*. Cambridge, MA: MIT Press.

Zeigler HP, Marler P, editors, 2004. *Behavioural Neurobiology of Birdsong*, volume 1016 of *Annals of the New York Academy of Sciences*. New York: New York Academy of Sciences.

Zielinski K, Jakubowska E, 1977. Auditory intensity generalization after CER differentiation training. *Acta Neurobiologiae Experimentalis* 37, 191–205.

Index